Clinical Applications of Spectral Photon Counting Computed Tomography Technology

Salim Aymeric Si-Mohamed
Krzysztof Kris Iniewski
Editors

Clinical Applications of Spectral Photon Counting Computed Tomography Technology

Springer

Editors
Salim Aymeric Si-Mohamed
Department of cardiovascular and thoracic imaging
CHU Louis Pradel, Hospices Civils de Lyon, Université Claude Bernard Lyon 1, INSA-Lyon, UJM Saint-Etienne, CNRS, Inserm, CREATIS UMR 5220, U1206
Lyon, France

Krzysztof Kris Iniewski
Redlen Technologies
Port Moody, BC, Canada

ISBN 978-3-031-85080-6 ISBN 978-3-031-85078-3 (eBook)
https://doi.org/10.1007/978-3-031-85078-3

This Springer imprint is published by the registered company Springer Nature Switzerland AG
The registered company address is: Gewerbestrasse 11, 6330 Cham, Switzerland

Preface

In recent decades, computed tomography (CT) has become an indispensable tool in modern medicine, revolutionizing how we diagnose and treat a vast array of medical conditions. CT ability to provide detailed cross-sectional images of the human body has profoundly enhanced the precision, speed, and reliability of medical diagnostics. Today, approximately 375 million CT scans are performed annually worldwide—a number that continues to rise by 3–4% each year. This exponential growth underscores the critical role of CT technology in patient care, from routine diagnostics to lifesaving interventions. The global computed tomography (CT) scanner market size is projected to grow from $7.5 billion in 2024 to $12 billion by 2032, at an annual growth rate of over 6%.

Spectral photon-counting computed tomography (SPCCT) represents the latest frontier in this transformative journey. Building on the foundational principles of CT, SPCCT offers unprecedented capabilities by harnessing photon counting detectors to differentiate X-ray photons by their energy levels. This innovation enables detailed tissue characterization, precise imaging of contrast agents, and potential for reducing radiation dose, marking a paradigm shift in diagnostic imaging.

This book is dedicated to exploring the full potential of SPCCT technology. It begins with a comprehensive overview of the technical principles behind the current SPCCT systems developed by various vendors. Next, radiology experts share their clinical experiences, providing valuable insights on utilizing this groundbreaking technology in diverse applications, including cardiovascular, thoracic, emergency, breast, musculoskeletal, and oncologic imaging. These advancements have significant implications for addressing chronic, acute, and emergency conditions with greater precision and efficiency. Finally, the book delves into innovative applications, such as color K-edge imaging, and examines how dedicated contrast agents are revolutionizing diagnostic possibilities.

A key focus of this book is to bridge the gap between research and clinical practice by presenting findings from phantom and animal studies, clinical comparison and validation studies, evidence-based literature medicine reviews, and real-world clinical cases. By highlighting emerging applications across diverse medical fields, the book aims to empower clinicians, researchers, and healthcare professionals to harness SPCCT's capabilities for improved patient outcomes. Whether enabling early disease detection, providing more detailed pathology characterization, or refining treatment planning, SPCCT is set to redefine the standards of medical imaging and elevate the quality of healthcare.

As you embark on this journey through the evolving landscape of SPCCT, we hope that this book serves as a valuable resource, illuminating the path to the next era of radiology and healthcare innovation.

Lyon, France — Philippe Douek
Vancouver, BC, Canada — Krzysztof Kris Iniewski
Lyon, France — Salim Aymeric Si-Mohamed

Contents

About the Editors

Salim Aymeric Si-Mohamed, (MD, PhD) is a full Professor of Radiology at the University of Lyon and a hospital practitioner at the Hospices Civils de Lyon, where he leads the Cardiothoracic Imaging Unit at Louis Pradel Hospital. He is also a senior researcher at the CREATIS imaging laboratory (CNRS/INSERM/INSA Lyon) and a faculty member at the Lyon Medical School. Pr. Si-Mohamed specializes notably in diagnostic and interventional imaging of interstitial lung diseases, cardiac diseases, and Rendu-Osler disease within national reference centers. His research focuses on spectral photon-counting and dual-energy spectral CT technologies, using advanced contrast agents for cardiothoracic imaging. He is actively involved in clinical and experimental work on a next-generation spectral photon-counting CT prototype and investigates cutting-edge methods in CT image reconstruction, spectral decomposition, ultra-/low-dose imaging protocols, and color K-edge imaging. He is the recipient of the 2024 ERC Starting Grant entitled "KOLOR SPCCT Imaging", dedicated to the simultaneous imaging of ventilation, perfusion, and inflammation in lung diseases through color K-edge imaging. Over the past 10 years, he has authored more than 130 peer-reviewed publications, patented contrast agents for CT imaging, and co-authored books on the clinical potential of spectral photon-counting CT. A recent metric analysis showed h index = 38, i10 index = 88, and number of citations = 4754.

Krzysztof Kris Iniewski is the Director of Detector Architecture and Applications at Redlen Technologies Inc. in British Columbia, Canada. During his 15 years at Redlen, he has managed development of highly integrated CdZnTe detector products in medical imaging and security applications. Prior to Redlen, Kris held various management and academic positions at PMC-Sierra, University of Alberta, SFU, UBC, and University of Toronto.

Dr. Iniewski has published over 150+ research papers in international journals and conferences. He holds 25+ international patents granted in the USA, Canada, France, Germany, and Japan. He has written and edited 75+ books for Wiley, Cambridge University Press, McGraw Hill, CRC Press, and Springer. He is a frequent invited speaker and has consulted for multiple organizations internationally. A recent metric analysis showed h index = 33, i10 index = 100, and number of citations = 4259.

Chapter 1
Development of Clinical Spectral Photon-Counting CT Systems: A Technical Review

Joël Greffier, Anaïs Viry, Antoine Robert, Mouad Khorsi, and Salim Aymeric Si-Mohamed

Contents

Joël Greffier and Anaïs Viry authors are equally contributed equally to this work.

J. Greffier (✉)
Department of Medical Imaging, Nîmes University Hospital, IMAGINE UR UM 103, Montpellier University, Nîmes, France
e-mail: joel.greffier@chu-nimes.fr

A. Viry · M. Khorsi
Institute of Radiation Physics, Lausanne University Hospital and University of Lausanne, Lausanne, Switzerland

A. Robert
University of Lyon, INSA-Lyon, University Claude Bernard Lyon 1, UJM-Saint Etienne, CNRS, Inserm, CREATIS UMR 5220, U1206, Villeurbanne, France

S. A. Si-Mohamed
University of Lyon, INSA-Lyon, University Claude Bernard Lyon 1, UJM-Saint Etienne, CNRS, Inserm, CREATIS UMR 5220, U1206, Villeurbanne, France

Department of Radiology, Louis Pradel Hospital, Hospices Civils de Lyon, Bron, France

S. A. Si-Mohamed, K. K. Iniewski (eds.), *Clinical Applications of Spectral Photon Counting Computed Tomography Technology*,
https://doi.org/10.1007/978-3-031-85078-3_1

Abbreviations

ASIC	Application-specific integrated circuit
CdTe	Cadmium Telluride
CT	Computed tomography
CZT	Cadmium Zinc Telluride
EID	Energy integrating detectors
PCCT	Photon-counting CT
PCD	Photon-counting detectors
Si	Silicon
SPCCT	Spectral photon-counting CT

1.1 Introduction

In recent years, Computed Tomography (CT) has undergone a number of developments to improve radiological care. These technological developments have been applied to acquisition (in particular by reducing rotation times, increasing the tube voltage (kV) and tube current (mA) ranges, and using additional filtration) [1, 2], detection (*e.g.*, reducting the size of z-axis detectors) [3] and reconstruction (e.g., iterative or deep-learning image reconstruction algorithms) [4–7]. These improvements have reduced the dose delivered to patients while improving image quality by reducing image noise and/or improving spatial resolution [8–18].

Another major advance has been the development of dual-energy tomography platforms [19–21]. These platforms are based on the acquisition or detection of two X-ray spectra at low and high energy, enabling the contribution of the photoelectric effect and the Compton effect to be assessed separately [19, 22]. Spectral imaging can be used to generate different types of images and the additional information contained in these images facilitates the detection and characterization of lesions [19–21].

The latest major innovation has been the development of photon counting detectors (PCD) [19, 23, 24, 25]. By comparison with the energy-integrating detectors (EIDs) traditionally used in CT, these detectors offer better dose efficiency, eliminate electronic noise, improve spatial resolution and have intrinsic spectral sensitivity [19, 23, 24]. These detectors also allow the energy of each photon to be counted, thus improving the sampling of the X-ray spectrum in several energy bins, to better distinguish between photoelectric and Compton attenuation coefficients, thus producing better spectral images and specific color K-edge images.

The purpose of this chapter was to make the reader more familiar with the basic principles and techniques of new photon-counting CT systems equipped with photon-counting detectors.

1.2 Technical Background

Spectral CT imaging is a generation of CT systems that uses the energy-dependent information present in the images to explore new applications in clinical routines with material characterization. For more than two decades, dual-energy CT systems have been used to perform spectral imaging by differentiating low- and high-energy photons [19, 26]. More recently, a newer generation of detectors, known as energy-resolving photon counting detectors (PCD), has emerged to replace conventional energy-integrating detectors (EID) previously present in both conventional and dual-energy CT systems [27].

Energy-resolving PCDs offer significant advantages over conventional EIDs [28–30]. The main difference between EIDs and PCDs is the detection principle. EIDs use scintillators to convert X-ray photons into visible light and then into electrical signals using photodiodes (Fig. 1.1a). EIDs collect and sum all the X-ray energy information without distinguishing between different energy levels. This loss of energy information degrades the spectral performances of dual-energy CT. On the opposite, PCDs use semiconductors to directly convert X-ray photons into electrical signals (Fig. 1.1b). PCDs are also so-called direct-conversion detectors. A bias voltage is applied on the semiconductor sensors; an X-ray photon creates electron/hole pair, resulting in a cloud of charges that is directly processed by an application-specific integrated circuit (ASIC) to produce an electric signal. This direct conversion enables PCDs to count each incoming photon individually and classify them based on their energy. Current photon-counting CT (PCCT) systems exhibit variations in detector technology, energy resolution, spatial resolution and the number of energy thresholds available. Despite these differences, PCCT systems can precisely discriminate various materials of interest, while improving spatial resolution and decreasing noise [23, 31–35].

1.2.1 Advantages of Energy-Resolving PCDs

1.2.1.1 Spatial Resolution

The spatial resolution in CT systems based on EIDs technology is currently limited by two factors. The first one is the minimal size of detector elements using scintillators and the second one is the use of septa to avoid light diffusion inside them, both of which create dead spaces. PCDs overcome both these limitations [23, 24]. Firstly, they are designed with pixel sizes at the isocenter ranging from 0.1×0.1 to

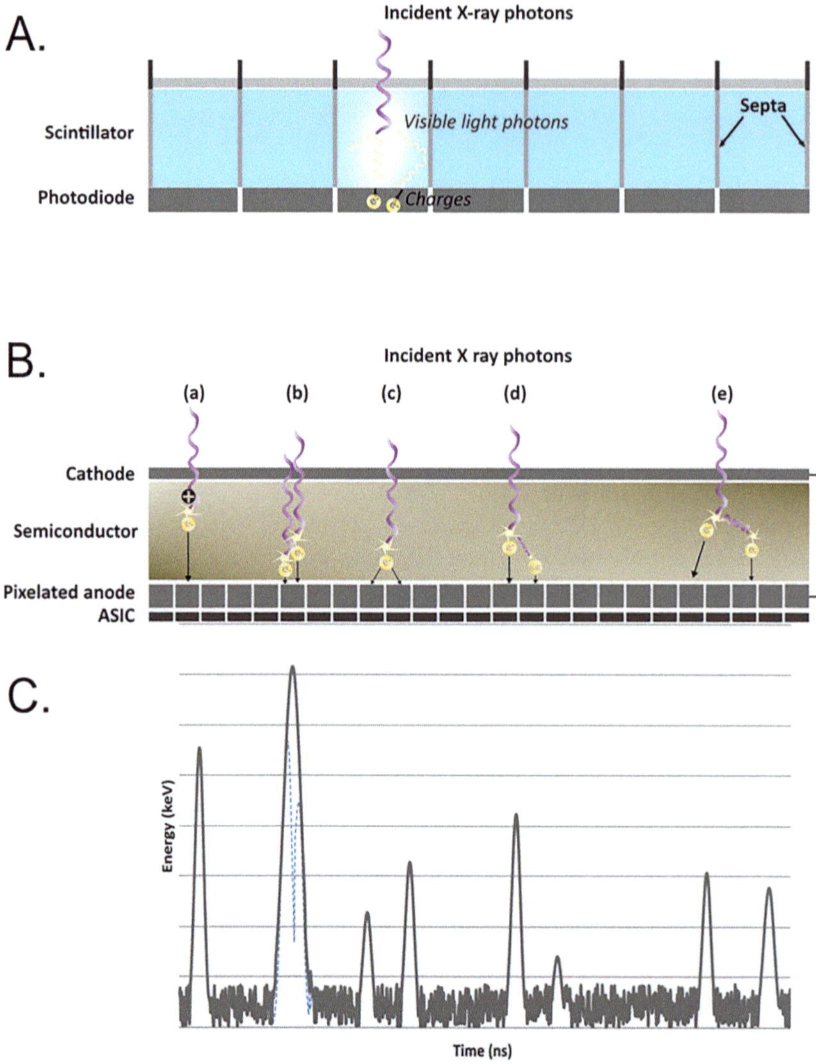

Fig. 1.1 Energy-integrating detector (EID) vs. photon-counting detector (PCD). (**A**) EID measures the total energy deposited by incoming X-rays. A scintillator transforms X-ray photons into visible light that is detected by a photodiode. The use of septa creates dead space that limits the spatial resolution; (**B**) (**a**) PCD counts individual photons by directly converting photon energy to an electric signal. (**b**) The pile-up effect occurs when multiple X-ray photons arrive at the detector within a short time, causing overlapping signals and compromising accurate energy measurement. (**c**) Charge-sharing effect arises when the charge generated by a single X-ray photon is distributed between two detector elements, leading to reduced spatial and energy resolution. (**d**) k-escape is seen when a new charge cloud is generated by the Kα X-ray fluorescence of the sensor materials (CdTe or Si) and registered by a pixel as two separate events. (**e**) Compton scattering effects occur when any secondary photons produced in the semiconductor material are registered by the same pixel or an adjacent pixel as two separate events, yielding to a lower record of the real energy value. (**C**) Registered electric signal for each type of interaction

0.4×0.4 mm², smaller than those used in EIDs. Secondly, the direct conversion between X-ray photons into electric signals avoids the use of reflecting septa. This is not only improves the spatial resolution but also the radiation dose efficiency and geometric effectiveness.

1.2.1.2 Noise Reduction and Contrast Improvement

Image noise arises from the statistical fluctuation of photons (the quantum noise) and electronic noise added by the electronic chain of EIDs [23, 24]. X-ray tube filtration (inherent and, in some cases, additional) generally filters low-energy photons below 20 keV. Therefore, the remaining low-energy signal detected with EIDs is due to electronic noise. In contrast, by implementing specific energy thresholds, PCDs can effectively filter out lower-energy noise, leading to a significant reduction in overall noise [23, 24].

In addition, a current limitation of EID is the higher resulting electric signal with high-energy photons by comparison with low-energy photons. This results in a loss of contrast because the photo-electric signal is dominant for low-energy photons. PCDs can register and count each individual incoming X-ray photon, accurately quantifying the X-ray signal and improving the contrast in the image. Therefore, the higher contrast-to-noise ratio of PCDs can be used to reduce radiation dose in many applications of PCCT.

1.2.1.3 Material Decomposition

Material decomposition relies on the energy dependence of the linear attenuation coefficient $\mu(\vec{x}, E)$ of materials in the diagnostic energy range (~15–150 keV). The linear attenuation coefficients can be modeled as the linear combination of energy-dependent and space-dependent function [36]:

$$\mu(\vec{x}, E) = \sum_{M}^{m=1} a_m(\vec{x}) f_m(E) \tag{1.1}$$

with $\vec{x}$ the 3D space position, E the energy, M the number of basis function, a_m the space-dependent functions and f_m the energy-dependent function. Knowing the energy functions f_m, the purpose of material decomposition is to obtain the volumes a_m.

For more than two decades, dual-energy CT has allowed the clinical use of spectral imaging. By measuring the X-ray attenuation in two different energy bins, material decomposition can be calculated using Eq. 1.1 with $M = 2$. Two types of basis functions have been proposed for human tissue. The first approach consists of using the energy dependence of both the photoelectric interaction f_{Ph} and the total cross-section for Compton scattering f_{Co}. Then, the volumes a_m describe the contributions of the photoelectric and Compton effects. As shown in the Alvarez et al.

study [36], the function f_{Ph} can be approximated by $\frac{1}{E^3}$ and f_{Co} by the Klein-Nishina function. Instead of using the photoelectric and Compton effect for the decomposition, the second approach uses the mass attenuation coefficient of two materials as energy-dependent basis functions. In that situation, the 3D space-dependent functions a_m correspond to the local density of the materials. Water and calcium (bone) can be used as base materials, but most frequently iodine is injected and therefore water and iodine are the materials used. Dual-energy CT combined with material decomposition algorithm helps differentiate materials, improves tissue contrast and reduces artifacts (such as metal or beam hardening artifacts) compared to conventional CT [37–39]. Many clinical applications have been proposed using both material decomposition maps (e.g., iodine maps or Z-effective) and virtual monoenergetic images [26].

Energy-resolving PCD could further advance spectral imaging [31, 40]. PCDs offer numerous advantages, one of which is to discriminate photons based on their energy with excellent resolution. This offers a greater resolution of photoelectric and Compton contribution/absorption coefficients that will enhance the current known spectral capabilities such as virtual monochromatic images.

Another key advantage of PDCs is the ability to record photons in more than two energy bins, allowing for the inclusion of additional materials in the spectral decomposition. This enables the use of contrast agents based on their K-edge energies (i.e., the binding energy between the K-shell and the nucleus). Indeed, at their K-absorption edges the atoms present a discontinuity in their attenuation spectra that can be used to discriminate them from other contributions to the global X-ray attenuation. This can be done by using Eq. 1.1 with $M = 3$ with the mass attenuation coefficient of the K-edge atom as an additional energy-dependent function basis. Then, the local density of the atom can be obtained after the decomposition.

This approach, referred to as color K-edge imaging [41, 42], allows the reconstruction of distinct images of both the contrast agent and the anatomy in a single CT acquisition. In this sense, color K-edge is similar to the nuclear imaging twin modality of positron emission tomography/CT in which low-resolution functional information on ^{18}F-fluorodeoxyglucose uptake is superimposed on high-resolution anatomical information. This opens up a completely new CT approach for functional, molecular or inflammation imaging and many other areas requiring exploration [43–48].

Figure 1.2 shows a schematic representation of two-material decomposition and K-edge imaging on a numerical phantom. The phantom consists of a water cylinder with a gadolinium insert (K-edge energy of 50.2 keV) on the left and an insert of iodine on the right. Two-material decomposition with water and iodine as basis materials distinguishes the iodine from the water but the gadolinium appears in the iodine map. By adding gadolinium as a third basis material, color K-edge imaging can also separate out iodine and gadolinium.

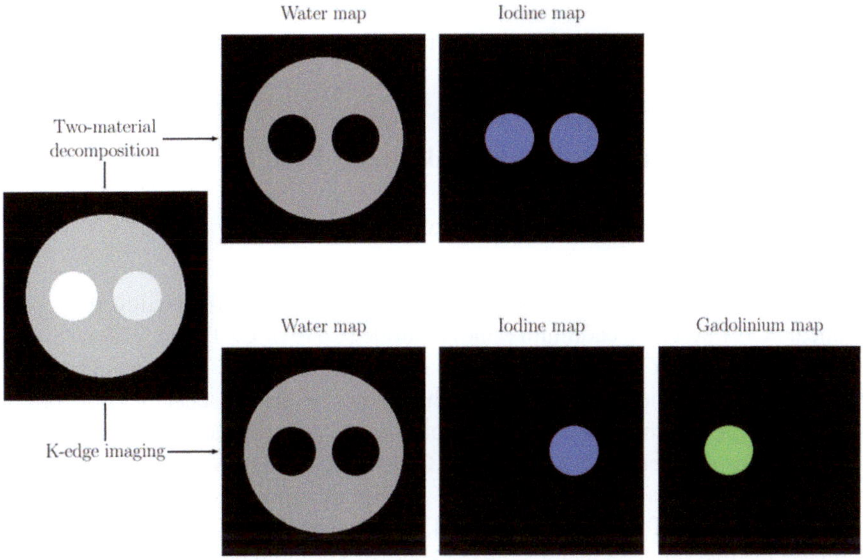

Fig. 1.2 Left: Conventional CT image of a numerical water phantom with one insert iodine (right) and one of gadolinium (left). Top row: Two-material decomposition using water and iodine as a basis function. Bottom row: color K-edge imaging done by adding gadolinium as a third material in the decomposition. For visualization, the iodine map is colored in blue and the gadolinium map in green

1.2.2 Technical Limitations: Charge Sharing, Pile-Up, Compton Scattering Effects and Energy Resolution

Despite having many advantages by comparison with EID-CT [49], PCCT still suffers from some limitations that could mitigate performance. Firstly, the choice of semiconductor materials influences the energy resolution capabilities of PCDs, which can affect material differentiation, particularly when dealing with very similar energy levels [50]. Second, some issues like charge sharing, Compton scattering, k-escape and pile-up effects can introduce errors in data interpretation [49]. These effects are described in Figure 1.1b. The charge-sharing effect occurs when a photon is counted by two adjacent pixels due to signal diffusion across adjacent detector elements. The simultaneous interactions are registered at lower energy, degrading both the spatial and energy resolutions. Similarly, Compton scattering effects occur when any secondary photons produced in the semiconductor material are registered by the same pixel or an adjacent pixel as two separate events, underestimating the real energy value. The K-escape effect is seen when a new charge cloud is generated by the K X-ray fluorescence of the sensor materials and detected by a pixel as two separate events. The pile-up effect occurs when two consecutive photons are counted at the same time, yielding a loss of counts and an incorrect energy registration [51]. This effect is always present and depends on the count rates and dead time of the detector [52]. The distortion in the measured energy of individual events impacts the accuracy of energy assessment and material differentiation capabilities.

1.3 Technological Differences Between PCCT Systems

1.3.1 Differences Between PCDs

Currently, there are two main semiconductor materials used in clinical or pre-clinical PCCT systems: cadmium (zinc) telluride (CdTe or CZT) and silicon (Si). In comparison with CZT, CdTe sensors are easier to produce in large quantities and have better homogeneity and reproducibility. However, the presence of zinc in CZT detectors improves the resistivity in comparison with CdTe detectors.

The major difference between Si and CdTe/CZT sensors is the stopping power ratio. For the same material thickness, the absorption efficiency is around 5% for Si and 90% for the CdTe/CZT detectors [53]. To obtain the same absorption efficiency between the two semiconductors, the thickness of Si should be equal to 55 mm [30]. As it is not feasible, the Si-semiconductor is mounted "edge on," parallel to the incident X-ray direction. In comparison, the thickness of CZT semiconductor is around 2 mm and is mounted "face on," perpendicular to the X-ray direction. The other difference between the two semiconductor materials comes from the different types of X-ray interaction in the semiconductor. With a high effective atomic number, CdTe or CZT have a higher probability of photo-electric absorption (84%) than Si (28%). X-ray interactions in Si are mainly the Compton effect, that cause a loss of spatial resolution. The Si fluorescence peak have a characteristic energy of 1.74 keV that is filtrated as electronic noise. The detector quantum efficiency is also lower for Si-based PCD, that require a higher radiation dose than CdTe or CZT PCDs for a similar detection task [24]. On the other hand, CdTe or CZT have a high probability of k-fluorescence emission between 23 and 27 keV, which could deform the energy spectrum (particularly for low energies) and also degrade the spatial resolution. The advantages and limitations of both semiconductor materials are summarized in Table 1.1 [24, 30, 40].

Another semiconductor technology is currently under development with a Gallium-Arsenide sensor. However, no PCCT systems have yet used this technology. Their performance will not be described here and more information can be found in E. Hamann et al. paper [54].

Table 1.1 Comparison of characteristics between cadmium-based and silicon-based detectors

Detector material	Cadmium-based PCD (CdTe or CZT)	Silicon-based PCD
Manufacturing complexity	Less complex	More complex
Thickness (orientation)	Face-on (1.4–2 mm)	Edge-on (30–60 mm)
Photo-electric effect/Compton effect	Photo-electric effect predominant	Compton effect predominant
k-fluorescence emission	Higher	Lower
Lowest-energy bin	20–25 keV	5–10 keV
Energy resolution (FWHM)	5–10 keV	3–5 keV
Spatial resolution	Higher	Lower
Energy-resolving capabilities	Lower	Higher

CZT cadmium zinc telluride, *CdTe* cadmium telluride, *FWHM* full width at half maximum, *PCD* photon-counting detectors

1.3.2 Technological Aspects of Spectral Photon-Counting CT Platforms

Various CT manufacturers have developed PCCT systems and two PCCTs have been FDA-approved (i.e., one wide-bore for all applications and one small-bore for head imaging) [19, 24]. The various systems developed differ in the choice of semiconductor material used, the gantry types and the application domain (Table 1.2). All these choices have an impact on the performance of the PCCTs and the image quality, particularly in terms of spatial resolution and spectral performance. PCD detectors are also under development in cone-beam CT equipment; their performance will not be detailed here.

1.3.2.1 Siemens Healthineers

Siemens Healthineers were the first to obtain FDA approval (2021) for their wide-bore NAEOTOM Alpha CT system, which can be used for different application domains, including head, body, cardiac and extremity [23, 24, 27]. This PCCT has passed through several prototypes (CounT and Count Plus). It is a third generation dual-source system consisting of two X-ray tube/detector pairs dephased by 95° and equipped with CdTe semiconductor-based photon counting detectors. The detector array of the first detector covers a 50 cm field-of-view (FOV), while the second detected has an effective width of 36 cm, limiting its acquisition and reconstruction FOV to 36 cm. CT acquisitions are usually carried out with a single X-ray tube, while two X-ray tubes are used for acquisitions requiring improved temporal resolution and/or high scan speeds (cardiac and high helical pitch scans with lowest rotation time of 0.25 s/rot and temporal resolution of 66 ms). With this system, acquisitions can be made for different kVp (70, 90, 120, 140, Sn100 and Sn140 kVp) with 144 × 0.4 mm beam collimation in standard mode or 120 × 0.2 mm in high-resolution mode, resulting in effective detector element sizes at the isocenter of

Table 1.2 Main characteristics of current prototype, pre-clinical or clinical photon-counting CT (PCCT) systems suitable for clinical imaging

Manufacturer	CT plateforms	PCD materials	Geometry	Beam collimation	Field-of-view	Energy threshold	Current status
Siemens Healthineers	NAEOTOM Alpha	CdTe	Dual-source	STD: 144 × 0.4 mm HR: 120 × 0.2 mm	50 cm; 36 cm used for cardiac scans; 36 cm for high helical pitch scans	4 energy thresholds mixed to generate a low and a high energy bins in routine modes	FDA-cleared in September 2021
Samsung Healthcare	OmniTom portable PCD head CT	CdTe	Mono-source	16 × 0.625 mm	30 cm	3 available	FDA-cleared in March 2022
MARS Bioimaging Limited (MBI)	MARS Extremity 5 × 120	CdZnTe	Mono-source	14 mm	12.4 cm	8 in total, 5 in "charge summing mode"	Pre-clinical PCCT
Canon Healthcare	Aquilion ONE ViSION CT	CdZnTe	Mono-source	STD: 16 × 0.62 mm HR: 48 × 0.21 mm	50 cm	6 available	Clinical prototype system
	Aquilion PRECISION	CdZnTe	Mono-source	STD: 64 × 0.6 mm HR: 192 × 0.2 mm	50 cm	5 available	Advanced clinical research PCCT
Philips Healthcare	iCT	CdZnTe	Mono-source	64 × 0.275 mm	50 cm	5 in standard mode; 5 in HR modes	Advanced research prototype system
GE Healthcare	LightSpeed, Revolution CT, Apex	Si	Mono-source	– – 5–80 mm for Revolution Apex CT	50 cm	8 available	Advanced research prototype system

Two CT platforms are designed with small bore gantry for head (Samsung Healthcare) and for extremities (MARS Bioimaging limited system) imaging and that two have been FDA-cleared (Siemens Healthineers and Samsung Healthcare)

The parameters listed are based on the status of the manufacturer's development at the time of publication and are expected to change in the near future

Cd cadmium, *Z* zinc, *Te* telluride, *Si* silicon, *FDA* Food and Drug Administration, *PCD* photon-counting detectors, *HR* high-resolution mode, *STD* standard

0.30 × 0.30 mm^2 and 0.15 × 0.15 mm^2, respectively. For spectral images, materials can be decomposed using the data from four energy thresholds mixed to generate a low- and high-energy image in routine modes.

1.3.2.2 Samsung Healthcare

Samsung Healthcare was the second to obtain FDA approval (2022) for their small-bore OmniTom Portable PCD Head CT system, which can be used only for head CT acquisitions [55]. This portable system is composed of a single X-ray tube/detector pair equipped with CdTe semiconductors. Acquisitions can be made for kV from 70 to 120 kVp with beam collimation of 16 rows of 0.625 mm and 30-cm FOV. The detector size at the isocenter is 0.703 × 0.707 mm^2 for standard mode and 0.117 × 0.141 mm^2 or 0.351 × 0.423 mm^2 for high-resolution modes. For spectral images, materials can be decomposed using three energy bins.

1.3.2.3 MARS Bioimaging Ltd (MBI)

MARS Bioimaging Litd has developed a pre-clinical PCCT specifically designed for the upper-extremities [56]. This system consists of a single X-ray tube/detector pair equipped with CdZnTe semiconductors. Unlike other CT systems, this system operates without a table, and the patient's upper extremity remains fixed while the pair (X-ray tube/detector) rotates. With this system, acquisitions can be performed at tube voltages ranging from 50 to 120 kVp, with a slice thickness of 0.09 mm, a 12.4 cm FOV, and a pixel size of 0.11 × 0.11 mm^2. For spectral images, materials can be decomposed using five energy bins in "charge summing" mode.

1.3.2.4 Canon Medical Systems

Canon Medical Systems has developed two PCCT versions. The first PCCT engineering prototype was developed on an Aquilion ONE ViSION CT platform [57] and the second is a clinical research PCCT (TSX-501R, Canon medical systems, Otawara, Japan) developed on an Aquilion PRECISION platform.

Both PCCTs are single-source scanners equipped with CdZnTe semiconductors. On both systems, acquisitions can be made at 120 kVp with a 50 cm FOV. For the first engineering prototype version, beam collimation is 16 × 0.62 mm in normal resolution mode (NR) and 48 × 0.21 mm in super-high-resolution mode (SHR) and 64 × 0.6 mm (NR) and 192 × 0.2 mm (SHR) on the Clinical Research PCCT version, respectively. The detector size at the isocenter is 0.2 × 0.2 mm^2 for both versions. For spectral images, materials can be decomposed using six energy bins for the first version and five for the second.

1.3.2.5 Philips Healthcare

Philips Healthcare has developed an advanced research spectral PCCT prototype (called Spectral Photon Counting CT, SPCCT) on the iCT platform, which could be used for all application domains. The SPCCT is equipped with a single X-ray tube and a CdZnTe semiconductor detector array (2-mm thick). Acquisitions can be made at 80, 100, 120 or 140 kVp with a 50 cm FOV. The lowest available rotation time is 0.33 s/rot (from 0.33 to 1 s/rot) and the greatest temporal resolution is 165 ms. The beam collimation is 64 × 0.275 mm (z-coverage in the isocenter 17.5 mm) and the detector size at the isocenter is 274 × 274 μm^2 [33]. For spectral images, materials can be decomposed using five energy bins for both standard and high-resolution modes (i.e., 512– or 1024–matrix sizes), with and without electrocardiogram-gating mode [58–61]. This system has been used in pre-clinical and clinical research studies to evaluate novel K-edge contrast agents and to investigate the potential of SPCCT in a clinical setting.

1.3.2.6 GE Healthcare

To our knowledge, GE Healthcare is the only company that has developed three advanced research single-source prototypes equipped with deep silicon semiconductor detectors [19, 24, 30, 62]. The first prototype was developed on a LightSpeed VCT platform, the second on a Revolution CT platform and the third on a Revolution Apex CT. With the third version of Si-PCCT prototype acquisitions are typically performed at 120 kVp using axial and helical collimations of between 5 mm and 80 mm, with modules focally aligned in the x-direction to provide a 50-cm FOV. With this system the rotation time ranges from 0.234 to 1 s/rot, the matrix size is 1024 × 1024 pixels and the pixel size is 0.2 × 0.2 mm^2 with a sensitive volume of 0.2 × 0.2 × 0.4 mm (width × height × depth). The system is capable of using up to eight energy bins for spectral imaging tasks such as material decomposition.

1.4 Conclusion

The introduction of photon-counting CT in the clinical field has opened a new era in medical imaging. The anticipated improvements in image quality are already being demonstrated for many applications, particularly those requiring better spatial resolution, contrast, and reduced radiation doses. In addition, this technology expands the capabilities of spectral CT for several applications that have not yet been fully exploited, especially with the development of color K-edge contrast agents. Overall, PCCT is expected to provide considerable benefits considerable benefits in the diagnosis, characterization, and staging of diseases, and to enable new approaches not previously with CT. However, PCCT systems are still scarce and open to technical development, that will facilitate the adoption of this technology.

Acknowledgments We wish to thank Sarah Kabani, a medical writer at the BESPIM, Nîmes University Hospital for her help in editing the manuscript. We also thank Kirsten Boedeker and Karim Haioun (Canon Medical Systems), Philippe Coulon (Philips Healthcare), Håkan Grundin and Dominic Crotty (GE Healthcare), and Sebastian Faby (Siemens Healthineers) for reviewing and validating the sections related to their PCCTs.

References

1. Greffier J, Fitton I, Ngoc Ty CV, Frandon J, Beregi JP, Dabli D. Impact of tin filter on the image quality of ultra-low dose chest CT: a phantom study on three CT systems. Diagn Interv Imaging. 2023;104:506–12.
2. Greffier J, Pereira F, Hamard A, Addala T, Beregi JP, Frandon J. Effect of tin filter-based spectral shaping CT on image quality and radiation dose for routine use on ultralow-dose CT protocols: a phantom study. Diagn Interv Imaging. 2020;101:373–81.
3. Greffier J, Viry A, Durand Q, Hajdu SD, Frandon J, Beregi JP, et al. Brain image quality according to beam collimation width and image reconstruction algorithm: a phantom study. Phys Med. 2023;108:102558.
4. Geyer LL, Schoepf UJ, Meinel FG, Nance JW Jr, Bastarrika G, Leipsic JA, et al. State of the art: iterative CT reconstruction techniques. Radiology. 2015;276:339–57.
5. Greffier J, Frandon J, Larbi A, Beregi JP, Pereira F. CT iterative reconstruction algorithms: a task-based image quality assessment. Eur Radiol. 2020;30:487–500.
6. Willemink MJ, de Jong PA, Leiner T, de Heer LM, Nievelstein RA, Budde RP, et al. Iterative reconstruction techniques for computed tomography part 1: technical principles. Eur Radiol. 2013;23:1623–31.
7. Willemink MJ, Noel PB. The evolution of image reconstruction for CT-from filtered back projection to artificial intelligence. Eur Radiol. 2019;29:2185–95.
8. Schwyzer M, Messerli M, Eberhard M, Skawran S, Martini K, Frauenfelder T. Impact of dose reduction and iterative reconstruction algorithm on the detectability of pulmonary nodules by artificial intelligence. Diagn Interv Imaging. 2022;103:273–80.
9. Akagi M, Nakamura Y, Higaki T, Narita K, Honda Y, Zhou J, et al. Deep learning reconstruction improves image quality of abdominal ultra-high-resolution CT. Eur Radiol. 2019;29:6163–71.
10. Greffier J, Boccalini S, Beregi JP, Vlassenbroek A, Vuillod A, Dupuis-Girod S, et al. CT dose optimization for the detection of pulmonary arteriovenous malformation (PAVM): a phantom study. Diagn Interv Imaging. 2020;101:289–97.
11. Greffier J, Macri F, Larbi A, Fernandez A, Khasanova E, Pereira F, et al. Dose reduction with iterative reconstruction: optimization of CT protocols in clinical practice. Diagn Interv Imaging. 2015;96:477–86.
12. Ichikawa Y, Kanii Y, Yamazaki A, Nagasawa N, Nagata M, Ishida M, et al. Deep learning image reconstruction for improvement of image quality of abdominal computed tomography: comparison with hybrid iterative reconstruction. Jpn J Radiol. 2021;39:598–604.
13. Jensen CT, Liu X, Tamm EP, Chandler AG, Sun J, Morani AC, et al. Image quality assessment of abdominal CT by use of new deep learning image reconstruction: initial experience. AJR Am J Roentgenol. 2020;215:50–7.
14. Nakamura Y, Higaki T, Tatsugami F, Zhou J, Yu Z, Akino N, et al. Deep learning-based CT image reconstruction: initial evaluation targeting Hypovascular hepatic metastases. Radiol Artif Intell. 2019;1:e180011.
15. Nakamura Y, Narita K, Higaki T, Akagi M, Honda Y, Awai K. Diagnostic value of deep learning reconstruction for radiation dose reduction at abdominal ultra-high-resolution CT. Eur Radiol. 2021;31:4700–9.

16. Greffier J, Hamard A, Pereira F, Barrau C, Pasquier H, Beregi JP, et al. Image quality and dose reduction opportunity of deep learning image reconstruction algorithm for CT: a phantom study. Eur Radiol. 2020;30:3951–9.
17. Greffier J, Frandon J, Durand Q, Kammoun T, Loisy M, Beregi JP, et al. Contribution of an artificial intelligence deep-learning reconstruction algorithm for dose optimization in lumbar spine CT examination: a phantom study. Diagn Interv Imaging. 2023;104:76–83.
18. Greffier J, Durand Q, Frandon J, Si-Mohamed S, Loisy M, de Oliveira F, et al. Improved image quality and dose reduction in abdominal CT with deep-learning reconstruction algorithm: a phantom study. Eur Radiol. 2023;33:699–710.
19. Greffier J, Villani N, Defez D, Dabli D, Si-Mohamed S. Spectral CT imaging: technical principles of dual-energy CT and multi-energy photon-counting CT. Diagn Interv Imaging. 2023;104:167–77.
20. Agostini A, Borgheresi A, Mari A, Floridi C, Bruno F, Carotti M, et al. Dual-energy CT: theoretical principles and clinical applications. Radiol Med. 2019;124:1281–95.
21. McCollough CH, Leng S, Yu L, Fletcher JG. Dual- and multi-energy CT: principles, technical approaches, and clinical applications. Radiology. 2015;276:637–53.
22. Greffier J, Si-Mohamed S, Dabli D, de Forges H, Hamard A, Douek P, et al. Performance of four dual-energy CT platforms for abdominal imaging: a task-based image quality assessment based on phantom data. Eur Radiol. 2021;31:5324–34.
23. Leng S, Bruesewitz M, Tao S, Rajendran K, Halaweish AF, Campeau NG, et al. Photon-counting detector CT: system design and clinical applications of an emerging technology. Radiographics. 2019;39:729–43.
24. Flohr T, Schmidt B. Technical basics and clinical benefits of photon-counting CT. Invest Radiol. 2023;58:441–50.
25. Greffier J, Viry A, Robert A, Khorsi M, Si-Mohamed S. Photon-counting CT systems: A technical review of current clinical possibilities. Diagn Interv Imaging. 2025;106(2):53–59. https://doi.org/10.1016/j.diii.2024.09.002. Epub 2024 Sep 20.
26. McCollough CH, Boedeker K, Cody D, Duan X, Flohr T, Halliburton SS, et al. Principles and applications of multienergy CT: report of AAPM task group 291. Med Phys. 2020;47:e881–912.
27. McCollough CH, Rajendran K, Leng S, Yu L, Fletcher JG, Stierstorfer K, et al. The technical development of photon-counting detector CT. Eur Radiol. 2023;33:5321–30.
28. Wehrse E, Klein L, Rotkopf LT, Wagner WL, Uhrig M, Heussel CP, et al. Photon-counting detectors in computed tomography: from quantum physics to clinical practice. Radiologe. 2021;61:1–10.
29. Sartoretti T, Wildberger JE, Flohr T, Alkadhi H. Photon-counting detector CT: early clinical experience review. Br J Radiol. 2023;96:20220544.
30. Danielsson M, Persson M, Sjolin M. Photon-counting x-ray detectors for CT. Phys Med Biol. 2021;66:03TR01.
31. Willemink MJ, Persson M, Pourmorteza A, Pelc NJ, Fleischmann D. Photon-counting CT: technical principles and clinical prospects. Radiology. 2018;289:293–312.
32. Nakamura Y, Higaki T, Kondo S, Kawashita I, Takahashi I, Awai K. An introduction to photon-counting detector CT (PCD CT) for radiologists. Jpn J Radiol. 2023;41:266–82.
33. Si-Mohamed SA, Boccalini S, Villien M, Yagil Y, Erhard K, Boussel L, et al. First experience with a whole-body spectral photon-counting CT clinical prototype. Invest Radiol. 2023;58:459–71.
34. Douek PC, Boccalini S, Oei EHG, Cormode DP, Pourmorteza A, Boussel L, et al. Clinical applications of photon-counting CT: a review of Pioneer studies and a glimpse into the future. Radiology. 2023;309:e222432.
35. Si-Mohamed SA, Miailhes J, Rodesch PA, Boccalini S, Lacombe H, Leitman V, et al. Spectral photon-counting CT technology in chest imaging. J Clin Med. 2021;10:10.
36. Alvarez RE, Macovski A. Energy-selective reconstructions in X-ray computerized tomography. Phys Med Biol. 1976;21:733–44.

37. Garnett R. A comprehensive review of dual-energy and multi-spectral computed tomography. Clin Imaging. 2020;67:160–9.
38. Omoumi P, Becce F, Racine D, Ott JG, Andreisek G, Verdun FR. Dual-energy CT: basic principles, technical approaches, and applications in musculoskeletal imaging (part 1). Semin Musculoskelet Radiol. 2015;19:431–7.
39. Tatsugami F, Higaki T, Nakamura Y, Honda Y, Awai K. Dual-energy CT: minimal essentials for radiologists. Jpn J Radiol. 2022;40:547–59.
40. Dane B, Froemming A, Schwartz FR, Toshav A, Ramirez-Giraldo JC, Ananthakrishnan L. Photon counting CT clinical adoption, integration, and workflow. Abdom Radiol (NY). 2024;49:4600. https://doi.org/10.1007/s00261-024-04503-5.
41. Nizovtseva IG, Galenko PK. Travelling-wave amplitudes as solutions of the phase-field crystal equation. Philos Trans A Math Phys Eng Sci. 2018;376:20170202.
42. Si-Mohamed S, Bar-Ness D, Sigovan M, Tatard-Leitman V, Cormode DP, Naha PC, Coulon P, Rascle L, Roessl E, Rokni M, Altman A, Yagil Y, Boussel L, Douek P. Multicolour imaging with spectral photon-counting CT: a phantom study. Eur Radiol Exp. 2018;2(1):34. https://doi.org/10.1186/s41747-018-0063-4. PMID: 30327898; PMCID: PMC6191405.
43. Si-Mohamed SA, Sigovan M, Hsu JC, Tatard-Leitman V, Chalabreysse L, Naha PC, et al. In vivo molecular K-edge imaging of atherosclerotic plaque using photon-counting CT. Radiology. 2021;300:98–107.
44. Boccalini S, Mayard C, Lacombe H, Villien M, Si-Mohamed S, Delahaye F, et al. Ultra-high-resolution and K-edge imaging of prosthetic heart valves with spectral photon-counting CT: a phantom study. Invest Radiol. 2024;59:589–98.
45. Cosset B, Sigovan M, Boccalini S, Farhat F, Douek P, Boussel L, et al. Bicolor K-edge spectral photon-counting CT imaging for the diagnosis of thoracic endoleaks: a dynamic phantom study. Diagn Interv Imaging. 2023;104:235–42.
46. Si-Mohamed S, Cormode DP, Bar-Ness D, Sigovan M, Naha PC, Langlois JB, et al. Evaluation of spectral photon counting computed tomography K-edge imaging for determination of gold nanoparticle biodistribution in vivo. Nanoscale. 2017;9:18246–57.
47. Si-Mohamed S, Tatard-Leitman V, Laugerette A, Sigovan M, Pfeiffer D, Rummeny EJ, et al. Spectral photon-counting computed tomography (SPCCT): in-vivo single-acquisition multi-phase liver imaging with a dual contrast agent protocol. Sci Rep. 2019;9:8458.
48. Si-Mohamed S, Thivolet A, Bonnot PE, Bar-Ness D, Kepenekian V, Cormode DP, et al. Improved peritoneal cavity and abdominal organ imaging using a biphasic contrast agent protocol and spectral photon counting computed tomography K-edge imaging. Invest Radiol. 2018;53:629–39.
49. Si-Mohamed S, Bar-Ness D, Sigovan M, Cormode DP, Coulon P, Coche E, et al. Review of an initial experience with an experimental spectral photon-counting computed tomography system. Nucl Instrum Methods Phys Res A Accel Spectrom Detect Assoc Equip. 2017;873:27–35.
50. Persson M, Wang A, Pelc NJ. Detective quantum efficiency of photon-counting CdTe and Si detectors for computed tomography: a simulation study. J Med Imaging (Bellingham). 2020;7:043501.
51. Wang AS, Harrison D, Lobastov V, Tkaczyk JE. Pulse pileup statistics for energy discriminating photon counting x-ray detectors. Med Phys. 2011;38:4265–75.
52. Hsieh SS, Rajbhandary PL, Pelc NJ. Spectral resolution and high-flux capability tradeoffs in CdTe detectors for clinical CT. Med Phys. 2018;45:1433–43.
53. Ballabriga R, Alozy J, Campbell M, Frojdh E, Heijne EHM, Koenig T, et al. Review of hybrid pixel detector readout ASICs for spectroscopic X-ray imaging. J Instrum. 2016;11:P01007.
54. Hamann E, Koenig T, Zuber M, Cecilia A, Tyazhev A, Tolbanov O, et al. Performance of a Medipix3RX spectroscopic pixel detector with a high resistivity gallium arsenide sensor. IEEE Trans Med Imaging. 2015;34:707–15.
55. Park SJ, Park J, Kim D, Lee D, Lee CL, Bechwati I, et al. The first mobile photon-counting detector CT: the human images and technical performance study. Phys Med Biol. 2023;68:095013.

56. Gallego Manzano L, Monnin P, Sayous Y, Becce F, Damet J, Viry A. Clinical commissioning of the first point-of-care spectral photon-counting CT for the upper extremities. Med Phys. 2023;50:2844–59.
57. Zhan X, Zhang R, Niu X, Hein I, Budden B, Wu S, et al. Comprehensive evaluations of a prototype full field-of-view photon counting CT system through phantom studies. Phys Med Biol. 2023;68:175007.
58. Si-Mohamed SA, Boccalini S, Lacombe H, Diaw A, Varasteh M, Rodesch PA, et al. Coronary CT angiography with photon-counting CT: first-in-human results. Radiology. 2022;303:303–13.
59. van der Werf NR, Si-Mohamed S, Rodesch PA, van Hamersvelt RW, Greuter MJW, Boccalini S, et al. Coronary calcium scoring potential of large field-of-view spectral photon-counting CT: a phantom study. Eur Radiol. 2022;32:152–62.
60. Greffier J, Si-Mohamed SA, Lacombe H, Labour J, Djabli D, Boccalini S, et al. Virtual monochromatic images for coronary artery imaging with a spectral photon-counting CT in comparison to dual-layer CT systems: a phantom and a preliminary human study. Eur Radiol. 2023;33:5476–88.
61. van der Werf NR, Rodesch PA, Si-Mohamed S, van Hamersvelt RW, Greuter MJW, Leiner T, et al. Improved coronary calcium detection and quantification with low-dose full field-of-view photon-counting CT: a phantom study. Eur Radiol. 2022;32:3447–57.
62. Almqvist H, Crotty D, Nyren S, Yu J, Arnberg-Sandor F, Brismar T, et al. Initial clinical images from a second-generation prototype silicon-based photon-counting computed tomography system. Acad Radiol. 2024;31:572–81.

Chapter 2
Spectral Photon-Counting Computed Tomography (SPCCT) for Cardiac and Aortic Imaging

Sara Boccalini, Salim Aymeric Si-Mohamed, Guillaume Fahrni, Benjamin Longere, Christos Gkizas, François Pontana, Aurelien Bustin, Hubert Cochet, and Philippe Douek

Contents

S. Boccalini (✉) · S. A. Si-Mohamed · P. Douek
Université Claude Bernard Lyon 1, Université Claude Bernard Lyon 1, Lyon, France

Hospices Civils de Lyon, Radiology, Lyon, France
e-mail: sara.boccalini@chu-lyon.fr

G. Fahrni
CHUV Lausanne, Radiology, Lausanne, Switzerland

B. Longere · C. Gkizas · F. Pontana
Université de Lille, Université de Lille, Lille, France

CHU de Lille, Radiology, Lille, France

A. Bustin
Université de Bordeaux, Université de Bordeaux, Bordeaux, France

H. Cochet
Université de Bordeaux, Université de Bordeaux, Bordeaux, France

CHU de Bordeaux, Radiology, Bordeaux, France

S. A. Si-Mohamed, K. K. Iniewski (eds.), *Clinical Applications of Spectral Photon Counting Computed Tomography Technology*,
https://doi.org/10.1007/978-3-031-85078-3_2

Abbreviations

CAC	Coronary artery calcium
ECV	Extracellular volume
PCCT	Photon counting CT
SPCCT	Spectral photon-counting CT
TAVI	Transcatheter aortic valve implantation
VMI	Virtual mono-energetic images

2.1 Coronary Artery Disease

2.1.1 Calcium Score

The benefits of PCCT for imaging of vascular calcifications have been studied in different in-vitro and ex-vivo settings. One of the reasons motivating these preliminary studies is the consequences, expected, likely and presumed, for coronary artery calcium (CAC) assessment [1]. CAC score based on the Agatston method is a well-established, relatively easy-to-perform and interpret CT exam that can provide an assessment of the cardiovascular risk of the person undertaking it. The calculation of this score is standardised and always performed on unenhanced prospectively ECCG-triggered CT acquisitions based on the density and volume of the calcifications [2].

The results of these investigations with PCCT confirmed what might have been expected in light of the established features of PCCT. Indeed with PCCT, the volume of calcifications is closer to the physical one as compared to results obtained with EID-CT, mainly due to improved spatial resolution and reduction of blooming artefacts [3, 4]. In addition, with PCCT smaller and less dense calcifications are detectable, CNR is improved and dense calcifications are visible at lower radiation doses as compared to EID-CT [4–6].

Although all the different aspects mentioned above are deemed to have a profound impact on CAC assessment, these are yet to be investigated in human cohorts. Surely, CAC is such a codified and established method to assess cardiovascular risk that robust studies in large populations will be necessary to change the status quo.

More recently, different groups have investigated CAC score assessment with PCCT in clinical studies, especially the possibility of computing this variable from injected coronary artery CTA acquisitions, that is, sparing patients the dedicated unenhanced acquisition. In fact, intrinsic spectral properties of PCCT can be used to derive CAC values from contrast-enhanced studies by virtually eliminating the contrast media. Human studies have shown that there is a good correlation (reported values of 0.98 [95% CI: 0.97, 0.99]) between results obtained with true noncontrast acquisitions (TNC) and images derived from injected scans [7]. Nevertheless, 26% of patients were reclassified with the virtual method as compared to the established one, TNC. Of particular concern are patients with CAC = 0 on TNC for whom virtually derived images identified calcifications in a significant number of cases, 37–46%, depending on the algorithm employed [8]. In addition, in the same study, different algorithms provided different results, not only for CAC = 0 patients but also for those with calcifications.

Summary/Conclusion *Detection and quantification of vascular calcifications yield different results with PCCT as compared to EID-CT. The impact of these differences on clinical practise is still unknown. Although PCCT allows for the calculation of calcium score from injected acquisitions by virtually removing iodine, results are different from what is obtained on true non-contrast images and should be taken with caution.*

2.2 Stenosis Assessment

Nowadays CT is regarded as the first-line imaging modality of choice to investigate stable coronary artery disease in patients with low to intermediate pre-test risk [9].

The first clinical study on coronary CTA with PCCT showed improved image quality, as compared to EID-CT, for arteries, plaques and, in fact, all other cardiac structures [10]. Additionally, as expected, with PCCT calcifications presented sharper borders and fewer blooming artefacts. Ever since these first findings, successive studies have explored the effects of this improved image quality and blooming artefact reduction for stenosis assessment. Indeed, once atherosclerosis is depicted on a CT coronary angiography, one of the most important aspects to be assessed is the degree of stenosis caused by the plaques since this parameter will guide the management of the patient [11].

Based on this first experience, few studies have recently explored the potential of PCCT for high- to very high-risk populations. A study focusing only on calcified plaques in patients with mild stenosis found that the degree of stenosis is reduced with PCCT as compared to EID-CT, for instance, from 34% to 25% [12]. Nevertheless, the correct estimation of stenosis with CT in patients with heavily calcified plaques not only is more challenging but yields more impactful clinical and management implications as well. In fact, the presence of blooming artefacts makes it impossible to visualize the real volume of calcifications and, thus, assess the patency of the lumen. Therefore, the use of coronary CTA for patients with heavily calcified arteries is even contraindicated by some guidelines [13, 14]. With PCCT this type of analysis has been shown to be easier and more reliable thanks to the above-mentioned features (Fig. 2.1). For instance, in high-risk patients

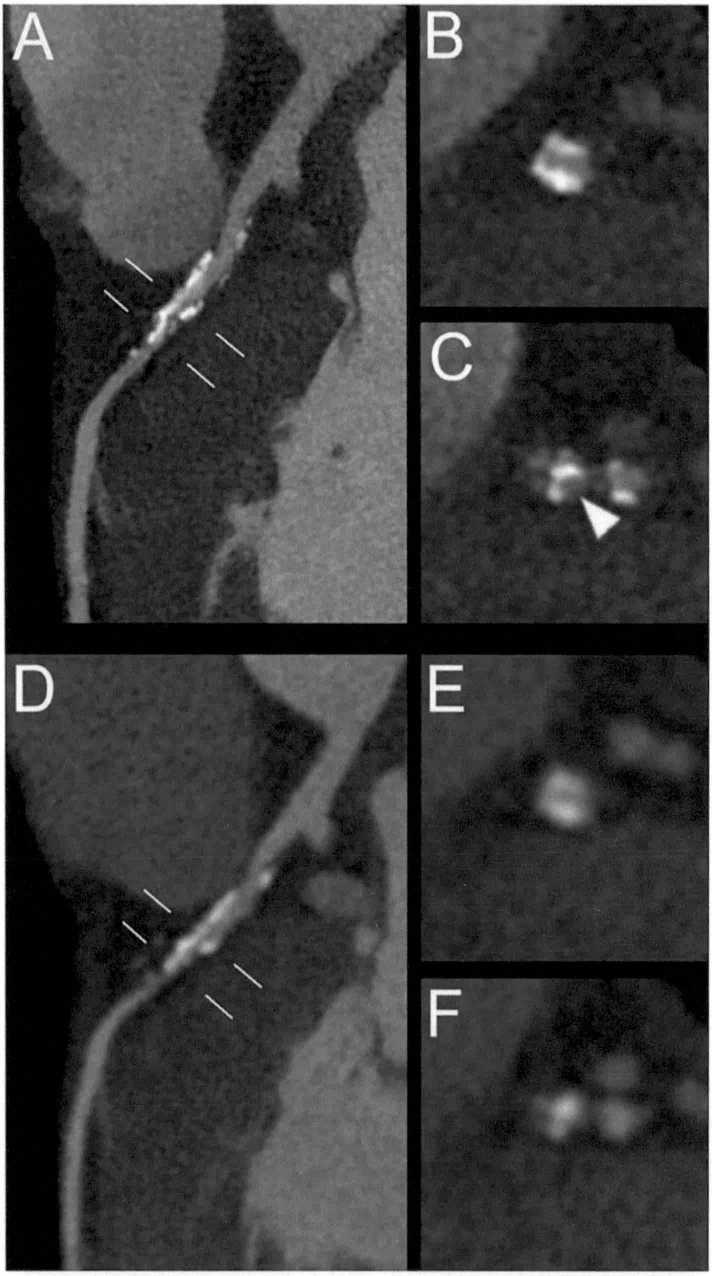

Fig. 2.1 Heavily calcified coronary artery and stenosis assessment with PCCT and EID-CT. (**a**–**c**) Photon counting CT ([PCCT], Philips Healthcare) and (**d**–**f**) dual-energy CT (DECT) images showing heavy calcifications of the left anterior descending coronary artery proximal and distal to the origin of the septal and diagonal branches of a 70-year-old woman that underwent coronary CT

undergoing CT examinations for pre-procedural planning before transcatheter aortic valve implantation (TAVI), PCCT showed very good performances for the assessment of the degree of coronary artery stenosis compared to invasive coronary angiography, with areas under the curve ≥0.92 for patient, vessel and segment based analysis [16]. In another population at high risk for coronary atherosclerosis [17], PCCT showed much better capacities for stenosis estimation than EID-CT, allowing for correct reclassification of 38% of cases (using invasive coronary angiography as the reference). Importantly, PCCT also increased the inter-reader variability from 0.65 of the intra-class correlation coefficient (ICC) with EID-CT to 0.94 with PCCT. In addition, a recent study has pushed the indication of CCTA for very high-risk patients following an acute myocardial infarction [17]. The present study has evaluated PCCT for ultra-high-resolution CCTA in very high-risk patients, comparing it to conventional CT using invasive coronary angiography (ICA) as the reference. Twenty-six patients with 26 coronary stenoses were assessed, 73% of which were >50%. PCCT demonstrated superior performance with a lower mean error (6% vs. 12%), higher sensitivity (100% vs. 75%), and greater specificity (90% vs. 50%) compared to conventional CT. PCCT reclassified 38% of stenoses, while conventional CT reclassified 4%. These findings show that UHR PCCT provides improved accuracy and diagnostic performance for quantifying coronary stenosis in very high-risk patients.

All these clinical studies have been conducted taking advantage of the improved spatial resolution of PCCT systems using the so-called "ultra-high resolution" mode, that is, with adapted FOVs, high matrix size, thin slice reconstructions resulting in almost isotropic voxels with dimensions in the order of 0.2–0.25 mm. Another important aspect to be taken into account is the choice of the kernel that should be adapted to the small size of these voxels. Indeed, using PCCT systems without ultra-high resolution or adapted kernels result in worse image quality, more blooming artefacts [18] and worse performances for coronary artery stenosis assessment [19].

to investigate the cause of typical chest pain. PCCT images were reconstructed with a matrix of 1024 and voxels of 0.29 × 0.29 × 0.25 mm whereas DECT images had a matrix of 512 with isotropic voxels of 0.67 mm. Curvilinear reconstructions of the (**a** and **d**) left main and left anterior descending coronary arteries show the two locations (white lines) where doubts arose regarding the presence of a significant stenosis. (**b** and **e**) The proximal view and (**c** and **f**) distal view cross-sectional images at these two locations show sharper, more defined and less bulky calcifications with (**b** and **c**) PCCT compared to (**e** and **f**) DECT. From a clinical point of view, this improved visualization of calcification confirmed the presence of (**b** and **e**) a significant proximal stenosis. Whereas, with (**e**) DECT the patency of the vessel was doubtful and the (**b**) PCCT images show the permeability of the vessel. More distally, after the origin of the branches, (**c**) PCCT allowed for the diagnosis of stenosis due to the calcifications and afforded good visualization of the patent lumen (arrow). On (**f**) DECT the distinction between calcifications and contrast material was not possible. (Clinical prototype from Philips Healthcare. With permission from Douek P and Boccalini S et al. Radiology 2023 [15])

Spectral properties of PCCT can be exploited in the attempt to virtually remove plaque calcifications. The results of algorithms trying to achieve this goal depend on their capacity to separate iodine and calcification spectra. However, this is very difficult, especially for low-density calcifications, since the spectra of these elements are quite similar. Recently, a clinical study has shown that the estimation of stenosis after virtual removal of calcifications is closer to results with invasive coronary angiography as compared to that calculated on VMI equivalent to conventional ones [20]. Nevertheless, very importantly, 12% of plaques had to be excluded from analysis since the algorithm failed to completely remove the calcifications, highlighting the limitations of this technique.

Another, seemingly more robust, way to use spectral images is to use low energetic VMI to increase the attenuation of iodinated contrast media and therefore reduce the dose of contrast media necessary to achieve diagnostic values in the lumen of the coronary arteries. Reducing the dose of 40% from a starting volume of 72–85.2 mL of 370 mg iodine/mL (the local standard injection protocol for coronary artery CTA) yielded lower attenuation values in the aorta but diagnostic image quality in all cases but one. In this single case, it was judged that motion artefacts hampered the results [21].

The next step of VMI using PCCT is to combine the UHR capabilities of this technology to boost the attenuation of the iodinated contrast agent. By combining spectral sensitivity with lower virtual mono-energetic imaging, spectral photon-counting CT (SPCCT) would enhance coronary artery assessment, potentially leading to more accurate diagnoses and better patient outcomes in cardiovascular imaging. As an example, a recent study demonstrated that UHR mode can be combined with 40 keV VMIs from a clinical SPCCT prototype [22]. In the present study, 40 keV outperformed 70 keV VMIs from SPCCT and both 40 and 70 keV VMIs from a dual-energy CT (DECT) for coronary stenosis evaluation. In addition, UHR 40 keV VMIs reduced measurement variability (6% limits of agreement) and provided superior image quality, whereas DECT showed significant variability (16% limits of agreement) and increased blooming artifacts, especially with calcified plaques. Hence, SPCCT's ultra-high resolution mitigated blooming artifacts and improved image noise scores. The findings support transitioning to UHR low-keV VMIs for better quality, with potential benefits like reduced contrast media use in clinical practice.

Summary/Conclusion *There is enough evidence today to state that PCCT allows for more accurate and reproducible coronary artery stenosis quantification as compared to EID-CT, even in patients at high risk of atherosclerosis and with heavily calcified coronary arteries. Contrast media reduction and use of low mono-energetic images to increase the attenuation of the lumen of coronary arteries with PCCT is feasible (as was already the case with dual-energy EID-CT). However, the separation of iodine and calcification with material decomposition remains a challenge also for PCCT.*

2.3 Plaque Composition Assessment

Besides the degree of stenosis, another important aspect of atherosclerosis of coronary arteries that has to be assessed is the composition of the plaque [11]. Indeed, the presence of specific components and their localisation in the plaque has been linked to a higher risk of plaque rupture and, thus, of myocardial infarction. In addition, it has been shown that, depending on the composition of the plaques, the response of atherosclerosis to medical therapy will be different [23]. Therefore, the need for a non-invasive method allowing to distinguish and quantify in a reliable way the different portions of the plaques as well as to monitor their changes with therapy is increasingly felt.

Ex-vivo studies have shown promising results for plaque component differentiation. For instance, an analysis of 23 plaques with PCCT compared to histological results showed that soft components (vessel wall, perivascular and intraplaque fat) as well as calcifications and iodine in the plaques could be differentiated by analysing photoelectric, Compton scatter and iodine maps [24].

More recently, for ex-vivo carotid arteries, it was demonstrated that different behaviours of the plaque components at progressive energetic levels of VMI allowed for differentiation of haemorrhage and thrombus [25]. It should be noted that, on the contrary, other tissues such as fat, fibrotic cap, lipid core and necrosis presented overlapping patterns and could not be distinguished.

Reconstruction parameters, including slice thickness, kernel [26] and, for spectral images, the level of energy of VMI [27], have an important impact on the relative quantification of plaque components. For instance, with ultra-high resolution and sharp kernels, the relative proportion of soft components was higher as compared to that of reconstructions with different parameters due to the reduction of the blooming artefacts of the calcifications. Nevertheless, these studies did not compare the PCCT results to histology or any other method, so their messages remain limited to alerting on the fact that these parameters need to be taken into account and further evaluated to identify the optimal ones.

Summary/Conclusion *Although promising studies on plaque characterisation with PCCT have been conducted, so far there is not enough evidence to indicate that the composition of coronary artery plaques should be assessed differently to what is currently done based on evidence with EID-CT.*

2.4 Coronary Stents

Together with calcifications, stents represent one of the major challenges for cardiac imaging. This is mainly due to their metal structure and consequent beam hardening and blooming artefacts. For small stents these constraints are even more important and their lumen is oftentimes entirely masked by artefacts. For these reasons, most guidelines state that stents with a diameter ≤3 mm cannot be evaluated with CT [13, 14].

Different coronary stents have been imaged in-vitro with PCCT in different experimental settings. All these studies reported the same main findings: with PCCT, the structure of the stent is better depicted, the dimensions of the struts are closer to reality, calcifications adjacent to the stent are easier to detect and, possibly most important, the artefacts inside the stents are reduced allowing for improved assessment of any intra-stent stenosis [28–32]. Phantom studies have also confirmed the importance of choosing the correct reconstruction parameters to exploit PCCT properties: high matrix, thin slice thickness and very sharp kernel [33, 34]. With these parameters, intrastent restenosis is easier to detect, notwithstanding the subsequent increase in noise. Additionally, specific kernels dedicated to stent analysis have been developed by vendors [35].

Most of these in-vitro results were confirmed in a small series of 16 stents in 18 patients in the first clinical study in 2021 [36] (Fig. 2.2). Indeed, blooming and intrastent artefacts were objectively reduced thanks to PCCT. Observers judged the image quality of the stents and the surrounding tissues obtained with PCCT better than that of EID-CT. Interestingly, calcifications of the plaques of the native coronary artery wall were identifiable with PCCT and distinguishable from the structure of the stents. However this and the other first clinical studies were mainly focused on image quality and did not investigate the performance of PCCT in the detection of intrastent restenosis, the most common long-term complication of coronary stent placement [37]. More recently, some data about the detection of intrastent restenosis has become available [16, 38, 39]. As an example, a recent clinical case of a 66-year-old man with left anterior descending stent implantation and suspected restenosis has illustrated how SPCCT allowed a comprehensive assessment of the coronary lesions in comparison to optical coherence tomography [40]. More in particular, the stent structure, a focal suboptimal expansion, neointimal hyperplasia, and a stent gap were detectable, explaining the clinical symptoms of the patient.

Overall, these works showed very good performances of PCCT as compared to invasive coronary angiography as a reference. In fact, accuracy values were included in a range of 87–93%. This is a very important improvement compared to the performances of EID-CT for which reported values of positive predictive values were of <50% even with last-generation, high-end CT systems [41]. Nevertheless, exams with insufficient image quality and complex stents were excluded from some of these analyses, limiting the generalisability of these findings.

Summary/Conclusion *PCCT offers an improved image quality of coronary stents as compared to EDI-CT. In particular, blooming and intrastent artefacts are reduced. Although the first results for intrastent restenosis detection are promising, more real-life data is needed to affirm the generalisability of these findings.*

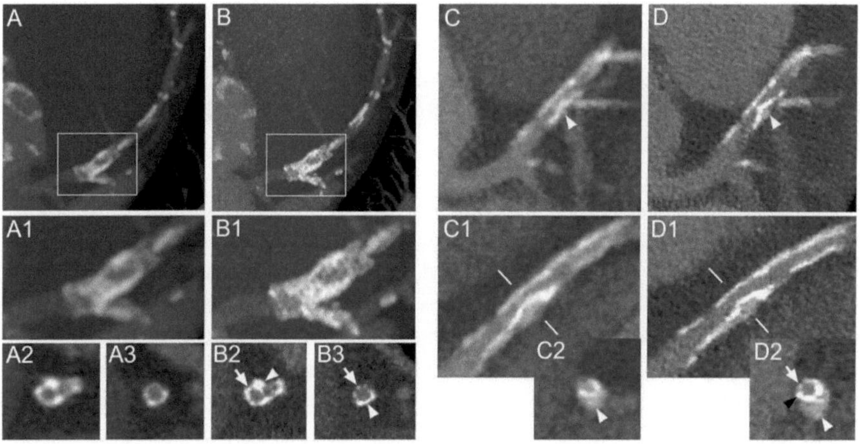

Fig. 2.2 Coronary stents and calcifications of the native coronary arteries. Panels (**a**) (EID-DLCT—XCD) and (**b**) (SPCCT—sharp) show a drug-eluting stent of the LAD, extending over the origin of the diagonal, of a 69-year-old man deployed in 2007. On MIP (**a** and **a1**; **a1** corresponds to the white box in **a**) and cross-axial (**a2** and **a3** at the level of the origin of the diagonal and a bit further downstream, respectively) reconstructions of the EID-DLCT some hyperdense images on the walls of the coronary images could represent either calcifications or struts since the stent structure is undistinguishable. Corresponding images on PCCT (MIP in **b** and **b1**; **b1** corresponds to the white box in **b**; cross-sectional in **b2** and **b3**) nicely show the stent struts (arrows) encrusted with calcifications (arrowheads). In panels (**c**) (EID-DLCT—XCD) and (**d**) (SPCCT—sharp) a 3.5 × 12 mm Synergy stent of the LAD, extending over the origin of the diagonal, was implanted 5 months prior to the CT scans. The stent had been placed over a rather voluminous calcification (white arrowheads) that is deforming the stent lumen as it is visible on EID-DLCT (**c**, **c1** and **c2**; **c2**: cross axial image at the level of the white lines on **c1**) as well as on corresponding SPCCT images (**d**, **d1** and **d2**). The lumen of the stent is hard to assess on EID-DLCT while it is clearly reduced in calibre but patent on SPCCT. Furthermore, on the SPCCT cross axiacl image (**d2**) the stent struts are nicely visible on one side (arrow) while on the other side the calcification has one peripheral less dense component (white arrowhead) and a more dense layer closer to the stent (black arrow). (Clinical prototype from Philips Healthcare. With permission from Boccalini S and Si-Mohamed S et al. Investigative Radiology 2021 [36])

2.5 Myocardial Tissue Characterization

Tissue characterization in cross-sectional imaging is considered to be the domain of CMR. However, the feasibility of studying myocardial late enhancement by CT (CT-MDE) with a semiology similar to that observed on CMR has been described for almost 20 years [42, 43]. Unfortunately, when compared to CMR as a gold standard, the detection of myocardial fibrosis/scar on late iodine-enhanced CT images is challenging due to a high level of noise and a low contrast between the lesions and the remote myocardium, as well as between the lesions and the blood pool. In addition, several acquisition and reconstruction parameters require further studies to identify the ones yielding optimal lesion detectability.

For instance, the amount of iodine to administrate for CT-MDE is not standardized and values ranging from 300 to 1000 mgI/kg have been reported [43]. In addition, there is no consensus on the optimal timing for CT-MDE acquisition on CT but a 3- to 10-min delay is the most frequently reported [43, 44]. In order to calculate the extracellular volume fraction (ECV), a 5- to 7-min delay appears to be a good compromise [44–46].

Although there is still very little published data, most being in the form of case reports [47], the impact of PCCT on myocardial late iodine enhancement image quality is expected to be relevant. Indeed, the reduction of electronic noise with PCCT makes it possible to explore the myocardium on spectral reconstructions at low energy and on iodine maps, while these spectral reconstructions were severely impacted by noise on conventional spectral CT systems [48]. PCCT can inherently provide spectral data, allowing for virtual mono-energetic images, iodine maps, iodine quantification, and improved tissue characterization [49, 50]. This is particularly advantageous in CT-MDE imaging, where iodine-based contrast agents highlight areas of myocardial damage through direct visualization or ECV maps (fibrosis, oedema) [51, 52].

The distribution space of iodinated contrast media (ICM) is represented by the ECV, that is, the intravascular sector without blood cells, estimated by the haematocrit and the interstitial space. The ECV fraction can be calculated from the equilibrium distribution of the ICM (5–10 min) according to the following equation [46]:

$$\mathrm{ECV} = \frac{\Delta \mathrm{HU}_{\text{myocardium}}}{\Delta \mathrm{HU}_{\text{blood pool}}} \times \left(1 - \text{Hematocrit}\right)$$

where ECV, $\Delta HU_{\text{myocardium}}$, and $\Delta HU_{\text{blood pool}}$ represent extracellular volume, change in attenuation of the myocardium, and change in attenuation of the blood pool, respectively.

PCCT allows for the direct calculation of iodine maps that represent the changes in attenuation (ΔHU) in both myocardium and blood since they are equivalent to the subtraction of CT-MDE and unenhanced series. Besides the dose and noise reductions achieved by PCCT, this spectral approach avoids the risk of misregistration during the subtraction process and thus yields a higher correlation with ECV values obtained from CMR [53].

If not available from blood tests, haematocrit can be calculated from virtual non-contrast series as follows [54]:

$$\text{Synthetic Hematocrit} = 0.0093 \times \mathrm{HU}_{\text{blood pool(VNC)}} - 0.053$$

where $HU_{\text{blood pool (VNC)}}$ represents the attenuation of the blood pool on virtual non-contrast series. Note that this formula is different from the one used with energy-integrating detectors CT [55].

2.6 Acute Pathologies: Myocardial Infarction and Myocarditis

In the setting of acute myocardial infarction, CT-MDE has been demonstrated to be accurate for the assessment of scar size and myocardial viability and for the prediction of major cardiovascular events over a follow-up of 2 years [56–59]. As for chronic myocardial infarction, poor sensitivity (53%) but excellent specificity (98%) has been reported for CT-MDE. Spectral features of PCCT offer the prospect of improving CT-MDE contrast resolution and thus increase its sensitivity. Cases of myocardial infarction scar assessment thanks to CT-MDE or ECV acquired with PCCT have been reported [51, 60–62]. In addition, the iodine map was reported to be helpful in depicting small left ventricular thrombi adhering to the myocardium due to the higher contrast between the thrombus and the blood pool as compared to conventional images [63].

In the framework of acute chest pain and troponin rise with no obstructive coronary artery disease (CAD), CT-MDE could identify the cause of symptoms in 81% of patients, most of the cases being acute myocarditis (Fig. 2.3) [64]. Indeed, CT-MDE obtained from spectral acquisition was reported to provide a 95% accuracy as compared to late gadolinium enhancement [65]. In addition to CT-MDE, the calculation of ECV is helpful for the identification of the aetiology of myocardial damage [64]. Based on a CMR study, regional ECV calculation within segments demonstrating myocardial delayed enhancement would help to differentiate myocarditis from cardiac sarcoidosis, the latter being responsible for higher ECV values (cut-off: 57%) [66]. Conversely, global ECV values did not differ between sarcoidosis and myocarditis.

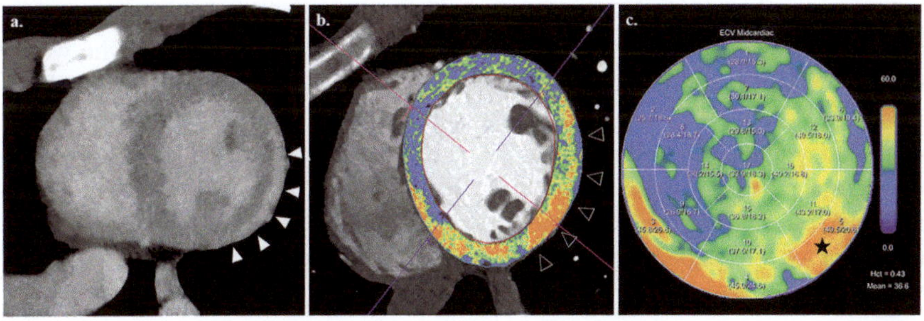

Fig. 2.3 Acute myocarditis. An example of a 32-year-old woman with acute myocarditis. Myocardial delayed enhancement (**a**) demonstrating subepicardial enhancement (white arrowheads). Extracellular volume (**b**) overlay demonstrating elevated values that confirm oedema (black arrowheads). The extracellular volume polar map (**c**) showing the same area of increased extracellular volume (star). (Naeotom Alpha, Siemens Healthineers)

2.7 Myocardial Ischemia

Stress CT myocardial perfusion imaging (CT-MPI) has been validated against reference standards such as invasive fractional flow reserve [9]. Stress is induced by pharmacological agents such as adenosine or regadenoson to add a functional dimension to the anatomical coronary CTA test [67]. Nevertheless, to date, no data regarding stress CT-MPI on PCCT has been published in the literature. However, the presence of areas of hypoattenuating myocardium and reduced iodine concentration in case of myocardial infarction have been described [62]. A similar approach can be used for the assessment of first-pass perfusion. Commercially available PCCT machines do not allow dynamic stress CT-MPI, and thus, only the static approach is possible. However, static CT-MPI is not inferior to the dynamic quantitative approach [68]. The optimal timing for CT-MPI acquisition should be at the peak of the myocardial enhancement curve [69]. Various delays and attenuation thresholds have been proposed for CT-MPI triggering [70–72]. The authors would recommend a 12-second delay after the beginning of ICM bolus injection and the use of iodine maps for iodine quantification (Fig. 2.4) [73].

2.8 Cardiomyopathies

It has been demonstrated that spectral acquisition of CT-MDE can adequately differentiate myocardial late enhancement patterns against CMR [74]. Intramyocardial enhancement of fibrotic scars benefits from the improved contrast allowed by PCCT and should be more conspicuous in the case of a thicker myocardial wall (Fig. 2.5), namely in the case of hypertrophic cardiomyopathy or concentric remodelling [47, 75].

Transthyretin cardiac amyloidosis has been reported to be prevalent in up to 16% of patients referred for transcatheter aortic valve replacement (TAVR), who would benefit from the additional acquisition for CT-MDE assessment [76]. Indeed, increased ECV on TAVR work-up CTA is associated with impaired recovery after treatment and underlines the prognostic value of this biomarker [75]. Moreover, in the case of cardiac amyloidosis, CT-MDE is accepted as an alternative to CMR, which gives access to the ECV calculation [77]. Indeed, an ECV > 37% is a strong argument for the diagnosis of cardiac amyloidosis and an ECV > 45% is associated with an increased risk of adverse events [78].

Summary/Conclusion *Even if CMR remains the reference method for tissue characterization, PCCT may facilitate the integration in future practice of CT-MDE and CT-derived ECV, which are powerful and promising biomarkers.*

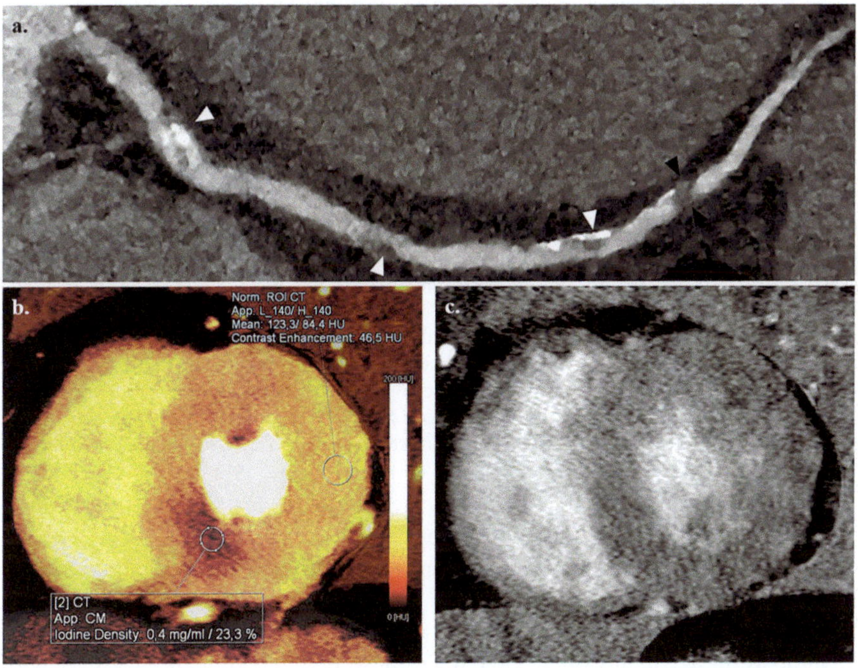

Fig. 2.4 Hypoperfused myocardium on first-pass perfusion CT imaging. A 66-year-old man with exertional dyspnoea, coronary CT angiography (**a**) demonstrates right coronary artery mixed plaques (white arrowheads) and focal total occlusion (black arrowheads). Iodine map (**b**) for quantitative assessment. Regadenoson first-pass myocardial perfusion demonstrates mid-inferior septal hypoattenuation with decreased iodine concentration (23% of that measured in remote myocardium). No myocardial infarction scar on iodine map (**c**) of the myocardial delayed enhancement acquisition. (Naeotom Alpha, Siemens Healthineers)

2.9 Arrhythmic Diseases

2.9.1 Detection and Diagnosis of Ventricular Arrhythmogenic Diseases

The role of cardiac computed tomography in the management of patients with ventricular arrhythmias has expanded over the last decade. The technique is employed to rule out obstructive coronary artery disease in patients with exercise-induced arrhythmias, to plan catheter ablation procedures and to detect and diagnose myocardial arrhythmogenic substrates in patients equipped with implantable cardioverter defibrillators (ICDs), or in those unable to undergo CMR. Myocardial arrhythmogenic substrates (of ischemic or non-ischemic origin) mostly consist of fibrosis and/or fat intermingled with surviving cardiomyocytes exhibiting slow conduction properties, exposing them to circuit re-entry and leading to ventricular tachycardia. These tissue changes are poorly visible on first-pass arterial-enhanced

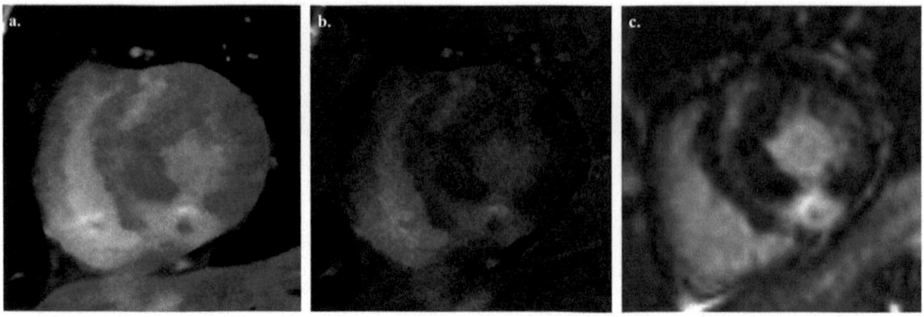

Fig. 2.5 Myocardial fibrosis in hypertrophic cardiomyopathy. A 52-year-old woman with hypertrophic cardiomyopathy. Septal myocardial fibrosis is conspicuous on a short axis view of a virtual mono-energetic image at 40 keV of myocardial delayed enhancement (**a**) The iodine map of the same plane (**b**) shows the same lesions. There is a good concordance with delayed gadolinium enhancement on cardiac magnetic resonance (**c**), short axis (same level). (Naeotom Alpha, Siemens Healthineers)

CT images and require the addition of late iodine-enhanced acquisitions. For the reasons already mentioned in the paragraphs above, PCCT is expected to increase the detectability of these arrhythmogenic substrates, unlocking the full benefit of spectral imaging. This may directly apply to the guidance of ventricular tachycardia ablation, as these patients often carry ICDs, preventing good-quality imaging with CMR. An example of 3D whole heart modelling from PCCT images before catheter ablation of ventricular tachycardia is shown in Fig. 2.6. Pending specific developments in right ventricular wall segmentation, the spatial resolution of PCCT systems should also allow for the assessment of right ventricular fibrosis, for instance, in arrhythmogenic right ventricular cardiomyopathy or cardiac sarcoid, potentially going beyond the reach of CMR, which is still limited by its spatial resolution for the assessment of the thin right ventricular wall. In addition, virtual mono-energetic images at high energy can be employed to improve image quality in the vicinity of cardiac metallic implants [79], often present in this population.

2.9.2 *Quantification of Myocardial Fibrosis*

In addition to the detection issue, the quantification of fibrosis and the objective assessment of its heterogeneity has been a long-standing conundrum in cardiac imaging. Fibrosis quantification is now recommended for the primary prevention of sudden cardiac death in non-ischemic diseases [80], but current late gadolinium-enhanced CMR techniques, being based on a relative contrast, fail to provide robust and standardized quantifications. Likewise, the detection of channels of grey zone within dense scar has been proposed to detect the isthmus of scar-related re-entrant

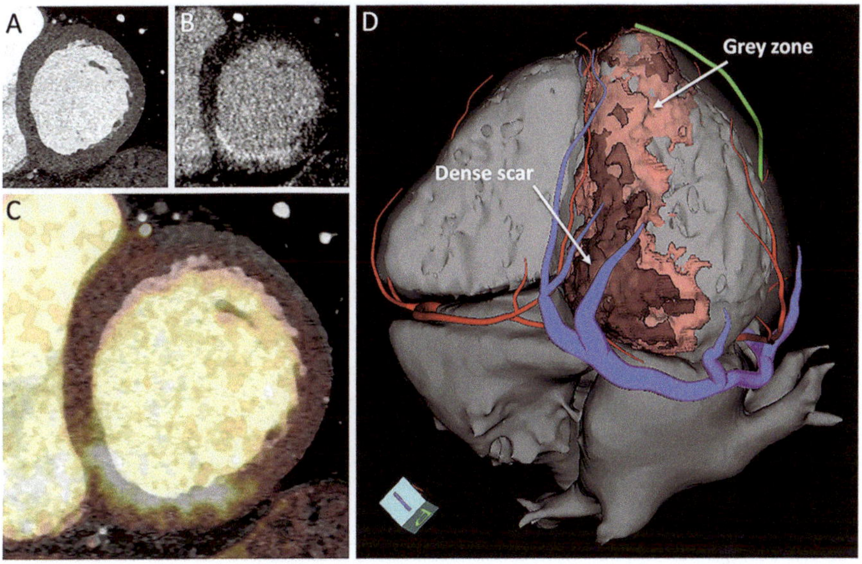

Fig. 2.6 3D whole heart modelling from PCCT images before catheter ablation of ventricular tachycardia. A 55-year-old man with post-infarction ventricular tachycardia referred for catheter ablation. (**a**) arterial-enhanced CT image. (**b**) late-enhanced CT image. (**c**) fused image. (**d**) 3-dimensional model used for catheter ablation guidance. (Naeotom Alpha, Siemens Healthineers)

tachycardia to be targeted by ablation [81], but the thresholds employed to characterize these tissue components on CMR are still largely debated. Furthermore, the spatial resolution of CMR sequences (1–2 mm) may be insufficient to detect smaller channels. PCCT allows for the reconstruction of ECV maps from single spectral acquisitions [51, 52]. This enables robust and reproducible quantification of fibrosis of the entire myocardium and might also allow for an objective definition of dense scar and grey zone at a regional level. Besides the quantification of fibrosis in chronic structural heart diseases, this approach could also be applied to quantify infarct size and microvascular obstruction after acute myocardial infarction [61], with important prognostic implications. Moreover, these ECV maps are reconstructed with parameters resulting in spatial resolution never achieved before in vivo, and approaching the microstructural scale of the myocardium (the length of a cardiomyocyte is around 100 μm, and the its thickness around 150 μm). This will dramatically improve our ability to detect arrhythmogenic substrates, with the potential to unravel novel disease entities (for instance, in patients experiencing sudden death with apparently normal hearts). Figure 2.7 presents a direct comparison between PCCT and CMR in a patient with ventricular tachycardia secondary to hypertrophic cardiomyopathy.

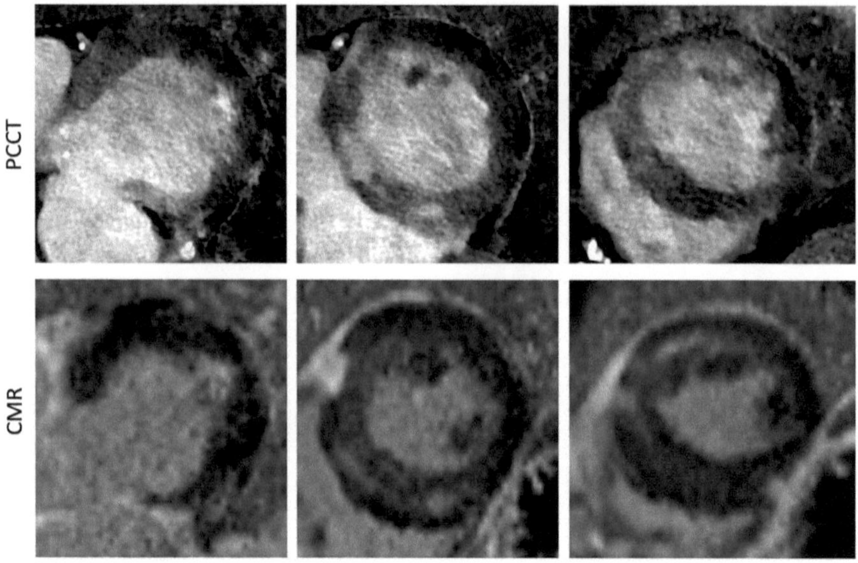

Fig. 2.7 Direct comparison between PCCT and CMR in a patient with ventricular tachycardia secondary to hypertrophic cardiomyopathy. Top row: iodine map acquired 6 min post-contrast using a PCCT system at the reconstructed resolution of 0.4 × 0.4 × 0.4 mm. Bottom Row: late gadolinium-enhanced images acquired using a 1.5T MR system (Magnetom Aera, Siemens Healthineers) using a free-breathing 3D gradient echo sequence at the resolution of 2.5 × 1.25 × 1.25 mm. (Naeotom Alpha, Siemens Healthineers)

2.9.3 Atrial Fibrillation

The role of CT in atrial fibrillation is currently limited to the ruling out of thrombus before cardioversion and to the obtention of detailed anatomy of the left atrium to plan catheter ablation. The structural remodelling of the atrium involves the development of interstitial fibrosis and intramural fat, both playing a major role in arrhythmia maintenance and in the risk of cardio-embolic stroke [82]. Unfortunately, this substrate is currently inaccessible to imaging due to the limited thickness of the atrial wall. Attempts to assess atrial fibrosis using late gadolinium-enhanced CMR failed to significantly improve patient outcomes [83]. PCCT systems could allow for robust measurements of atrial ECV and robust assessment of atrial fat once appropriate and specific segmentation tools adapted to the left atrium will be developed. In addition, high-resolution arterial-enhanced PCCT images may also be leveraged to generate maps of atrial thickness, which can be employed to guide catheter ablation in areas more amenable to transmural ablation [84]. These innovations will likely expand the use of CT in atrial fibrillation and the impact of therapies on patient outcomes.

Summary/Conclusion *Overall, the advent of PCCT technology is likely to deepen our understanding of ventricular and atrial arrhythmias and further expand the*

Table 2.1 Potential PCCT indications in the context of arrhythmias

Potential PCCT indications in the context of arrhythmias	Existing CT indication	New CT indication	Rationale
Ruling out CAD in exercise-induced ventricular arrhythmias	X		– Higher accuracy for the detection of CAD
Pre-procedural planning before catheter ablation of ventricular arrhythmias	X		– Better analysis of fibrosis distribution – More accurate definition of ablation targets
Diagnosing the underlying substrate in ventricular arrhythmias of unknown etiology		X	– Improvements in fibrosis/fat detection
Primary prevention of sudden cardiac death with ICDs in chronic structural heart disease		X	– Robust quantification of myocardial fibrosis – Greater availability of CT over CMR for screening
Primary prevention of sudden death with wearable defibrillators in acute infarction		X	– Quantification of infarct size and microvascular obstruction during hospitalization – Greater availability of CT over CMR for screening
Pre-procedural planning before catheter ablation of atrial fibrillation	X		– Atrial tissue characterization to assist catheter ablation (wall thickness, fat, fibrosis)
Primary prevention of stroke in patients with atrial fibrillation		X	– Atrial tissue characterization (wall thickness, fat, fibrosis) to stratify stroke risk

CAD coronary artery disease, *CMR* cardiac magnetic resonance, *ICD* implantable cardioverter defibrillator, *PCCT* photon counting computed tomography

indications for cardiac CT in this area of cardiology. An overview of potential key indications for cardiac PCCT in patients presenting with (or at risk of) cardiac arrhythmias is shown in Table 2.1.

2.10 Cardiac Valves

CT has been acquiring a more and more central place in the management of patients with cardiac valve pathologies including diagnostic, pre-procedural, pre-surgical, follow up and post-interventional assessment.

In the first paper on cardiac CT with PCCT, the authors also analysed the image quality of cardiac valves and found that it was better than that obtained with EID-CT for the same patients [10]. Recently, some cases were reported showing very good image quality of both the biological [15] (Fig. 2.8) and the metallic components of prosthetic heart valves [49].

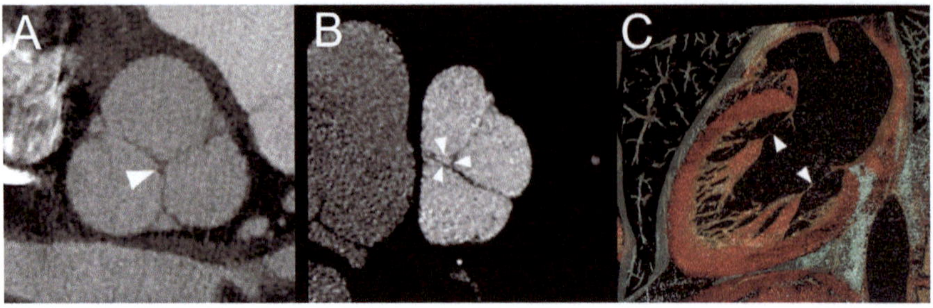

Fig. 2.8 Examples of Cardiac valves imaged with Photon counting CT. (**a**) An aortic valve in a multiplanar reconstruction with a well-depicted nodule of Arantius of the non-coronary cusp (arrowhead) of a 45-year-old man with a history of myocardial infarction in the territory of the left anterior descending. (**b**) A pulmonary valve showing a small orifice of regurgitation (arrowheads) of the same patient whose aortic valve is imaged in panel (**a**). (**c**) A volume rendering showing the mitral valve apparatus, including the chordae (arrowheads) and the trabeculae of the left ventricle (white dotted circle) of a 47-year-old woman with a history of acute coronary artery disease. (Clinical prototype from Philips Healthcare. With permission from Douek P and Boccalini S et al. Radiology 2023 [15])

Nevertheless, so far, the only systematic evaluation of heart valves with PCCT has been performed in vitro [85]. This study demonstrated that PCCT allowed detecting more valve components as compared to EID-CT. In addition, almost all parts of both biological and metallic valves presented better conspicuity and sharpness with PCCT. Metallic and blooming artefacts were less important with PCCT in both subjective and objective analyses. Most importantly, small lesions were better detectable with PCCT, which, if confirmed in vivo, might allow for earlier, easier and more accurate diagnosis of complications of heart valve prosthesis, that are often lethal pathologies.

Other clinical studies with PCCT focused more on pre-interventional planning exams rather than on diagnostic ones. The main goal of these exams is to provide reliable measures of the structures of interest and information about the access route. PCCT seems like a promising technique for these exams also in light of the frailty of some of these patients [86]. CT scans of patients undergoing pre-TAVI assessment with either PCCT or EID-CT systems but with otherwise similar acquisition protocols showed higher CNR and SNR with EID-CT than with PCCT but higher subjective image quality with the latter [87]. In particular, the femoral access route and the three hinge points of the aortic cusps could be assessed with PCCT in 99% of cases against 85% with EID-CT. Between retrospectively ECG-gated acquisitions and ECG-triggered systolic high-pitch acquisitions with PCCT, the first showed better image quality at the expenses of higher radiation doses (DLP (mGy*cm) 902 [736–1071] 292 [262–365]; $P < 0.01$) [88].

Another aim of pre-TAVI exams is to help establish the severity of the aortic stenosis, especially in patients with low flow state [89]. This is normally achieved with quantification of aortic valve calcifications on non-enhanced prospectively triggered acquisitions, as also done for the assessment of coronary artery calcifications. Also in this case, the spectral properties of PCCT might help avoid the acquisition by deriving data on aortic calcifications directly from arterial acquisitions. This was partially confirmed by a recent study conducted on 56 patients with aortic calcifications [90]. The results showed that the correlation between calcium scores of the aortic valve on non-enhanced and arterial acquisitions was good, and in only one case the patient was classified differently in severe or non-sever aortic stenosis. Nevertheless, in eight cases where valvular calcifications were evident on the unenhanced CT, the images with virtually removed iodine did not show any.

2.11 Aortic Diseases

Similar to what is already detailed for coronary arteries, low-energy VMI can help reduce the dose of contrast media for aortic CT scans. For instance, reducing the volume of contrast media by 25% and using 50 keV VMI resulted in non-inferior image quality as compared to conventional images at full contrast dose in patients undergoing an examination of the entire aorta and of the iliac arteries with PCCT [91].

Although there is no systematic data available yet, it is likely that photon counting will allow for a better depiction of prosthetic materials, in particular those including metal. This would possibly allow for easier detection of material corruption and also small lesions of the native wall of the aorta adjacent to the metal and small endoleaks (Fig. 2.9).

Another theoretical advantage of PCCT could be the possibility to reduce the number of acquisitions per patient by deriving non-enhanced images from injected acquisitions thanks to the virtual removal of iodine, as already proposed for dual-energy CT [92]. Nevertheless, complete removal of iodine is difficult to achieve especially using arterial acquisitions as the base. With PCCT, differences of approximately 10 HU were found between TNC and virtual noncontrast images of the aorta derived from portal acquisitions [93]. Interestingly, in the same study, differences were also significant for virtual noncontrast images derived from unenhanced acquisitions, showing the limits of this type of reconstruction. Therefore, given the fundamental role of CT for the diagnosis of acute aortic syndrome, great caution should be exerted when relying on these virtual images.

Another possible application of spectral imaging for aortic imaging is the use of virtual noncontrast reconstructions to distinguish surgical material from iodinated contrast material on post-injection acquisitions. This feature is particularly interesting during follow-up after open surgery or interventions when no previous CT scans are available [94]. Unfortunately, up to date, limited data is available on the subject. One study investigating the removal of EVAR components with different types of virtual removal found that the best one was judged to provide images that could substitute a true non-contrast acquisition in clinical practise in only 75% of cases [95].

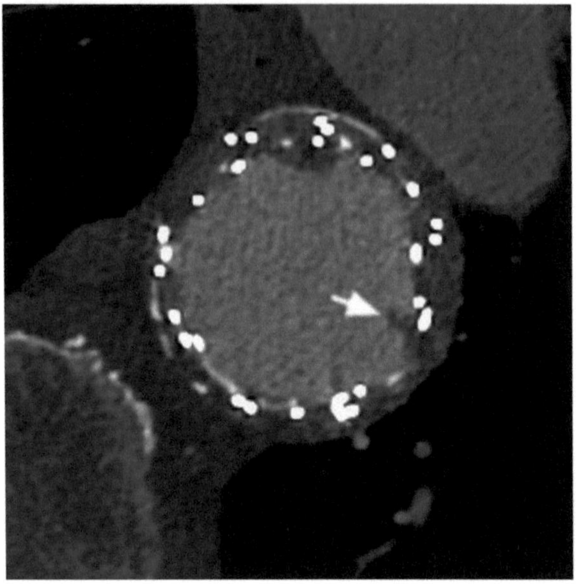

Fig. 2.9 An example of image quality of aortic endoprosthesis of the descending aorta. This axial image passes through a plane where two prostheses overlap. The metallic struts of the two layers are clearly distinguishable. This allows for accurate assessment of the malpositioning of certain parts of the prosthesis. In addition, a fold of the graft of the internal one is clearly depicted (arrow) with PCCT. (Prototype from Philips Healthcare)

2.12 Future Developments

2.12.1 Spectral Imaging

The evaluation of spectral imaging applications for photon counting CT is currently undergoing and the results of some studies on the subject have been reported above in the paragraphs dedicated to specific diseases. As a general consideration, all the applications of spectral imaging that have been reported and investigated with dual-energy CT should also be possible with photon counting [96]. Some of them have already been mentioned in the other specific paragraphs of this chapter.

In addition to all these applications, photon counting might offer some further advantages. For sure, it will be possible to combine ultra-high spatial resolution and spectral properties. As of today, these features can be combined only with some systems and they haven't been studied systematically [10].

2.12.2 K-Edge Imaging

Notwithstanding the possible advancements in spectral imaging with SPCCT, the real revolution in CT is more likely to derive from color K-edge imaging and contrast media development. In fact, possible applications of color K-edge imaging in cardiovascular radiology range from implant imaging to the use of one or more extracellular K-edge contrast media to molecular imaging.

Visualization of implants or their components is the easiest clinical application of K-edge. This would mean exploiting the K-edge of exogenous materials already present in the body of patients, namely metallic implants, to image them. This was possible for stents [29] and for prosthetic heart valves [85] in-vitro. It has been hypothesized that the structure of the stents might be easier to assess and to distinguish from dense calcifications thanks to K-edge. For valves, it has been shown that K-edge images show less metallic and blooming artefact as compared to conventional images with K-edge. Surely, the clinical benefits of this type of imaging in clinical routine still need to be proven in vivo.

Using an extracellular contrast media assessable with K-edge could help solve most problems related to the quantification of the stenosis of arteries, coronary arteries or others. In fact, with color k-edge imaging the contrast media and only the contrast media would be imaged, and if the acquisition was performed during the arterial phase, only the patent lumen of the vessel would fill with contrast, resulting in a purely angiographic image. This was shown to be feasible with K-edge of gadolinium in rabbits with atherosclerosis [97]. A single acquisition allowed reconstructing both conventional and K-edge images, on both of which aortic diameters were measurable without significant differences between reconstructions with similar parameters. On K-edge images, only the lumen of the arteries, even the very small ones of <1 mm, and parenchyma of organs that are vascularized precociously were visible and no atherosclerotic plaques were detectable. Unfortunately, to obtain these results, the dose of contrast media that was administered was higher than what is currently allowed in humans. Nevertheless, research in contrast materials and improvements in the K-edge reconstruction chain are both providing reasons for believing these limitations will be overcome. Recently, the first exemple of gadolinum K-edge in human has demonstrated not only that this technique is feasible but also that it might be possible with commercially available contrast media and with clinically approved doses [98].

Another potential application of K-edge imaging for cardiovascular imaging would be using two contrast materials injected at different times and one single acquisition to highlight their different distribution. This could be interesting, for instance, to distinguish endoleaks after injection of gadolinium and iodine as shown in a phantom study [99]. Surely, in this case gadolinium-based contrast material can be imaged with K-edge and iodinated contrast media could be vibe thanks to the spectral properties of PCCT since its K-edge falls below the threshold of exploitable energies.

Furthermore, multi-material reconstructions allow for the combination of multiple contrast agents (e.g. iodine and gadolinium) [100]. As an example, this may be employed to suppress the blood pool of the left ventricle, thereby facilitating the detection of subendocardial scars.

Another fascinating field for color K-edge applications is molecular imaging. One example of application in this field is the depiction of inflammatory plaques of atherosclerosis. For these experiments, gold nanoparticles have been used since they can be actively phagocytized by macrophages that are found in large numbers inside inflammatory plaques [101, 102]. A few days after the injection of the nanoparticles, K-edge imaging allowed for their visualization in the wall of the

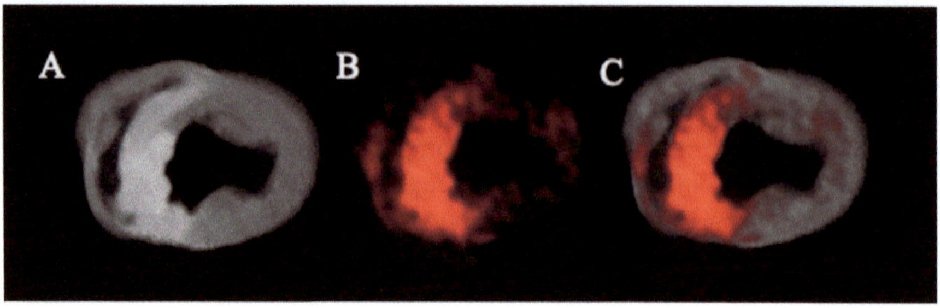

Fig. 2.10 Spectral photon-counting CT imaging of a myocardial infarction in a model of ischemia-reperfusion in pig, using a phase 2 gadolinium-based nanodrug (AGuIX). Contrary to conventional images (**a**) K-edge imaging (**b**) revealed the specific area of the myocardial infarction in the territory of the left anterior descending coronary artery. Overlay image (**c**) showed the specific capability provided by color K-edge imaging despite the underlying healthy myocardial tissue. (Clinical prototype from Philips Healthcare. Under the framework of ANR ATHERSPECTRAL CT)

aorta of the rabbits, indicating the presence of atherosclerosis. Unfortunately, the elimination of gold from human bodies is incomplete which, added to the fact that it is an expensive arterial, will make its use in humans difficult. Another nanoparticle that might be of interest and that is already approved in Europe for human use in Hafnium although literature on the subject is still lacking. Surely, ideas about potential targets are numerous. For instance, in the context of cardiac arrhythmias, the targeted tissues may include adipocytes and collagene (type I and type III) in structural heart diseases, as well as Purkinje cells in "idiopathic" ventricular fibrillation.

Finally, following these exploratory approaches, repurposing human-tested nanodrugs for diagnostic applications represents a way to accelerate the development of color K-edge imaging. For instance, AGuIX®, a phase 2 gadolinium-based nanodrug, shows potential for advancing K-edge imaging of myocardial infarction due to its well-documented distribution in pathological tissues via the enhanced permeability and retention effect [103]. This nanodrug, initially meant to improve the efficacy of radiotherapy of cancer, may hold promise as a theranostic imaging tool, such as for pharmacological postconditioning in myocardial reperfusion injury (Fig. 2.10) [104].

References

1. Willemink MJ, van der Werf NR, Nieman K, Greuter MJW, Koweek LM, Fleischmann D. Coronary artery calcium: a technical argument for a new scoring method. J Cardiovasc Comput Tomogr. 2019;13:347–52. https://doi.org/10.1016/j.jcct.2018.10.014.
2. Agatston AS, Janowitz WR, Hildner FJ, Hildner FJ, Zusmer NR, Viamonte M Jr, et al. Quantification of coronary artery calcium using ultrafast computed tomography. J Am Coll Cardiol. 1990;15:827–32. https://doi.org/10.1016/0735-1097(90)90282-T.

3. Sandstedt M, Marsh J, Rajendran K, Gong H, Tao S, Persson A, et al. Improved coronary calcification quantification using photon-counting-detector CT: an ex vivo study in cadaveric specimens. Eur Radiol. 2021;31:6621–30. https://doi.org/10.1007/s00330-021-07780-6.
4. van der Werf NR, Si-Mohamed S, Rodesch PA, van Hamersvelt RW, Greuter MJW, Boccalini S, et al. Coronary calcium scoring potential of large field-of-view spectral photon-counting CT: a phantom study. Eur Radiol. 2022;32:152–62. https://doi.org/10.1007/s00330-021-08152-w.
5. van der Werf NR, van Gent M, Booij R, van der Lugt A, Budde RPJ, Greuter MJW, et al. Dose reduction in coronary artery calcium scoring using mono-energetic images from reduced tube voltage dual-source photon-counting CT data: a dynamic phantom study. Diagnostics. 2021;11 https://doi.org/10.3390/diagnostics11122192.
6. Symons R, Sandfort V, Mallek M, Ulzheimer S, Pourmorteza A. Coronary artery calcium scoring with photon-counting CT: first in vivo human experience. Int J Cardiovasc Imaging. 2019;35:733–9. https://doi.org/10.1007/s10554-018-1499-6.
7. Haag NP, Michael AE, Lennartz S, Panknin C, Niehoff JH, Borggrefe J, et al. Coronary artery calcium scoring using virtual versus true noncontrast images from photon-counting coronary CT angiography. Radiology. 2024;310:e230545. https://doi.org/10.1148/radiol.230545.
8. Sharma SP, van der Bie J, van Straten M, Hirsch A, Bos D, Dijkshoorn ML, et al. Coronary calcium scoring on virtual non-contrast and virtual non-iodine reconstructions compared to true non-contrast images using photon-counting computed tomography. Eur Radiol. 2024;34:3699–707. https://doi.org/10.1007/s00330-023-10402-y.
9. Vrints C, Andreotti F, Koskinas KC, Rossello X, Adamo M, Ainslie J, et al. 2024 ESC guidelines for the management of chronic coronary syndromes. Eur Heart J. 2024;45:3415–537. https://doi.org/10.1093/eurheartj/ehae177.
10. Si-Mohamed SA, Boccalini S, Lacombe H, Diaw A, Varasteh M, Rodesch PA, et al. Coronary CT angiography with photon-counting CT: first-in-human results. Radiology. 2022;303:303–13. https://doi.org/10.1148/RADIOL.211780.
11. Cury RC, Leipsic J, Abbara S, Achenbach S, Berman D, Bittencourt M, et al. CAD-RADS™ 2.0 - 2022 coronary artery disease-reporting and data system: an expert consensus document of the Society of Cardiovascular Computed Tomography (SCCT), the American College of Cardiology (ACC), the American College of Radiology (ACR), and the North America Society of Cardiovascular Imaging (NASCI). J Cardiovasc Comput Tomogr. 2022;16:536–57. https://doi.org/10.1016/j.jcct.2022.07.002.
12. Koons EK, Rajiah PS, Thorne JE, Weber NM, Kasten HJ, Shanblatt ER, et al. Coronary artery stenosis quantification in patients with dense calcifications using ultra-high-resolution photon-counting-detector computed tomography. J Cardiovasc Comput Tomogr. 2024;18:56–61. https://doi.org/10.1016/j.jcct.2023.10.009.
13. Knuuti J, Wijns W, Saraste A, Capodanno D, Barbato E, Funck-Brentano C, et al. 2019 ESC guidelines for the diagnosis and management of chronic coronary syndromes. Eur Heart J. 2020;41:407–77. https://doi.org/10.1093/eurheartj/ehz425.
14. Collet JP, Thiele H, Barbato E, Barthélémy O, Bauersachs J, Bhatt DL, et al. 2020 ESC guidelines for the management of acute coronary syndromes in patients presenting without persistent ST-segment elevation. Eur Heart J. 2021;42:1289–367.
15. Douek PC, Boccalini S, Oei EHG, Cormode DP, Pourmorteza A, Boussel L, et al. Clinical applications of photon-counting CT: a review of Pioneer studies and a glimpse into the future. Radiology. 2023;309:e222432. https://doi.org/10.1148/radiol.222432.
16. Hagar MT, Soschynski M, Saffar R, Rau A, Taron J, Weiss J, et al. Accuracy of ultrahigh-resolution photon-counting CT for detecting coronary artery disease in a high-risk population. Radiology. 2023;307:e223305. https://doi.org/10.1148/radiol.223305.
17. Fahrni G, Boccalini S, Mahmoudi A, Lacombe H, Houmeau A, Elbaz M, et al. Quantification of coronary artery stenosis in very-high-risk patients using ultra-high resolution spectral photon-counting CT. Invest Radiol. 2024;60:114. https://doi.org/10.1097/RLI.0000000000001109.

18. Mergen V, Sartoretti T, Baer-Beck M, Schmidt B, Petersilka M, Wildberger JE, et al. Ultra-high-resolution coronary CT angiography with photon-counting detector CT. Invest Radiol. 2022;57:780–8. https://doi.org/10.1097/rli.0000000000000897.
19. Halfmann MC, Bockius S, Emrich T, Hell M, Schoepf UJ, Laux GS, et al. Ultrahigh-spatial-resolution photon-counting detector CT angiography of coronary artery disease for stenosis assessment. Radiology. 2024;310:e231956. https://doi.org/10.1148/radiol.231956.
20. Mergen V, Rusek S, Civaia F, Rossi P, Rajagopal R, Bättig E, et al. Virtual calcium removal in calcified coronary arteries with photon-counting detector CT—first in-vivo experience. Front Cardiovasc Med. 2024;11:1–10. https://doi.org/10.3389/fcvm.2024.1367463.
21. Cundari G, Deilmann P, Mergen V, et al. Saving contrast media in coronary ct angiography with photon-counting detector CT. Acad Radiol. 2024;31(1):212–20. https://doi.org/10.1016/j.acra.2023.06.025.
22. Fahrni G, Boccalini S, Lacombe H, de Oliveira F, Houmeau A, Francart F, et al. Ultra-high-resolution 40 keV virtual monoenergetic imaging using spectral photon-counting CT in high-risk patients for coronary stenoses. Eur Radiol. 2024; https://doi.org/10.1007/s00330-024-11237-x.
23. Lee SE, Sung JM, Andreini D, Al-Mallah MH, Budoff MJ, Cademartiri F, et al. Differences in progression to obstructive lesions per high-risk plaque features and plaque volumes with CCTA. JACC Cardiovasc Imaging. 2020;13:1409–17. https://doi.org/10.1016/j.jcmg.2019.09.011.
24. Boussel L, Coulon P, Thran A, Martens G, Sigovan M, Douek P. Photon counting spectral CT component analysis of coronary artery atherosclerotic plaque samples. Br J Radiol. 2014;87:20130798. https://doi.org/10.1259/bjr.20130798.
25. Shami A, Sun J, Gialeli C, Markstad H, Edsfeldt A, Aurumskjöld M-L, et al. Atherosclerotic plaque features relevant to rupture-risk detected by clinical photon-counting CT ex vivo: a proof-of-concept study. Eur Radiol Exp. 2024;8:14. https://doi.org/10.1186/s41747-023-00410-4.
26. Mergen V, Eberhard M, Manka R, Euler A, Alkadhi H. First in-human quantitative plaque characterization with ultra-high resolution coronary photon-counting CT angiography. Front Cardiovasc Med. 2022;9:1–9. https://doi.org/10.3389/fcvm.2022.981012.
27. Vattay B, Szilveszter B, Boussoussou M, Vecsey-Nagy M, Lin A, Konkoly G, et al. Impact of virtual monoenergetic levels on coronary plaque volume components using photon-counting computed tomography. Eur Radiol. 2023;33:8528–39. https://doi.org/10.1007/s00330-023-09876-7.
28. Symons R, De BY, Roosen J, Van Camp L, Cork TE, Kappler S, et al. Journal of cardiovascular computed tomography quarter-millimeter spectral coronary stent imaging with photon-counting CT: initial experience ☆. J Cardiovasc Comput Tomogr. 2018;12:509–15. https://doi.org/10.1016/j.jcct.2018.10.008.
29. Sigovan M, Si-Mohamed S, Bar-Ness D, Bar-Ness D, Mitchell J, Langlois J-B, et al. Feasibility of improving vascular imaging in the presence of metallic stents using spectral photon counting CT and K-edge imaging. Sci Rep. 2019;9:1–9. https://doi.org/10.1038/s41598-019-56427-6.
30. Bratke G, Hickethier T, Bar-Ness D, Bunck AC, Maintz D, Pahn G, et al. Spectral photon-counting computed tomography for coronary stent imaging: evaluation of the potential clinical impact for the delineation of in-stent restenosis. Invest Radiol. 2020;55:61–7. https://doi.org/10.1097/RLI.0000000000000610.
31. Mannil M, Hickethier T, Von Spiczak J, Baer M, Henning A, Hertel M, et al. Photon-counting CT: high-resolution imaging of coronary stents. Invest Radiol. 2018;53:143–9. https://doi.org/10.1097/RLI.0000000000000420.
32. Petritsch B, Petri N, Weng AM, Andreas M, Petersilka M, Allmendinger T, et al. Photon-counting computed tomography for coronary stent imaging. Invest Radiol. 2021;56:653. https://doi.org/10.1097/rli.0000000000000787.

33. Decker JA, O'Doherty J, Schoepf UJ, Todoran TM, Aquino GJ, Brandt V, et al. Stent imaging on a clinical dual-source photon-counting detector CT system—impact of luminal attenuation and sharp kernels on lumen visibility. Eur Radiol. 2023;33:2469–77. https://doi.org/10.1007/s00330-022-09283-4.
34. Stein T, Taron J, Verloh N, Doppler M, Rau A, Hagar MT, et al. Photon-counting computed tomography of coronary and peripheral artery stents: a phantom study. Sci Rep. 2023;13:1–11. https://doi.org/10.1038/s41598-023-41854-3.
35. Elias Michael A, Schoenbeck D, Becker-Assmann J, Niehoff JH, Flohr T, Schmidt B, et al. Coronary stent imaging in photon counting computed tomography: optimization of reconstruction kernels in a phantom. Eur J Radiol. 2023;166:110983. https://doi.org/10.1016/j.ejrad.2023.110983.
36. Boccalini S, Si-Mohamed SA, Lacombe H, Diaw A, Varasteh M, Rodesch P-A, et al. First in-human results of computed tomography angiography for coronary stent assessment with a spectral photon counting computed tomography. Invest Radiol. 2021;57:212. https://doi.org/10.1097/rli.0000000000000835.
37. Shlofmitz E, Iantorno M, Waksman R. Restenosis of drug-eluting stents. Circ Cardiovasc Interv. 2019;12:1–8. https://doi.org/10.1161/CIRCINTERVENTIONS.118.007023.
38. Hagar MT, Soschynski M, Saffar R, Molina-Fuentes MF, Weiss J, Rau A, et al. Ultra-high-resolution photon-counting detector CT in evaluating coronary stent patency: a comparison to invasive coronary angiography. Eur Radiol. 2024;34:4273–83. https://doi.org/10.1007/s00330-023-10516-3.
39. Qin L, Zhou S, Dong H, Li J, Zhang R, Yang C, et al. Improvement of coronary stent visualization using ultra-high-resolution photon-counting detector CT. Eur Radiol. 2024;34:6568–77. https://doi.org/10.1007/s00330-024-10760-1.
40. Fahrni G, Boccalini S, Prieur C, Douek P, Si-Mohamed SA. Comprehensive imaging of coronary stent using ultra-high resolution spectral photon counting CT: a multimodality validation. JACC Cardiovasc Interv. 2023;16:2466–8. https://doi.org/10.1016/j.jcin.2023.07.035.
41. de Graaf FR, Schuijf JD, van Velzen JE, Kroft LJ, de Roos A, Reiber JHC, et al. Diagnostic accuracy of 320-row multidetector computed tomography coronary angiography to noninvasively assess in-stent restenosis. Invest Radiol. 2010;45:331–40. https://doi.org/10.1097/RLI.0b013e3181dfa312.
42. Gerber BL, Belge B, Legros GJ, Lim P, Poncelet A, Pasquet A, et al. Characterization of acute and chronic myocardial infarcts by multidetector computed tomography: comparison with contrast enhanced magnetic resonance. Circulation. 2006;113:823–33. https://doi.org/10.1161/CIRCULATIONAHA.104.529511.
43. Rodriguez-Granillo GA. Delayed enhancement cardiac computed tomography for the assessment of myocardial infarction: from bench to bedside. Cardiovasc Diagn Ther. 2017;7:159–70. https://doi.org/10.21037/cdt.2017.03.16.
44. Oyama-Manabe N, Oda S, Ohta Y, Takagi H, Kitagawa K, Jinzaki M. Myocardial late enhancement and extracellular volume with single-energy, dual-energy, and photon-counting computed tomography. J Cardiovasc Comput Tomogr. 2024;18:3–10. https://doi.org/10.1016/j.jcct.2023.12.006.
45. Hamdy A, Kitagawa K, Goto Y, Yamada A, Nakamura S, Takafuji M, et al. Comparison of the different imaging time points in delayed phase cardiac CT for myocardial scar assessment and extracellular volume fraction estimation in patients with old myocardial infarction. Int J Cardiovasc Imaging. 2019;35:917–26. https://doi.org/10.1007/s10554-018-1513-z.
46. Nacif MS, Kawel N, Lee JJ, Chen X, Yao J, Zavodni A, et al. Interstitial myocardial fibrosis assessed as extracellular volume fraction with low-radiation-dose cardiac CT. Radiology. 2012;264:876–83. https://doi.org/10.1148/radiol.12112458.
47. Nishihara T, Miyoshi T, Akagi N, Nakashima M, Nakamura K, Ito H. Myocardial late iodine enhancement using photon-counting computed tomography in patients with hypertrophic cardiomyopathy. Circ Rep. 2023;5:269–70. https://doi.org/10.1253/circrep.cr-23-0034.

48. Sartoretti T, Wildberger JE, Flohr T, Alkadhi H. Photon-counting detector CT: early clinical experience review. Br J Radiol. 2023:1–11. https://doi.org/10.1259/bjr.20220544.
49. van der Bie J, Sharma SP, van Straten M, Bos D, Hirsch A, Dijkshoorn ML, et al. Photon-counting detector CT in patients pre- and post-transcatheter aortic valve replacement. Radiol Cardiothorac Imaging. 2023;5:2–6. https://doi.org/10.1148/ryct.220318.
50. Willemink MJ, Persson M, Pourmorteza A, Pelc NJ, Fleischmann D. Photon-counting CT: technical principles and clinical prospects. Radiology. 2018;289:293–312. https://doi.org/10.1148/radiol.2018172656.
51. Mergen V, Sartoretti T, Klotz E, Schmidt B, Jungblut L, Higashigaito K, et al. Extracellular volume quantification with cardiac late enhancement scanning using dual-source photon-counting detector CT. Invest Radiol. 2022;57:406–11. https://doi.org/10.1097/RLI.0000000000000851.
52. Aquino GJ, O'Doherty J, Schoepf UJ, Ellison B, Byrne J, Fink N, et al. Myocardial characterization with extracellular volume mapping with a first-generation photon-counting detector CT with MRI reference. Radiology. 2023;307:e222030. https://doi.org/10.1148/radiol.222030.
53. Han D, Lin A, Kuronuma K, Gransar H, Dey D, Friedman JD, et al. Cardiac computed tomography for quantification of myocardial extracellular volume fraction: a systematic review and meta-analysis. JACC Cardiovasc Imaging. 2023;16:1306–17. https://doi.org/10.1016/j.jcmg.2023.03.021.
54. Mergen V, Ehrbar N, Moser LJ, Harmes JC, Manka R, Alkadhi H, et al. Synthetic hematocrit from virtual non-contrast images for myocardial extracellular volume evaluation with photon-counting detector CT. Eur Radiol. 2024:7845–55. https://doi.org/10.1007/s00330-024-10865-7.
55. Kim NY, Im DJ, Youn JC, Hong YJ, Choi BW, Kang S-M, et al. Synthetic extracellular volume fraction derived using virtual unenhanced attenuation of blood on contrast-enhanced cardiac dual-energy CT in nonischemic cardiomyopathy. AJR Am J Roentgenol. 2022;218:454–61. https://doi.org/10.2214/AJR.21.26654.
56. Mahnken AH, Buecker A, Wildberger JE, Ruebben A, Stanzel S, Vogt F, et al. Coronary artery stents in multislice computed tomography. Invest Radiol. 2004;39:27–33. https://doi.org/10.1097/01.rli.0000095471.91575.18.
57. Rodriguez-Granillo GA, Rosales MA, Baum S, Rennes P, Rodriguez-Pagani C, Curotto V, et al. Early assessment of myocardial viability by the use of delayed enhancement computed tomography after primary percutaneous coronary intervention. JACC Cardiovasc Imaging. 2009;2:1072–81. https://doi.org/10.1016/j.jcmg.2009.03.023.
58. Jacquier A, Boussel L, Amabile N, Bartoli JM, Douek P, Moulin G, et al. Multidetector computed tomography in reperfused acute myocardial infarction. Assessment of infarct size and no-reflow in comparison with cardiac magnetic resonance imaging. Invest Radiol. 2008;43:773–81. https://doi.org/10.1097/rli.0b013e318181c8dd.
59. Sato A, Nozato T, Hikita H, Akiyama D, Nishina H, Hoshi T, et al. Prognostic value of myocardial contrast delayed enhancement with 64-slice multidetector computed tomography after acute myocardial infarction. J Am Coll Cardiol. 2012;59:730–8. https://doi.org/10.1016/j.jacc.2011.10.890.
60. Lemmens MJDK, Heuts S, Bidar E, Wildberger JE, Mihl C, Smulders MW. Integrating coronary artery assessment and myocardial late enhancement imaging with photon-counting detector CT: visualizing the invisible. Circ Cardiovasc Imaging. 2024:1–4. https://doi.org/10.1161/CIRCIMAGING.124.017238.
61. Kotronias RA, Raman B, Ferreira V, Neubauer S, Antoniades C. Photon-counting computed tomography: 'one-stop shop' for coronary stenosis, inflammation, and myocardial assessment in ST-segment elevation acute coronary syndrome. Eur Heart J Cardiovasc Imaging. 2024;25:e165. https://doi.org/10.1093/ehjci/jeae003.

62. Polacin M, Templin C, Manka R, Alkadhi H. Photon-counting computed tomography for the diagnosis of myocardial infarction with non-obstructive coronary artery disease. Eur Heart J Case Rep. 2022;6:1–2. https://doi.org/10.1093/ehjcr/ytac028.
63. Kisohara M, Kitera N, Itoh T, Murai K, Hiwatashi A, Kawai T. Identification of a small thrombus in the left ventricle identified on iodine maps derived from dual-source photon-counting detector CT. Radiol Case Rep. 2024;19:1404–8. https://doi.org/10.1016/j.radcr.2024.01.020.
64. Palmisano A, Vignale D, Tadic M, Moroni F, De Stefano D, Gatti M, et al. Myocardial late contrast enhancement CT in troponin-positive acute chest pain syndrome. Radiology. 2022;302:545–53. https://doi.org/10.1148/radiol.211288.
65. Bouleti C, Baudry G, Iung B, Arangalage D, Abtan J, Ducrocq G, et al. Usefulness of late iodine enhancement on spectral CT in acute myocarditis. JACC Cardiovasc Imaging. 2017;10:826–7. https://doi.org/10.1016/j.jcmg.2016.09.013.
66. Treiber J, Novak D, Fischer-Rasokat U, Wolter JS, Kriechbaum S, Weferling M, et al. Regional extracellular volume within late gadolinium enhancement-positive myocardium to differentiate cardiac sarcoidosis from myocarditis of other etiology: a cardiovascular magnetic resonance study. J Cardiovasc Magn Reson. 2023;25:1–12. https://doi.org/10.1186/s12968-023-00918-z.
67. Le Ven F, Pontana F, Barone-Rochette G, Macron L, Garot J, Genée O, et al. Position paper on stress cardiac MRI in chronic coronary syndrome: Endorsed by the Société Française de Radiologie (SFR) the Société Française d'Imagerie CardioVasculaire (SFICV) and the Société Française de Cardiologie (SFC). Diagn Interv Imaging. 2021;102:337–45. https://doi.org/10.1016/j.diii.2021.02.005.
68. Danad I, Szymonifka J, Schulman-Marcus J, Min JK. Static and dynamic assessment of myocardial perfusion by computed tomography. Eur Heart J Cardiovasc Imaging. 2016;17:836–44. https://doi.org/10.1093/ehjci/jew044.
69. van Assen M, Vonder M, Pelgrim GJ, Doeberitz PLVK, Vliegenthart R. Computed tomography for myocardial characterization in ischemic heart disease: a state-of-the-art review. Eur Radiol Exp. 2020;4:1–13. https://doi.org/10.1186/s41747-020-00158-1.
70. Pelgrim GJ, Nieuwenhuis ER, Duguay TM, van der Geest RJ, Varga-Szemes A, Slump CH, et al. Optimal timing of image acquisition for arterial first pass CT myocardial perfusion imaging. Eur J Radiol. 2017;86:227–33. https://doi.org/10.1016/j.ejrad.2016.11.024.
71. Bischoff B, Bamberg F, Marcus R, Schwarz F, Becker H-C, Becker A, et al. Optimal timing for first-pass stress CT myocardial perfusion imaging. Int J Cardiovasc Imaging. 2013;29:435–42. https://doi.org/10.1007/s10554-012-0080-y.
72. Huber AM, Leber V, Gramer BM, Muenzel D, Leber A, Rieber J, et al. Myocardium: Dynamic versus single-shot CT perfusion imaging. Radiology. 2013;269:378–86. https://doi.org/10.1148/radiology.13121441.
73. Delgado Sánchez-Gracián C, Oca Pernas R, Trinidad López C, Armentia ES, Liste AV, Caamaño MV, et al. Quantitative myocardial perfusion with stress dual-energy CT: iodine concentration differences between normal and ischemic or necrotic myocardium. Initial experience. Eur Radiol. 2016;26:3199–207. https://doi.org/10.1007/s00330-015-4128-y.
74. Ohta Y, Kitao S, Yunaga H, Fujii S, Mukai N, Yamamoto K, et al. Myocardial delayed enhancement CT for the evaluation of heart failure: comparison to MRI. Radiology. 2018;288:682–91. https://doi.org/10.1148/radiol.2018172523.
75. Han D, Tamarappoo B, Klein E, Tyler J, Chakravarty T, Otaki Y, et al. Computed tomography angiography-derived extracellular volume fraction predicts early recovery of left ventricular systolic function after transcatheter aortic valve replacement. Eur Heart J Cardiovasc Imaging. 2021;22:179–85. https://doi.org/10.1093/ehjci/jeaa310.
76. Castano A, Narotsky DL, Hamid N, Khalique OK, Morgenstern R, DeLuca A, et al. Unveiling transthyretin cardiac amyloidosis and its predictors among elderly patients with severe aortic stenosis undergoing transcatheter aortic valve replacement. Eur Heart J. 2017;38:2879–87. https://doi.org/10.1093/eurheartj/ehx350.

77. Kitaoka H, Izumi C, Izumiya Y, Inomata T, Ueda M, Kubo T, et al. JCS 2020 guideline on diagnosis and treatment of cardiac amyloidosis. Circ J. 2020;84:1610–71. https://doi.org/10.1253/circj.CJ-20-0110.
78. Pan JA, Kerwin MJ, Salerno M. Native T1 mapping, extracellular volume mapping, and late gadolinium enhancement in cardiac amyloidosis: a meta-analysis. J Am Coll Cardiol Imaging. 2020;13(6):1299–310. https://doi.org/10.1016/j.jcmg.2020.03.010.
79. Konst B, Ohlsson L, Henriksson L, Sandstedt M, Persson A, Ebbers T. Optimization of photon counting CT for cardiac imaging in patients with left ventricular assist devices: an in-depth assessment of metal artifacts. J Appl Clin Med Phys. 2024;25:1–20. https://doi.org/10.1002/acm2.14386.
80. Zeppenfeld K, Tfelt-Hansen J, De Riva M, Winkel BG, Behr ER, Blom NA, et al. 2022 ESC guidelines for the management of patients with ventricular arrhythmias and the prevention of sudden cardiac death: developed by the task force for the management of patients with ventricular arrhythmias and the prevention of sudden cardiac death. Eur Heart J. 2022;43:3997–4126.
81. Soto-Iglesias D, Penela D, Jáuregui B, Acosta J, Fernández-Armenta J, Linhart M, et al. Cardiac magnetic resonance-guided ventricular tachycardia substrate ablation. JACC Clin Electrophysiol. 2020;6:436–47. https://doi.org/10.1016/j.jacep.2019.11.004.
82. Schotten U, Goette A, Verheule S. Translation of pathophysiological mechanisms of atrial fibrosis into new diagnostic and therapeutic approaches. Nat Rev Cardiol. 2024; https://doi.org/10.1038/s41569-024-01088-w.
83. Marrouche NF, Wazni O, McGann C, Greene T, Dean JM, Dagher L, et al. Effect of MRI-guided fibrosis ablation vs conventional catheter ablation on atrial arrhythmia recurrence in patients with persistent atrial fibrillation: the DECAAF II randomized clinical trial. JAMA. 2022;327:2296–305. https://doi.org/10.1001/jama.2022.8831.
84. Alderete J, Fernández-Armenta J, Zucchelli G, Sommer P, Nazarian S, Falasconi G, et al. The ablate-by-LAWT multicentre prospective study: personalized paroxysmal atrial fibrillation ablation with ablation index adapted to local left atrial wall thickness. J Interv Card Electrophysiol. 2024;67:2089. https://doi.org/10.1007/s10840-024-01871-2.
85. Boccalini S, Mayard C, Lacombe H, Villien M, Si-Mohamed S, Delahaye F, et al. Ultra-high-resolution and K-edge imaging of prosthetic heart valves with spectral photon-counting CT. Invest Radiol. 2024;00:1–10. https://doi.org/10.1097/rli.0000000000001068.
86. Francone M, Budde RPJ, Bremerich J, Dacher JN, Loewe C, Wolf F, et al. CT and MR imaging prior to transcatheter aortic valve implantation: standardisation of scanning protocols, measurements and reporting—a consensus document by the European Society of Cardiovascular Radiology (ESCR). Eur Radiol. 2020;30:2627–50. https://doi.org/10.1007/s00330-019-06357-8.
87. Dirrichs T, Schröder J, Frick M, Huppertz M, Iwa R, Allmendinger T, et al. Photon-counting versus dual-source CT for Transcatheter aortic valve implantation planning. Acad Radiol. 2024:1–10. https://doi.org/10.1016/j.acra.2024.06.014.
88. Hagar MT, Kluemper T, Hein M, von Zur Muhlen C, Faby S, Capilli F, et al. Photon-counting CT-angiography in pre-TAVR aortic annulus assessment: effects of retrospective vs. prospective ECG-synchronization on prosthesis valve selection. Int J Cardiovasc Imaging. 2024;40:811–20. https://doi.org/10.1007/s10554-024-03050-w.
89. Pawade T, Sheth T, Guzzetti E, Dweck MR. Why and how to measure aortic valve calcification in patients with aortic stenosis. JACC Cardiovasc Imaging. 2019;12 https://doi.org/10.1016/j.jcmg.2019.01.045.
90. Feldle P, Scheuber M, Grunz JP, Heidenreich JF, Pannenbecker P, Nora C, et al. Virtual non-iodine photon-counting CT-angiography for aortic valve calcification scoring. Sci Rep. 2024;14:1–7. https://doi.org/10.1038/s41598-024-54918-9.
91. Higashigaito K, Mergen V, Eberhard M, et al. CT angiography of the aorta using photon-counting detector CT with reduced contrast media volume. Radiol Cardiothorac Imaging. 2023;5 https://doi.org/10.1148/ryct.220140.

92. Si-Mohamed S, Dupuis N, Tatard-Leitman V, Rotzinger D, Boccalini S, Dion M, et al. Virtual versus true non-contrast dual-energy CT imaging for the diagnosis of aortic intramural hematoma. Eur Radiol. 2019;29:6762–71. https://doi.org/10.1007/s00330-019-06322-5.
93. Schoenbeck D, Pauline Haag N, Elias Michael A, Woeltjen MM, Boriesosdick J, Saeed S, et al. Dedicated virtual non-contrast images adapted for liver tissue in clinical photon counting CT improve virtual non-contrast imaging in various organs beyond the liver. Eur J Radiol. 2023;167:111031. https://doi.org/10.1016/j.ejrad.2023.111031.
94. Boccalini S, Swart LE, Bekkers JA, Nieman K, Krestin GP, Bogers AJ, et al. CT angiography for depiction of complications after the Bentall procedure. Br J Radiol. 2018;20180226 https://doi.org/10.1259/bjr.20180226.
95. Decker JA, Bette S, Scheurig-Muenkler C, Jehs B, Risch F, Woźnicki P, et al. Virtual non-contrast reconstructions of photon-counting detector CT angiography datasets as substitutes for true non-contrast acquisitions in patients after EVAR—Performance of a novel calcium-preserving reconstruction algorithm. Diagnostics. 2022;12 https://doi.org/10.3390/diagnostics12030558.
96. Boccalini S, Si-Mohamed S. Spectral photon counting CT: not just a pimped-up new version of dual-energy CT. Diagn Interv Imaging. 2022;000:2–3. https://doi.org/10.1016/j.diii.2022.10.009.
97. Boccalini S, Dessouky R, Rodesch PA, Lacombe H, Yagil Y, Lahoud E, et al. Gadolinium K-edge angiography with a spectral photon counting CT in atherosclerotic rabbits. Diagn Interv Imaging. 2023;104:490. https://doi.org/10.1016/j.diii.2023.05.002.
98. Boccalini S, Si-Mohamed S, Erhard K, Bordet M, Yagil Y, Douek P. First in-human gadolinium K-edge imaging with spectral photon counting CT. Diagn Interv Imaging. 2025; https://doi.org/10.1016/j.diii.2024.12.004. S2211-5684(24)00279-1
99. Cosset B, Sigovan M, Boccalini S, Farhat F, Douek P, Boussel L, et al. Bicolor K-edge spectral photon-counting CT imaging for the diagnosis of thoracic endoleaks: a dynamic phantom study. Diagn Interv Imaging. 2023;104:235–42. https://doi.org/10.1016/j.diii.2022.12.003.
100. Symons R, Cork TE, Lakshmanan MN, Evers R, Davies-Venn C, Rice KA, et al. Dual-contrast agent photon-counting computed tomography of the heart: initial experience. Int J Cardiovasc Imaging. 2017;33:1253–61. https://doi.org/10.1007/s10554-017-1104-4.
101. Cormode DP, Si-Mohamed S, Bar-Ness D, Sigovan M, Naha PC, Balegamire J, et al. Multicolor spectral photon-counting computed tomography: in vivo dual contrast imaging with a high count rate scanner. Sci Rep. 2017;7:1–11. https://doi.org/10.1038/s41598-017-04659-9.
102. Si Mohamed SA, Sigovan M, Hsu JC, Tatard-Leitman V, Chalabreysse L, Naha PC, et al. In vivo molecular K-edge imaging of atherosclerotic plaque using photon-counting CT. Radiology. 2021;300:98–107. https://doi.org/10.1148/radiol.2021203968.
103. Bennett S, Verry C, Kaza E, Miao X, Dufort S, Boux F, et al. Quantifying gadolinium-based nanoparticle uptake distributions in brain metastases via magnetic resonance imaging. Sci Rep. 2024;14:11959. https://doi.org/10.1038/s41598-024-62389-1.
104. Si-Mohamed S, Tillement O, Lux F, et al. Methods for imaging using spcct; 2024. WO2024083937A2.

Chapter 3
Lung Applications of Spectral Photon-Counting Computed Tomography (SPCCT)

Salim Aymeric Si-Mohamed, Alexandre Bleunven, Sara Boccalini, Alexandra Braillon, Ségolène Turquier, Julie Traclet, Rémi Diesler, François Lestelle, Loic Boussel, Vincent Cottin, and Philippe Douek

Contents

S. A. Si-Mohamed (✉) · S. Boccalini · A. Braillon · L. Boussel · P. Douek
Thoracic and Cardiovascular Radiology Department, Hospices Civils de Lyon, CHU Louis Pradel, Hospices Civils de Lyon, Lyon, France

Université Claude Bernard Lyon 1, INSA-Lyon, UJM Saint-Etienne, CNRS, Inserm, CREATIS UMR 5220, U1206, Lyon, France
e-mail: salim.si-mohamed@chu-lyon.fr

A. Bleunven
Thoracic and Cardiovascular Radiology Department, Hospices Civils de Lyon, CHU Louis Pradel, Hospices Civils de Lyon, Lyon, France

S. Turquier · J. Traclet · R. Diesler · F. Lestelle · V. Cottin
Centre National de Référence des Maladies Pulmonaires Rares, CHU Louis Pradel, Hospices Civils de Lyon, Université Claude Bernard Lyon 1, UMR 754, Service de Pneumologie, ERN-LUNG, Lyon, France

S. A. Si-Mohamed, K. K. Iniewski (eds.), *Clinical Applications of Spectral Photon Counting Computed Tomography Technology*,
https://doi.org/10.1007/978-3-031-85078-3_3

3.1 Introduction

Lung imaging is constantly evolving and the demand for diagnostic performance is growing in many areas such as lung cancer, detection of pulmonary nodules, chronic obstructive pulmonary disease, as well as evaluation and follow-up of interstitial lung diseases (ILD). While CT is the most powerful tool currently available for this purpose, it has many limitations such as the fact that accurate diagnosis often requires additional histological analyses or other non-radiological examinations.

SPCCT provides new diagnostic perspectives compared to the conventional EID found in current CT systems. Studies have therefore been conducted in recent years, mainly limited to non-contrast lung imaging, to determine the potential of SPCCT as an alternative to conventional CT. In the present chapter, we will discuss these studies and explore the perspectives of SPCCT for lung imaging.

3.2 Pre-clinical and Clinical Applications

3.2.1 Parenchyma Imaging

Despite the recent availability of high-resolution CT (HR–CT), evaluation of lung parenchyma integrity is a daily challenge for the radiologist. The principal reason is that radiological evaluation is sometimes limited by the fine semiology of the anatomical structures of the lung lobule [1]; pathological analysis therefore remains the technique of choice in case of diagnostic uncertainty [2]. However, due to its high spatial resolution, SPCCT systems could revolutionize the evaluation of the lung. Accordingly, numerous studies have investigated SPCCT imaging of the parenchyma by studying the influence of the reconstruction parameters such as the matrix size, the slice thickness, and the iterative algorithms recently made available for this new technology.

For instance, Bartlett et al. [3] demonstrated in patients the value of using a 1024 matrix with a high-frequency filter (Q65) with a limited field-of-view (FOV) SPCCT system, in comparison with a 512 and 1024 matrix and standard detailed filter (B46) with SPCCT and EID-CT. These reconstruction parameters were used to convey the high spatial resolution of the SPCCT system, which improved the depiction of the lung parenchyma structures, such as the bronchi and morphological features of lesions (e.g., nodule, ground-glass nodules). Recently, on a different SPCCT system [4], Si-Mohamed et al. demonstrated in human volunteers with a large FOV and a 1024 matrix that lung structures, such as fissures, distal airways and vessels, were of greater conspicuity and sharpness with better overall image quality than standard CT, despite a 23% flux reduction (Fig. 3.1). Interestingly, the noise was also better rated by three experienced lung radiologists who noted a particular texture that characterizes high-frequency noise. In this study, Si-Mohamed et al. showed that the UHR parameters allowed to depict a 178 μm line width on a line pair phantom,

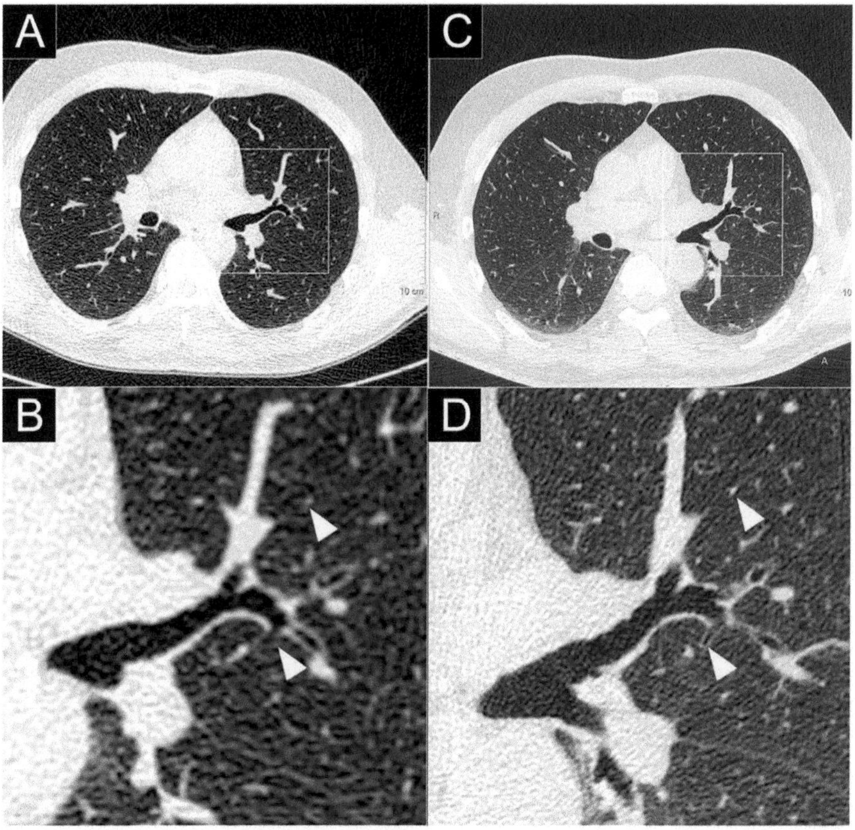

Fig. 3.1 Comparison of high-resolution lung imaging between a conventional CT (Brilliance 64; Philips Haifa, Israel; **a**, **b**), a clinical prototype SPCCT (SPCCT; Philips; **c**, **d**) and in a human volunteer. Close-up views on the left pulmonary hilum found a greater overall image quality as well as better depiction and a greater number of small structures (bronchial wall, vessels; white arrows). (Re-use after permission from Si-Mohamed et al. [6])

which was enabled by a spatial resolution at 22.3 l p/cm at 10% modulation transfer function. Similarly, Leng et al. [5] reported in patients with a different SPCCT system the value of a UHR mode for small lung structures using a sharp filter with a spatial frequency cutoff of 32.4 lp/cm, a 1024 matrix and 0.25 slice thickness, in comparison to a standard (Macro) and Sharp mode. The Sharp and UHR modes showed similar modulation transfer functions, both better than for Macro mode; 10% modulation transfer function value was 9.48 lp/cm for Macro, 16.05 lp/cm for Sharp, and 17.69 lp/cm for UHR modes.

Taken together, these studies show the potential for more accurate parenchyma evaluation in vitro as well as in human volunteers and patients with the UHR mode of the SPCCT technology that requires confirmation in a prospective clinical study.

3.2.2 Nodule Imaging

Lung nodule imaging is a challenging task that necessitates an accurate characterization of the morphological features of a nodule [7]. In recent years, several studies have proven the performance of SPCCT in the detection and morphological characterization of pulmonary nodules, thus demonstrating its potential for this public health problem. For example, Zhou et al. [8] studied the impact of UHR mode for lung nodule volume evaluation and shape characterization using a SPCCT system. They used a 512 matrix with a small FOV (11 cm) resulting in a pixel size of 215 µm and compared the UHR mode to the macro mode using two reconstruction algorithms (s80f and b 46f). The UHR mode used an effective detector pixel size of 0.25 mm by 0.25 mm at the iso-center, while the conventional macro mode was limited to 0.5 mm by 0.5 mm. For small nodules (diameter ≤ 5 mm), UHR mode was able to achieve more accurate volume measurements than macro mode due to higher spatial resolution, which is an advantage for routine use. Moreover, all acquisitions showed comparatively lower bias in the evaluation of the volume of spherical nodules. Additionally, for irregularly shaped (star) nodules, particularly of small size (diameter ≤ 5 mm), UHR acquisitions with the sharp kernel provided substantially better accuracy in volume measurements compared to the other three acquisition mode/reconstruction kernel combinations. Furthermore, receiver-operating characteristics analyses showed a clear benefit of the UHR mode, demonstrating an increased ability to differentiate small, smooth, spherical nodules from small, irregular-shaped nodules. These finding are particularly important, as studies have shown that nodules with irregular or spiculated margins, particularly with distortion of adjacent vessels, are likely to be malignant.

In a second study, Zhou et al. [9] performed a similar study in phantoms with similar parameters and proved that for all nodules the volume estimation was more accurate as there was a lower mean absolute percent error (6.5%) compared with macro mode (11.1% to 12.9%). The improvement of volume measurement with the UHR mode was more obvious for small nodule sizes (3 mm, 5 mm) or star-shape nodules. Thus, the authors demonstrated that the UHR mode of the evaluated SPCCT system was able to improve measurement accuracy for nodule volume and nodule shape characterization.

In line with these studies, Kopp et al. [10] demonstrated a better morphological and volumetric evaluation of different nodule phantoms presenting spikes with a different SPCCT system. The volume estimation indicated better accuracy for SPCCT compared to CT and HR-CT (with a root mean squared error of 21.3 mm^3 for SPCCT, 28.5 mm^3 for CT, and 26.4 mm^3 for HR-CT). The Dice similarity coefficient was greater for SPCCT considering all nodule shapes and sizes (mean, SPCCT 0.90; CT: 0.85; HR-CT: 0.85). These findings were explained by the higher spatial resolution of the system as demonstrated by a higher modulation transfer function (10% modulation transfer function at 21.7 lp/cm).

Recently, using a clinical prototype [11], Si-Mohamed et al. demonstrated the enhanced performances of SPCCT for solid and ground glass lung nodule detectability in a phantom study using a task-based model observer assessment and a subjective image quality evaluation. The comparison between a dual-layer CT and the SPCCT system from the same manufacturer with optimized parameters for each platform found a noise reduction with SPCCT as demonstrated by the noise power spectrum analysis and an improved spatial resolution as demonstrated by the task transfer function. As a consequence, the task-based model observer (d') evaluation demonstrated a greater detectability for solid and ground-glass nodules. It is noteworthy that the difference in detectability between CT systems was more pronounced for the ground-glass nodules as confirmed by volumetric quantification (Fig. 3.2), which are known to be more difficult to detect due to their low contrast with conventional CT systems. Finally, experienced lung radiologists determined that the best compromise for both optimal image quality, noise reduction and increased spatial resolution was reached by using the iterative reconstruction at level iDose 6, which may enable clinical use.

Finally, Jungblut et al. [12] evaluated the image quality of pulmonary nodules at different low radiation doses for SPCCT and EID-CT systems. Their analysis was based on a subjective evaluation by observers and on a commercial artificial intelligence-based computer-aided detection system. They analyzed an anthropomorphic lung phantom containing 14 pulmonary nodules of different sizes with three vendor-specific scanning modes at decreasing matched radiation dose levels, from which an HR mode. By comparing SPCCT to EID-CT images, they found that using the HR mode resulted in better image quality with SPCCT in the subjective

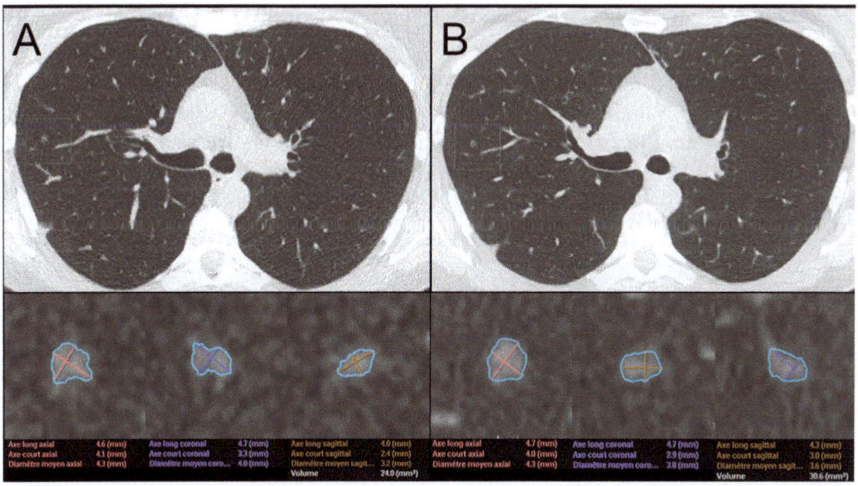

Fig. 3.2 Comparison of high-resolution CT lung imaging of ground glass nodule (GGN) between a dual-layer EID-CT (iQon; Philips, Haifa, Israel) (**a**) and a clinical prototype SPCCT (SPCCT; Philips) (**b**) in a 42-year-old female patient using a low-dose protocol (120 kVp, 10 mAs). GGN was automatically measured by a clinical workstation (InteliSpace Portal, Philips Healthcare) at 24 mm^3 on EID-CT images and at 30.6 mm^3 on SPCCT images, i.e., an underestimation of 27.5% using EID-CT

analysis as well as the lowest image noise in favor of a better objective analysis. Despite the improved performance, the artificial intelligence-based computer-aided detection system delivered comparable results for lung nodule detection and volumetry between SPCCT and dose-matched EID-CT. Nevertheless, the mean sensitivity of SPCCT images was higher than with EID-CT with, for example, a rate of 95% versus 86% for dose-matched EID-CT with a volume CT dose index of 0.41 mGy.

Taken together, these studies show the potential for more accurate lung nodule evaluation in vitro and in vivo with SPCCT technology that requires confirmation in a prospective clinical study.

3.2.3 Lung Cancer Screening Imaging

Lung cancer is the leading cause of cancer deaths worldwide [13] and despite the conclusions of the National Lung Screening Trial that CT screening significantly reduces mortality associated with lung cancer, among other diseases [14], to date, there is no worldwide agreement on the screening recommendations. This can, in part, be explained by the fact that the diagnostic benefits of lung cancer screening have to be balanced with the inherent risks of ionizing radiation, as well as overdiagnosis and work-up of false positive findings [15].

Symons et al. [16] studied the impact of a low-dose acquisition with SPCCT and a conventional CT with a 512 matrix on 2 phantoms by comparing attenuation accuracy, noise, and CNR values between different dose settings. Hounsfield unit stability and accuracy for lung, ground glass nodules, and emphysema were greater with SPCCT than EID-CT, particularly at lower doses where the attenuation decreased significantly with EID-CT. In addition, a reduction of noise was consistently found for SPCCT at 80, 100, 120 kVp in comparison to EID-CT. As a result, CNR were improved compared to EID for ground glass nodules and emphysema, particularly for 100 and 80 kVp, without significant difference for other objects (acrylic, water, lung foam, air, …). These results support the technical benefits of PCD making SPCCT a promising tool in radiation dose optimization that is critical in further improving the risk-benefit ratio of CT lung cancer screening; however, they require confirmation in a prospective clinical study.

3.2.4 Low and Ultra-low Dose Imaging

Ultra-low and low-dose CT has been increasingly used in low-radiation risk assessment of lung cancer, but this is not the only pulmonary disease where the risk-benefit ratio seems positive; for instance the follow-up of fibrosing disease at low doses of radiation could be of great interest because of the iterative CT scans. However, the main limitation for use of these protocols is the impairment of the objective and subjective image quality because of the photon starvation and increase in artefacts such as beam hardening using EID-CT. With SPCCT, numerous studies have suggested an improvement of the image quality. For instance, Symons et al. [16] have proven on phantoms using a 512 matrix the capacity of PCD technology to be more stable in measuring the parenchyma attenuation at low dose. This may also provide better reproducibility and accuracy for quantitative imaging in the future such as performed for lung density. The latter is necessary to monitor disease such as alpha-1 antitrypsin deficiency, emphysema, and idiopathic pulmonary fibrosis but can also be used for the treatment strategy such as performed in alpha-1 antitrypsin deficiency [17]. More recently the same group reported in 30 healthy volunteers the feasibility of a CT lung cancer screening protocol (120 kVp, 20 mAs) with a 512 matrix and an iterative reconstruction algorithm [18]. They showed that experienced readers identified the SPCCT images as having better diagnostic quality for assessment of lung tissue, lung nodules, soft tissue, and bone with lower subjective image noise when compared to those obtained with the EID system. They also found a better image quality in areas with known beam hardening, e.g., around the vertebrae and in the apical lobes. Quantitative measurements showed that SPCCT images had 15.2–16.8% lower noise at two different dose levels, with 21.0% higher lung nodule CNR. In addition, Si-Mohamed et al. [4] demonstrated the feasibility of using low-dose (CTDI: 1.11 mGy) and ultra-low-dose (CTDI: 0.10 mGy) protocols in lung nodule evaluation with phantoms and one human volunteer while preserving UHR parameters. The authors used a 1024 matrix, a large FOV and slice thickness of 0.25 mm and found satisfactory detection with a low X-ray dose protocol for solid, mixed, and ground-glass nodules. Using ultra-low-dose parameters (80 kVp, 3 mAs), ground-glass nodules were not clearly defined because of increasing noise. However, iterative reconstruction (i.e., iDose 5 and 9) enabled a satisfactory nodule visualization. Moreover, improved image quality was found with low-dose SPCCT images compared to standard-dose EID-CT for lung imaging in a human volunteer despite 87% of flux reduction. Additionally, the use of newly developed iterative reconstruction for SPCCT data led to a decrease in noise by up to 68% when using iDose reconstruction algorithm (idose 9). In conclusion, this study showed the feasibility of significant dose reduction with SPCCT with or without iterative reconstruction while conserving the high-resolution parameters for lung analysis. More recently, Lacombe et al. [19] from the same group have compared the lung image quality of ultra-high resolution SPCCT with dual-layer CT at both standard and low radiation doses using phantoms. The authors showed that SPCCT images had 29.1% less noise than DLCT images at 10 mGy and 8.9% less noise even at a lower dose of 6 mGy, indicating better image clarity and quality. SPCCT also demonstrated higher noise power spectrum (NPS) spatial frequency and task-based transfer function (TTF), enhancing the visibility of fine lung details compared to DLCT (Figs. 3.3 and 3.4) [19]. Overall, SPCCT provided superior image quality while allowing for a 40% reduction in radiation exposure.

Taken together, these studies highlight the feasibility of SPCCT imaging for improving lung nodule detection with a good compromise between increased spatial resolution, decreased dose, and equivalent or reduced noise.

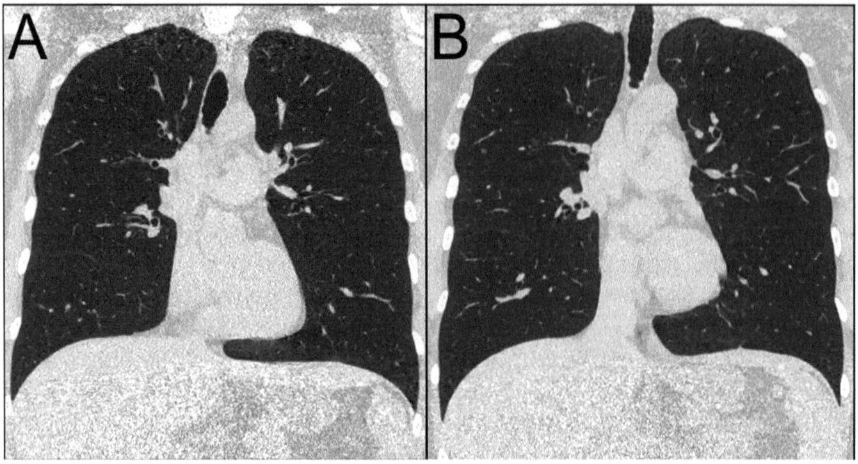

Fig. 3.3 Comparison of high-resolution low-dose lung imaging (120 kVp, 10 mAs) between a dual-layer EID-CT (CT7500, Philips, Haifa, Israel; **a**: filter YB) and a clinical prototype SPCCT (Philips; **b**: filter detailed 1) in a 42-year-old female patient. Both 1024 matrix and field-of-view of 350 mm were matched. Beam hardening was greatly reduced on the SPCCT images, greatly improving the image quality

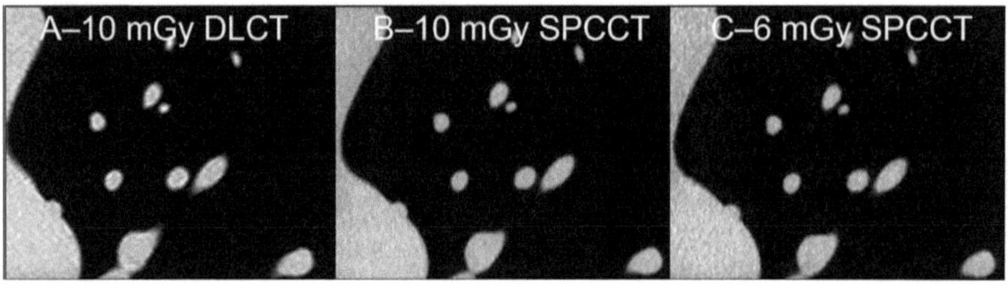

Fig. 3.4 Comparison of high-resolution standard (a, b: 120 kVp, 10 mGy) and low-dose lung imaging (c: 120 kVp, 10 mAs) between a dual-layer EID-CT (CT7500, Philips, Haifa, Israel; **a**: filter YB) and a clinical prototype SPCCT (Philips; b, **c**: filter detailed 1) in an anthroporphic lung phantom (Kyoto Kagaku N1 "LUNGMAN"). Both 1024 matrix and field-of-view of 300 mm were matched. Sharpness was greatly improved on the SPCCT images either on 10 or 6 mGy, in comparison to EID-CT

3.2.5 Interstitial Lung Disease Imaging

Interstitial lung diseases are complex pathologies and HR-CT evaluation is essential for accurate diagnosis and determination of the best treatment. However, this requires very high spatial resolution because of the microsemiology of their key signs, e.g., intralobular reticulations, bronchiectasis and honeycombing, which are frequently indeterminate and may lead to invasive procedures such as biopsies for histological evaluation [20]. In addition, current low-dose protocols, i.e., lower than 1 mSv, are not recommended due to the impaired image quality that may lead to miss-classification of the interstitial lung disease [2]. Therefore, PCD, because of its high resolution and noise reduction capacity, could be a promising tool for ILD evaluation.

In an attempt to evaluate the feasibility of SPCCT in interstitial lung diseases, a recent study conducted by Ferda et al. [21] with a clinical prototype large FOV SPCCT system identified a few cases in a cohort of 60 patients addressed for various reasons. They investigated the diagnostic quality of SPCCT lung imaging using a standard and a low-dose protocol by subjectively assessing critical structures such as bronchi and bronchial walls. When SPCCT images were compared to those obtained with a similar dose using EID-CT, the background noise was significantly lower and the signal-to-noise ratios were significantly higher in SPCCT images. An important point is that the subjective quality score between full and low-dose lung imaging was comparable. In patients with interstitial lung disease, the SPCCT demonstrated a better visualization of higher-order bronchi and third–/fourth–/five-order bronchial walls, such as previously suggested with a SPCCT system from a different manufacturer [4]. However, the authors did not evaluate the diagnostic performance of SPCCT compared to EID-CT for key signs of fibrosis. Nevertheless, these initial findings indicated an important potential for further radiation dose reduction in interstitial lung disease imaging diagnosis and follow-up. Finally, Gaillandre et al. [22] have compared lung parenchyma imaging of interstitial lung disease (ILD) between SPCCT and high-resolution EID-CT. SPCCT images had higher objective noise but offered superior visualization of distal bronchial divisions, bronchial wall sharpness, and ILD features, which led to a reclassification of some ILD cases. Additionally, SPCCT allowed for clearer detection of micronodules, linear opacities, and reticulation, enhancing diagnostic accuracy. SPCCT also achieved a significant reduction in radiation dose by 27–32% compared to EID-CT (Figs. 3.5 and 3.6).

3.2.6 Distal Airways and Bronchial Imaging

Bronchial diseases are difficult to evaluate with conventional CT imaging. For patients with chronic obstructive pulmonary disease, the initial morphological evaluation of the bronchial tubes is difficult as well as their follow-up under treatment due to the lack of spatial resolution of conventional scanners. This limitation is also observed for interstitial lung disease, where the differentiation between traction bronchiolectasis and honeycomb leads to numerous controversies in the literature [20], leaving room for image quality improvement using SPCCT.

For example, Bartlett et al. [3] included 22 patients with pulmonary conditions (pneumonia, pulmonary nodule…) and showed improvement in the detection of higher-order bronchi compared with the EID-CT system, irrespective of the image reconstruction kernel used (readers detected more seventh- and eighth-order bronchi). The addition of a sharp reconstruction kernel designed to convey the higher spatial resolution of the SPCCT system (Q65) further improved the visualization of

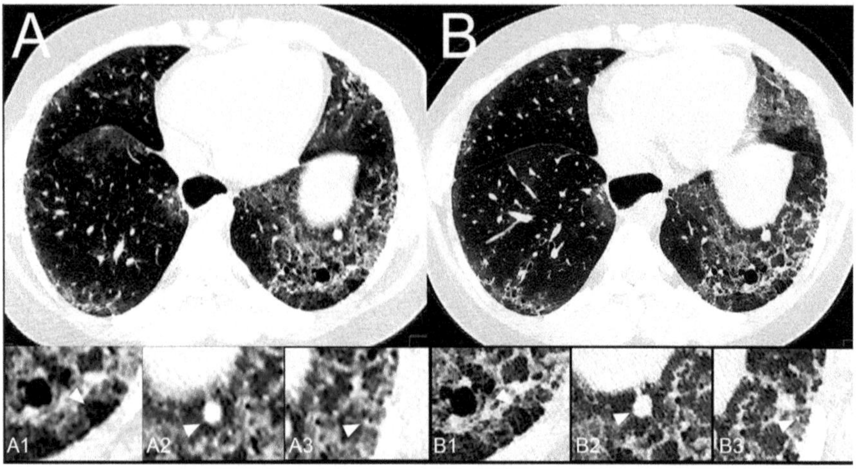

Fig. 3.5 Comparison of high-resolution lung imaging (120 kVp, 40 mAs) between a dual-layer EID-CT (**a**: CT7500, Philips, Haifa, Israel; filter YB) and a clinical prototype SPCCT (**b**: Philips; filter detailed 2) in a 25-year-old female patient with fibrotic ILD associated with systemic sclerosis. Zoomed views highlight greater image quality of fibrotic features such as reticular opacities as well as associated lesions such as solid nodules and pulmonary dendritic ossifications

small bronchi and bronchial walls (walls of the third- and fourth-order bronchi in both lungs). The results of this study demonstrated the potential benefit of SPCCT lung HR-CT, particularly in assessing airway diseases. Similarly, Kopp et al. [10] compared in vivo images of a rabbit and a patient acquired with SPCCT and HR-CT. With HR-CT, bronchi and bronchioles down to a diameter of 1.5–2 mm could be identified. Comparing images of HR-CT and SPCCT adjusted to the same size, vessels and walls of bronchioles could be visualized more distinctly with SPCCT. In addition, Si-Mohamed et al. showed better conspicuity and sharpness of the distal airways as rated by 3 experienced lung radiologists on a human volunteer, which allowed seeing the origin of a 0.5 mm diameter bronchiole that was not visible on an EID-CT [4]. Taken together, these studies support the potential of SPCCT to propose a new evaluation of bronchial diseases (Fig. 3.7).

3.2.7 Pulmonary Vascularization Imaging

Pulmonary vascular involvement is highly prevalent in various lung diseases such as in chronic obstructive pulmonary disease, emphysema, and chronic thromboembolic pulmonary hypertension, as well as infectious diseases such as COVID-19 [23–27]. Current standard CT can allow the diagnosis of vessel structure abnormalities such as thrombus or caliper disparity, while DECT enables the specific imaging of the iodine distribution in the lung, allowing microcirculation imaging, known as a surrogate marker of lung perfusion [28]. Considering the probable improved

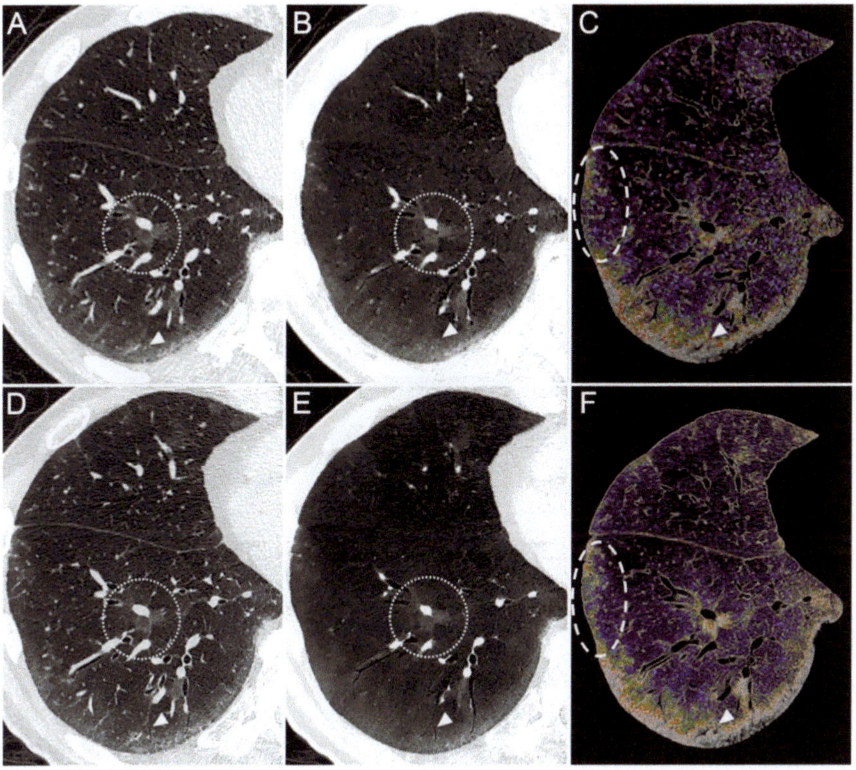

Fig. 3.6 Comparison of ultra-high-resolution lung imaging (120 kVp, 40 mAs) between a dual-layer EID-CT (**a**–**c**: CT7500, Philips, Haifa, Israel; filter YB) and a clinical prototype SPCCT (**d**–**f**: Philips; filter detailed 2) in an 83-year-old male patient with fibrotic ILD associated with systemic sclerosis. Ultra high resolution images (voxel size 250 μm^3) allow improved detection of ground glass, more precise contours and extension (dotted white circle) and improved detection of fibrosis lesions such as distal subpleural bronchiectasis and those co-located with ground glass (arrowheads). The improved detection of ground glass and bronchiectasis can also be seen in the minimal-intensity projection images (images **b** and **e**) and the volume rendering images (images **c** and **f**)

performance of SPCCT compared to EID-CT, there is hope for better monitoring of the proximal and distal pulmonary vascular involvement in the diseases cited above (Fig. 3.8). For example, this would be of great interest in chronic thromboembolic pulmonary hypertension for the characterization of the involvement of microvasculopathy in the sub-pleural area that might contribute to severe hemodynamics, as suggested by Onishi et al. [29]. In addition, the virtual monoenergetic capabilities of SPCCT would allow the use of low monoenergetic images, e.g., 40 or 50 keV, to boost iodine attenuation that would help decrease the amount of iodinated contrast agent for a pulmonary CT angiography.

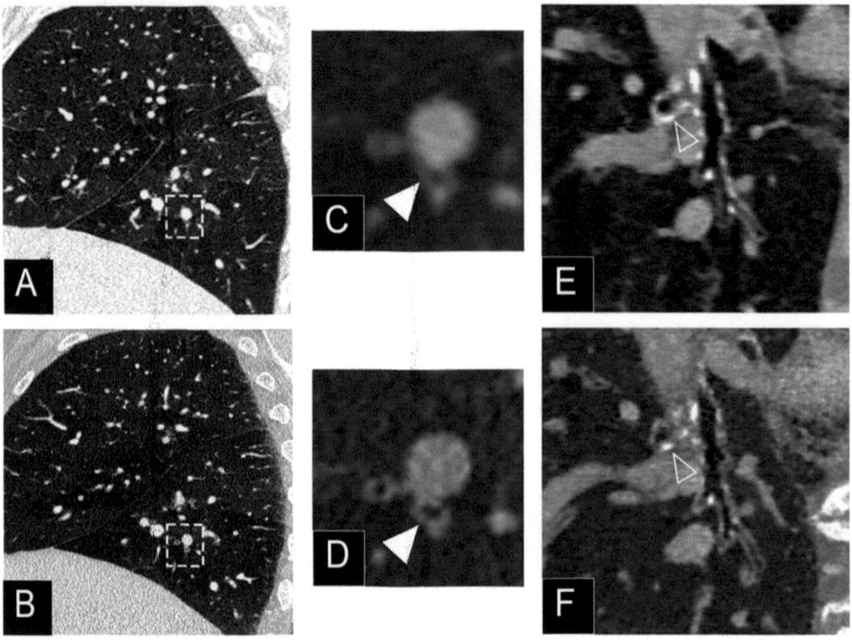

Fig. 3.7 Comparison of the airways imaging in a 72-year-old patient with respiratory bronchiolitis, which was the same day on a clinical prototype SPCCT (first row; Philips, Haifa, Israel) and a dual-layer EID-CT (second row; iQon, Philips). Sagittal views (**a**, **b**) demonstrated thickening of the bronchial wall as well as mosaicism probably due to air trapping. Close-up views of the air-spaces (**c**, **d**) showed a better sharpness and conspicuity of the bronchial wall (white arrowheads) with SPCCT (**d**). Greater sharpness of the bronchial calcifications was noticed on SPCCT images (**f**) compared to EID-CT images (**e**; white empty arrowheads). (Re-use after permission from Si-Mohamed et al. [6])

3.2.8 Color K-edge Imaging

Through its multi-energy CT capabilities, PCD technology enables spectral tissue characterization and differentiation with or without the use of contrast agents, explaining the use of the term spectral photon-counting CT (SPCCT) in recent studies [30–47]. Among these capabilities, color K-edge imaging is one of the new features of SPCCT enabled by the energy-resolving characteristics of PCD. It is based on the recognition of the specific K-edge signature of an atom, i.e., the binding energy between the inner electron shell and the atom [48]. Indeed, by dividing the spectrum into well-chosen energy-based datasets, it is possible to detect multiple elements such as gadolinium, gold, and other atoms that have a K-edge within the relevant energy range of the X-ray spectrum used (e.g., ~40–100 keV; Fig. 3.8) [49]. The main advantage of this technique is to permit specific and quantitative imaging of the agent without any signal arising from the background, which would allow an improved contrast of the tissue as compared to conventional CT imaging.

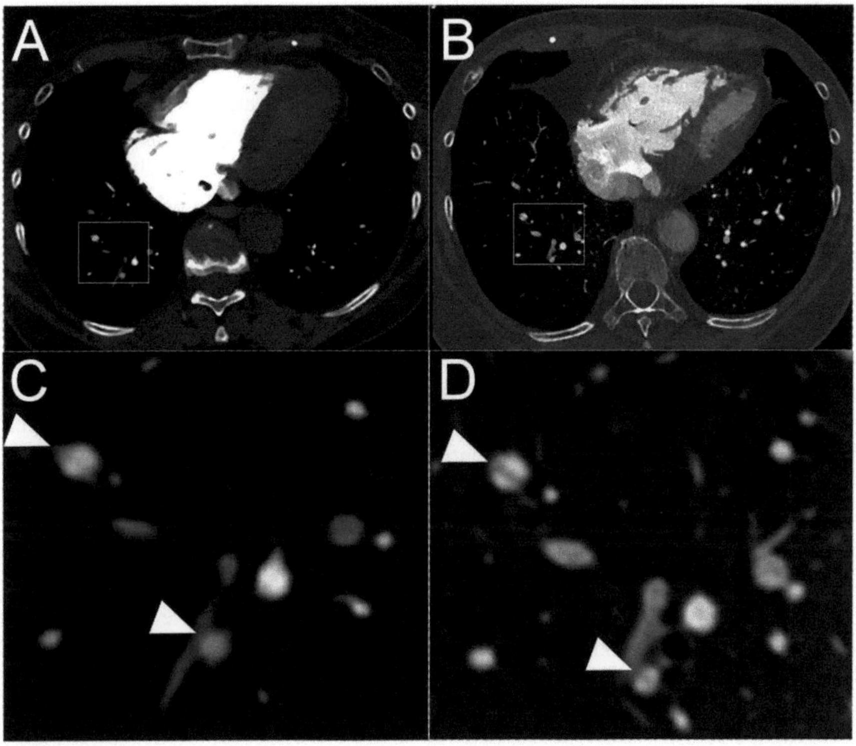

Fig. 3.8 Comparison of the distal arterial pulmonary tree with a dual-layer EID-CT (iQon, Philips, Haifa, Israel; **a**, zoom magnification; c) and a clinical prototype SPCCT (Philips; **b**, zoom magnification; d) in a patient with a chronic thromboembolic pulmonary hypertension, after the injection of iodinated contrast agent. The improvement in quality is visible, notably in the distal vessel lumen where arterial web can be seen only on SPCCT images due to chronic pulmonary clots

This type of imaging requires the use of contrast agents adapted for SPCCT, which opens the door for research and development. Among the candidate contrast agents for K-edge imaging nanoparticles are becoming the agents of choice because of their numerous advantages such as the high payload of atoms, the potential for blood pool effect, tunable physicochemical properties, the potential for targeting imaging, cell tracking, and theranostic applications [31, 50, 51].

In 2018, Si-Mohamed et al. [32] showed in vitro the feasibility of monocolor and bicolor imaging in order to detect and quantify different mixed contrast agents. Three contrast agents (iodine, gadolinium, and gold-based) distributed in tubes at varying proportions were tested by reconstructing the specific images of each material, i.e., based on a two-basis material decomposition (MD) for iodine, three-basis MD for gadolinium or gold with or without iodine, and four-basis MD for gadolinium and gold. The mixtures were prepared such that the solutions could not be differentiated in conventional images. However, a distinction was observed in the material images within the same samples, and the measured and prepared concentrations were strongly correlated, confirming that SPCCT enables multicolor quantitative imaging in vitro.

Additionally, in 2017, Cormode and Si-Mohamed et al. [37] used a first-of-its-kind system with high-count rate capabilities to demonstrate in vivo in rabbits the feasibility of a dynamic monocolor K-edge imaging of a blood pool agent using PEGylated gold nanoparticles. In the organs analyzed, the authors showed persistent visualization of lung vascularization during the first hours after injection. In addition, the authors presented for the first time the feasibility of bicolor K-edge imaging in animals, i.e., simultaneous differentiation of two different contrast agents, using gold nanoparticles and an iodinated contrast agent. Quantification of these contrast agents in the lungs demonstrated persistent concentration for gold while for iodine an arterial first-pass was noticed with rapid clearance in the lungs as expected. More recently, under the framework of a Franch grant, Si-Mohamed et al. [52] have shown the feasibility of a theranostic agent (AGuIX® (NH TherAguix; France) after increasing its own concentration of gadolinium load from 10 to 45 mg/mL, which led to a patent [52]. This enabled the performance of a color K-edge first-pass lung arterial phase as a substitute marker of lung perfusion while taking benefit from an agent that could be functionalized and embedded with therapeutic drugs as well as other atoms. Taken together, all of these first proofs-of-concept may potentially allow a new form of functional imaging, where multiple contrast agents with different pharmacokinetics are used simultaneously in the same biological system for exploring a new form of diagnostic and theranostic imaging (Figs. 3.9 and 3.10).

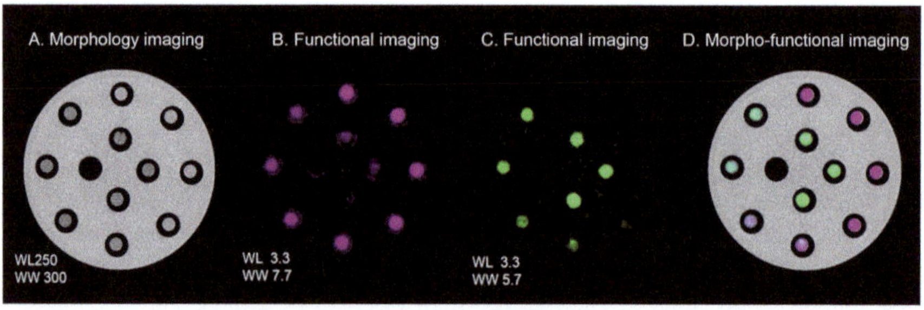

Fig. 3.9 Color K-edge image enabling images of a phantom containing tubes with clockwise decrease concentrations of iodine mixed with increased concentrations of gadolinium using a clinical prototype (SPCCT; Philips, Haifa, Israel) (adapted from Si-Mohamed et al. [32]). Conventional image (**a**) represents morphology imaging. Each color image represents a functional image as they identify specifically the signal of each contrast agent (**b**: iodine image, **c**: gadolinium image). Altogether, color K-edge imaging enabled a potential morpho-functional imaging of the lung (d)

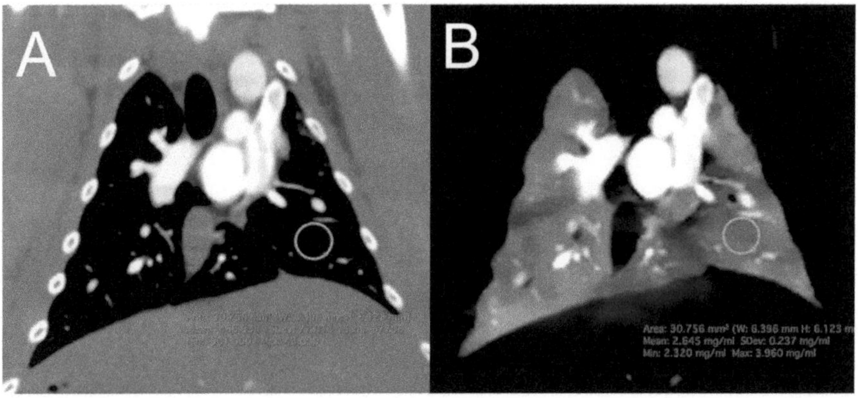

Fig. 3.10 Color K-edge imaging of the lung perfusion blood volume in a rabbit after injection of a theranostic contrast agent (AGuIX® (NH TherAguix; France) using a spectral photon-counting CT (SPCCT; Philips, Haifa, Israel). Conventional image (**a**) showed the enhancement of the lung vessels as well as the underlying tissue while the color K-edge image showed only the signal of gadolinium (**b**)

3.3 Limitations and Perspectives

The studies cited above provide insight into the future potential of SPCCT imaging, particularly in vivo studies that confirm the improvement in spatial resolution and noise management. However, they remain preliminary and larger patient cohort studies are still needed to identify all the clinical benefits of this new technology. Despite this it is already possible to consider the clinical potential of SPCCT thanks to studies such as that reported by Yanagawa et al. [53] who proved the value of studying lung adenocarcinomas with a high spatial resolution (2048 matrix on a conventional CT) to define local invasion of the tumor before the surgery.

With respect to the Color K-edge imaging feature of SPCCT, there are still many hurdles to overcome before considering a clinical translation. These are related to the detection chain and spectral model performance but also to the current contrast agents. It should be noted that, for example, the gadolinium doses needed for in vivo imaging are far higher than currently used and recommended in MR and other contrast agents are in an experimental stage that do not allow any kind of clinical human use in the near future.

In addition, the current technical limitations of the SPCCT systems, which depend on the progress made by manufacturers, should be noted, such as the detector array number, the rotation time, and the energy-resolving capabilities. However, these limitations are expected to be addressed in the near future.

3.4 Conclusion

Spectral photon-counting detector computed tomography (SPCCT) represents an emerging medical imaging modality. The key features are an improved spatial and contrast resolution and a significant noise reduction in comparison to the standard energy-integrating detectors CT (EID-CT) as well as spectral capabilities such a scolor K-edge imaging that may contribute to the improvement of current CT practice in lung imaging.

References

1. Webb WR. Thin-section CT of the secondary pulmonary lobule: anatomy and the image—the 2004 Fleischner lecture. Radiology. 2006;239:322–38. https://doi.org/10.1148/radiol.2392041968.
2. Raghu G, Collard HR, Egan JJ, et al. An official ATS/ERS/JRS/ALAT statement: idiopathic pulmonary fibrosis: evidence-based guidelines for diagnosis and management. Am J Respir Crit Care Med. 2011;183:788–824. https://doi.org/10.1164/rccm.2009-040GL.
3. Bartlett DJ, Koo CW, Bartholmai BJ, Rajendran K, Weaver JM, Halaweish AF, et al. High-resolution lung computed tomography imaging of the lungs: impact of 1024 matrix reconstruction and photon-counting detector computed tomography. Invest Radiol. 2019;54:129–37. https://doi.org/10.1097/RLI.0000000000000524.
4. Si-Mohamed S, Boccalini S, Rodesch P-A, Dessouky R, Lahoud E, Broussaud T, et al. Feasibility of lung imaging with a large field-of-view spectral photon-counting CT system. Diagn Interv Imaging. 2021;102:305–12. https://doi.org/10.1016/j.diii.2021.01.001.
5. Leng S, Rajendran K, Gong H, Zhou W, Halaweish AF, Henning A, et al. 150-μm spatial resolution using photon-counting detector computed tomography technology: technical performance and first patient images. Invest Radiol. 2018;53:655–62. https://doi.org/10.1097/RLI.0000000000000488.
6. Si-Mohamed SA, Miailhes J, Rodesch P-A, Boccalini S, Lacombe H, Leitman V, et al. Spectral photon-counting CT technology in chest imaging. J Clin Med. 2021;10:5757. https://doi.org/10.3390/jcm10245757.
7. MacMahon H, Naidich DP, Goo JM, Lee KS, Leung ANC, Mayo JR, et al. Guidelines for management of incidental pulmonary nodules detected on CT images: from the Fleischner society 2017. Radiology. 2017;284:228–43. https://doi.org/10.1148/radiol.2017161659.
8. Zhou W, Montoya J, Gutjahr R, Ferrero A, Halaweish A, Kappler S, et al. Lung nodule volume quantification and shape differentiation with an ultra-high resolution technique on a photon-counting detector computed tomography system. J Med Imaging Bellingham Wash. 2017;4:043502. https://doi.org/10.1117/1.JMI.4.4.043502.
9. Zhou W, Montoya J, Gutjahr R, Ferrero A, Halaweish A, Kappler S, et al. Lung nodule volume quantification and shape differentiation with an ultra-high resolution technique on a photon counting detector CT system. J Med Imaging (Bellingham). 2017;10132 https://doi.org/10.1117/12.2255736.
10. Kopp FK, Daerr H, Si-Mohamed S, Sauter AP, Ehn S, Fingerle AA, et al. Evaluation of a preclinical photon-counting CT prototype for pulmonary imaging. Sci Rep. 2018;8:17386. https://doi.org/10.1038/s41598-018-35888-1.
11. Si-Mohamed SA, Greffier J, Miailhes J, Boccalini S, Rodesch P-A, Vuillod A, et al. Comparison of image quality between spectral photon-counting CT and dual-layer CT for

the evaluation of lung nodules: a phantom study. Eur Radiol. 2022;32:524–32. https://doi.org/10.1007/s00330-021-08103-5.

12. Jungblut L, Blüthgen P, Messerli M, Schmidt B, Euler A, et al. First performance evaluation of an artificial intelligence-based computer-aided detection system for pulmonary nodule evaluation in dual-source photon-counting detector CT at different low-dose levels. Invest Radiol. 2021;57:108. https://doi.org/10.1097/RLI.0000000000000814.
13. Jemal A, Bray F, Center MM, Ferlay J, Ward E, Forman D. Global cancer statistics. CA Cancer J Clin. 2011;61:69–90. https://doi.org/10.3322/caac.20107.
14. National Lung Screening Trial Research Team, Aberle DR, Adams AM, et al. Reduced lung-cancer mortality with low-dose computed tomographic screening. N Engl J Med. 2011;365:395–409. https://doi.org/10.1056/NEJMoa1102873.
15. Mayo JR, Aldrich J, Muller NL, Fleischner Society. Radiation exposure at chest CT: a statement of the Fleischner society. Radiology. 2003;228:15–21. https://doi.org/10.1148/radiol.2281020874.
16. Symons R, Cork TE, Sahbaee P, Fuld MK, Kappler S, Folio LR, et al. Low-dose lung cancer screening with photon-counting CT: a feasibility study. Physics in Medicine and Biology. 2017;62:202–13. https://doi.org/10.1088/1361-6560/62/1/202.
17. Green CE, Parr DG, Edgar RG, Stockley RA, Turner AM. Lung density associates with survival in alpha 1 antitrypsin deficient patients. Respir Med. 2016;112:81–7. https://doi.org/10.1016/j.rmed.2016.01.007.
18. Symons R, Pourmorteza A, Sandfort V, Ahlman MA, Cropper T, Mallek M, et al. Feasibility of dose-reduced chest CT with photon-counting detectors: initial results in humans. Radiology. 2017;285:980–9. https://doi.org/10.1148/radiol.2017162587.
19. Lacombe H, Labour J, de Oliveira F, Robert A, Houmeau A, Villien M, et al. Ultra-high resolution spectral photon-counting CT outperforms dual layer CT for lung imaging: results of a phantom study. Diagn Interv Imaging. 2024;S2211-5684(24):00208–0. https://doi.org/10.1016/j.diii.2024.09.011.
20. Watadani T, Sakai F, Johkoh T, Noma S, Akira M, Fujimoto K, et al. Interobserver variability in the CT assessment of honeycombing in the lungs. Radiology. 2013;266:936–44. https://doi.org/10.1148/radiol.12112516.
21. Ferda J, Vendiš T, Flohr T, Schmidt B, Henning A, Ulzheimer S, et al. Computed tomography with a full FOV photon-counting detector in a clinical setting, the first experience. Eur J Radiol. 2021;137:109614. https://doi.org/10.1016/j.ejrad.2021.109614.
22. Gaillandre Y, Duhamel A, Flohr T, Faivre J-B, Khung S, Hutt A, et al. Ultra-high resolution CT imaging of interstitial lung disease: impact of photon-counting CT in 112 patients. Eur Radiol. 2023;33:5528–39. https://doi.org/10.1007/s00330-023-09616-x.
23. Si-Mohamed S, Chebib N, Sigovan M, Zumbihl L, Turquier S, Boccalini S, et al. In vivo demonstration of pulmonary microvascular involvement in COVID-19 using dual-energy computed tomography. Eur Respir J. 2020;56:2002608. https://doi.org/10.1183/13993003.02608-2020.
24. Simonneau G, Torbicki A, Dorfmüller P, Kim N. The pathophysiology of chronic thromboembolic pulmonary hypertension. Eur Respir Rev. 2017;26:160112. https://doi.org/10.1183/16000617.0112-2016.
25. Kovacs G, Agusti A, Barberà JA, Celli B, Criner G, Humbert M, et al. Pulmonary vascular involvement in chronic obstructive pulmonary disease. Is there a pulmonary vascular phenotype? Am J Respir Crit Care Med. 2018;198:1000–11. https://doi.org/10.1164/rccm.201801-0095PP.
26. Peinado VI, Pizarro S, Barberà JA. Pulmonary vascular involvement in COPD. Chest. 2008;134:808–14. https://doi.org/10.1378/chest.08-0820.
27. Iyer KS, Newell JD, Jin D, Fuld MK, Saha PK, Hansdottir S, et al. Quantitative dual-energy computed tomography supports a vascular etiology of smoking-induced inflammatory lung disease. Am J Respir Crit Care Med. 2016;193:652–61. https://doi.org/10.1164/rccm.201506-1196OC.

28. Si-Mohamed S, Moreau-Triby C, Tylski P, Tatard-Leitman V, Wdowik Q, Boccalini S, et al. Head-to-head comparison of lung perfusion with dual-energy CT and SPECT-CT. Diagn Interv Imaging. 2020;101:299–310. https://doi.org/10.1016/j.diii.2020.02.006.
29. Onishi H, Taniguchi Y, Matsuoka Y, Yanaka K, Izawa Y, Tsuboi Y, et al. Evaluation of microvasculopathy using dual-energy computed tomography in patients with chronic thromboembolic pulmonary hypertension. Pulm Circ. 2021;11:2045894020983162. https://doi.org/10.1177/2045894020983162.
30. Si-Mohamed S, Bar-Ness D, Sigovan M, Cormode DP, Coulon P, Coche E, et al. Review of an initial experience with an experimental spectral photon-counting computed tomography system. Nucl Instrum Methods Phys Res Sect Accel Spectrometers Detect Assoc Equip. 2017;873:27–35. https://doi.org/10.1016/j.nima.2017.04.014.
31. Kim J, Bar-Ness D, Si-Mohamed S, Coulon P, Blevis I, Douek P, et al. Assessment of candidate elements for development of spectral photon-counting CT specific contrast agents. Sci Rep. 2018;8:12119. https://doi.org/10.1038/s41598-018-30570-y.
32. Si-Mohamed S, Bar-Ness D, Sigovan M, Tatard-Leitman V, Cormode DP, Naha PC, et al. Multicolour imaging with spectral photon-counting CT: a phantom study. Eur Radiol Exp. 2018;2:34. https://doi.org/10.1186/s41747-018-0063-4.
33. Si-Mohamed S, Cormode DP, Bar-Ness D, Sigovan M, Naha PC, Langlois J-B, et al. Evaluation of spectral photon counting computed tomography K-edge imaging for determination of gold nanoparticle biodistribution in vivo. Nanoscale. 2017;9:18246. https://doi.org/10.1039/c7nr01153a.
34. Si-Mohamed S, Thivolet A, Bonnot P-E, Bar-Ness D, Képénékian V, Cormode DP, et al. Improved peritoneal cavity and abdominal organ imaging using a biphasic contrast agent protocol and spectral photon counting computed tomography K-edge imaging. Invest Radiol. 2018;53:629–39. https://doi.org/10.1097/RLI.0000000000000483.
35. Si-Mohamed S, Tatard-Leitman V, Laugerette A, Sigovan M, Pfeiffer D, Rummeny EJ, et al. Spectral photon-counting computed tomography (SPCCT): in-vivo single-acquisition multiphase liver imaging with a dual contrast agent protocol. Sci Rep. 2019;9:8458. https://doi.org/10.1038/s41598-019-44821-z.
36. Cormode DP, Roessl E, Thran A, et al. Atherosclerotic plaque composition: analysis with multicolor CT and targeted gold nanoparticles. Radiology. 2010;256:774–82. https://doi.org/10.1148/radiol.10092473.
37. Cormode DP, Si-Mohamed S, Bar-Ness D, et al. Multicolor spectral photon-counting computed tomography: in vivo dual contrast imaging with a high count rate scanner. Sci Rep. 2017;7:4784. https://doi.org/10.1038/s41598-017-04659-9.
38. de Vries A, Roessl E, Kneepkens E, Brendel B, Martens G, Proska R, et al. Quantitative spectral K-edge imaging in preclinical photon-counting x-ray computed tomography. Invest Radiol. 2015;50:297–304. https://doi.org/10.1097/RLI.0000000000000126.
39. Muenzel D, Bar-Ness D, Roessl E, Blevis I, Bartels M, Fingerle AA, et al. Spectral photon-counting CT: initial experience with dual-contrast agent K-edge colonography. Radiology. 2017;283:723–8. https://doi.org/10.1148/radiol.2016160890.
40. Riederer I, Bar-Ness D, Kimm MA, Si-Mohamed S, Noël PB, Rummeny EJ, et al. Liquid embolic agents in spectral X-ray photon-counting computed tomography using tantalum K-edge imaging. Sci Rep. 2019;9:5268. https://doi.org/10.1038/s41598-019-41737-6.
41. Sigovan M, Si-Mohamed S, Bar-Ness D, Mitchell J, Langlois J-B, Coulon P, et al. Feasibility of improving vascular imaging in the presence of metallic stents using spectral photon counting CT and K-edge imaging. Sci Rep. 2019;9:19850. https://doi.org/10.1038/s41598-019-56427-6.
42. Thivolet A, Si-Mohamed S, Bonnot P-E, Blanchet C, Képénékian V, Boussel L, et al. Spectral photon-counting CT imaging of colorectal peritoneal metastases: initial experience in rats. Sci Rep. 2020;10:13394. https://doi.org/10.1038/s41598-020-70282-w.
43. Bratke G, Hickethier T, Bar-Ness D, Bunck AC, Maintz D, Pahn G, et al. Spectral photon-counting computed tomography for coronary stent imaging: evaluation of the potential clinical

impact for the delineation of in-stent restenosis. Invest Radiol. 2020;55:61–7. https://doi.org/10.1097/RLI.0000000000000610.
44. Boussel L, Coulon P, Thran A, Roessl E, Martens G, Sigovan M, et al. Photon counting spectral CT component analysis of coronary artery atherosclerotic plaque samples. Br J Radiol. 2014;87:20130798. https://doi.org/10.1259/bjr.20130798.
45. Halttunen N, Lerouge F, Chaput F, Vandamme M, Karpati S, Si-Mohamed S, et al. Hybrid Nano-GdF3 contrast media allows pre-clinical in vivo element-specific K-edge imaging and quantification. Sci Rep. 2019;9:12090. https://doi.org/10.1038/s41598-019-48641-z.
46. Riederer I, Si-Mohamed S, Ehn S, Bar-Ness D, Noël PB, Fingerle AA, et al. Differentiation between blood and iodine in a bovine brain-initial experience with spectral photon-counting computed tomography (SPCCT). PLoS One. 2019;14:e0212679. https://doi.org/10.1371/journal.pone.0212679.
47. Si-Mohamed SA, Sigovan M, Hsu JC, Tatard-Leitman V, Chalabreysse L, Naha PC, et al. In vivo molecular K-edge imaging of atherosclerotic plaque using photon-counting CT. Radiology. 2021;300:98–107. https://doi.org/10.1148/radiol.2021203968.
48. Roessl E, Brendel B, Engel K-J, Schlomka J-P, Thran A, Proksa R. Sensitivity of photon-counting based K-edge imaging in X-ray computed tomography. IEEE Trans Med Imaging. 2011;30:1678–90. https://doi.org/10.1109/TMI.2011.2142188.
49. Roessl E, Proksa R. K-edge imaging in x-ray computed tomography using multi-bin photon counting detectors. Phys Med Biol. 2007;52:4679–96. https://doi.org/10.1088/0031-9155/52/15/020.
50. Hsu JC, Nieves LM, Betzer O, Sadan T, Noël PB, Popovtzer R, et al. Nanoparticle contrast agents for X-ray imaging applications. Wiley Interdiscip Rev Nanomed Nanobiotechnol. 2020;12:e1642. https://doi.org/10.1002/wnan.1642.
51. Bouché M, Hsu JC, Dong YC, Kim J, Taing K, Cormode DP. Recent advances in molecular imaging with gold nanoparticles. Bioconjug Chem. 2020;31:303–14. https://doi.org/10.1021/acs.bioconjchem.9b00669.
52. Si-Mohamed S, Douek, Philippe, Boussel L, et al Methods for imaging using SPCCT. WO2024083937A2. Patent.
53. Yanagawa M, Tsubamoto M, Satoh Y, Hata A, Miyata T, Yoshida Y, et al. Lung adenocarcinoma at CT with 0.25-mm section thickness and a 2048 matrix: high-spatial-resolution imaging for predicting invasiveness. Radiology. 2020;297:462–71. https://doi.org/10.1148/radiol.2020201911.

Chapter 4
Novel Contrast Agents for Spectral Photon Counting Computed Tomography (SPCCT)

Derick N. Rosario-Berríos, Mahima Gupta, Agnieszka Gutwinska, Amanda Y. Pang, and David P. Cormode

Contents

D. N. Rosario-Berríos
Department of Biochemistry and Molecular Biophysics, University of Pennsylvania, Philadelphia, PA, USA

M. Gupta · A. Y. Pang
Department of Radiology, University of Pennsylvania, Philadelphia, PA, USA

A. Gutwinska
University of Lyon, INSA-Lyon, University Claude Bernard Lyon 1, UJM-Saint Etienne, CNRS, Inserm, CREATIS UMR 5220, U1206, Villeurbanne, France

D. P. Cormode (✉)
Department of Radiology, University of Pennsylvania, Philadelphia, PA, USA

Department of Bioengineering, University of Pennsylvania, Philadelphia, PA, USA
e-mail: david.cormode@pennmedicine.upenn.edu

S. A. Si-Mohamed, K. K. Iniewski (eds.), *Clinical Applications of Spectral Photon Counting Computed Tomography Technology*,
https://doi.org/10.1007/978-3-031-85078-3_4

4.1 Introduction

Computed tomography (CT), first introduced into clinical practice in 1971, is one of the most commonly used non-invasive imaging tools such that more than 80 million scans are now performed annually in the USA [1]. It utilizes X-rays and can be used for a wide range of diagnostic purposes as it generates detailed cross-sectional images of the body's internal structures [2]. CT scans are considered superior to conventional X-ray images due to the three-dimensional information they produce of tissues and organs [1]. As such, they are widely used in emergency medicine, for monitoring disorders like gastrointestinal and cardiovascular issues [3–6], and in detecting various conditions such as bone fractures, cancer, tumors, and infectious diseases [7–9]. The fundamental principle of CT lies in its ability to differentiate between tissues of varying densities. When X-rays pass through tissues, they are absorbed at different rates depending on the tissue density. For example, highly attenuating materials absorb more X-rays and are typically displayed as white or light grey. On the other hand, low attenuating materials absorb fewer X-ray beams and are usually displayed as dark grey or black.

CT imaging is a sophisticated process involving several steps to produce three-dimensional anatomical information about the body's internal structures. Before a scan, the patient is positioned on a motorized table that slides into the CT scanner, with the area of interest centered within the scanner. The CT scanner then emits X-ray beams as the X-ray source, and the detector rotates around the patient. Typically, collimation is used to reduce scattered radiation, thereby enhancing image quality. Detectors opposite to the X-ray source absorb some of the transmitted photons that are not absorbed by the patient's body. The intensity of the X-ray beam reduces as it passes through the body, a phenomenon known as X-ray attenuation. As different tissues absorb X-rays at varying rates, the strength of the transmitted X-ray beams varies, which is measured by the detectors. This information is sent to a computer, which processes the data to construct cross-sectional slices of the body. These slices can then be combined to form a 3D image, providing information about structure and conditions involving the patient's tissues, organs, and bones [10].

Conventional CT has several limitations, including its spatial and contrast resolutions. For instance, it cannot differentiate gold-labeled cells from bones or microcalcifications if found in close proximity [11]. As a result, while CT technology has reached a point of widespread adoption and technical refinement, it can still be improved upon. An example of an advancement in CT technology that seeks to address the issues of conventional CT is spectral photon-counting computed tomography (SPCCT).

SPCCT is a cutting-edge imaging technology that has rapidly evolved in the last decade, representing a significant advancement in diagnostic capabilities over traditional CT systems [12]. The key factor that differentiates SPCCT from conventional CT is its use of photon-counting detectors (PCDs) that are made of semiconductor materials like cadmium zinc telluride or silicon [13]. These detectors capture X-ray photons and convert them into electrical pulses, enabling superior spatial resolution and the categorization of photons based on energy [12]. This innovation results in spectral imaging, a feature lacking in traditional CT systems, that lowers noise levels, improves contrast-to-noise ratio, enhances spectral separation, and increases dose efficiency (Fig. 4.1) [12, 14–16].

SPCCT imaging allows for the specific and absolute quantification of exogenous contrast agents in vivo, which can facilitate targeted or molecular imaging [15]. SPCCT has proven its capability to differentiate materials within biological structures, making it a valuable tool in both small and large animal studies [13]. Clinical studies also underscore the technology's comparable image quality to conventional CT scanners, solidifying its acceptance in the medical field [17, 18]. Ongoing research focuses on refining image reconstruction methods and detector modules to further improve overall imaging performance.

Although imaging modalities, including SPCCT, serve as non-invasive, cost-effective diagnostic medical tools, contrast agents are often needed in conjunction to enhance the visibility of anatomical structures, provide dynamic information, and identify pathologies and phenotype disease [19]. While substantial progress has been made in the area of detectors and other facets of CT technology, there remains a significant amount of work to be done in the domain of contrast agents. The impact of SPCCT would be enhanced by the development of specialized contrast agents tailored to augment its capabilities. Currently, the contrast agents employed, i.e., iodinated small molecules, fall short of fully exploiting the potential of SPCCT due to the predominant absorption of X-ray light by the patient, rather than the contrast agent. This absorption pattern poses a significant challenge, resulting in insufficient data for generating element-specific images or "K-edge" images for elements with low K-edge energy elements like iodine. Despite these limitations of existing contrast agents, no new contrast agents have been approved for more than two decades. However, with the advent of a new technology, i.e., SPCCT, there is renewed

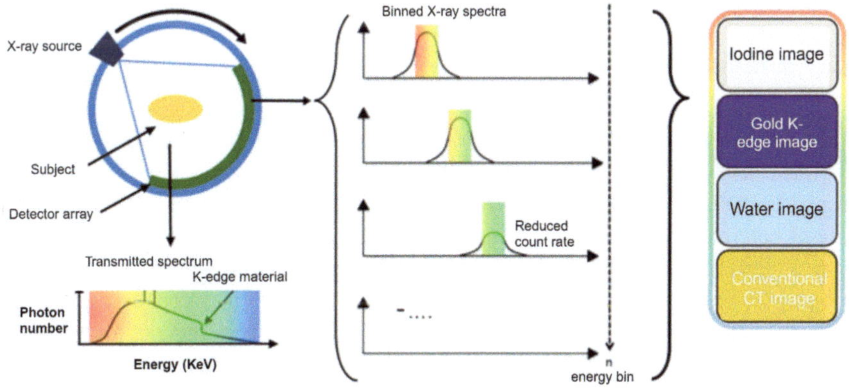

Fig. 4.1 Photon-counting CT systems differentiate energy spectra with specific algorithms based on material concentration. (Figure reproduced with permission from reference [16])

interest in novel X-ray contrast agents, with researchers actively engaged in focusing on elements whose K-edges are at higher energy levels than iodine's.

The most commonly used intravenous contrast agents for SPCCT currently are the same iodinated aromatic small molecules that are used for various other X-ray imaging modalities [15]. While these iodinated small molecules are effective and valuable for many purposes, they have a number of drawbacks. Their contrast generation is comparatively low, particularly at high tube voltages. This is due to the low K-edge energy of iodine (33.2 kV), which limits K-edge imaging in clinical SPCCT [15]. In addition to the limited contrast generation, iodinated small molecule contrast agents have also been associated with contrast-induced allergies and nephropathies [20]. As a result, there has been an increasing push towards the development of novel SPCCT contrast agents that not only improve contrast generation but minimize toxicity. To meet these ends, the field has focused on elements whose K-edge is significantly higher than iodine's (K-edges that are approximately in the range of 40–90 keV) and thereby allow K-edge imaging [15, 21]. This has led to experimentation with heavy metal elements, including gold, ytterbium, tantalum, bismuth, tungsten, and gadolinium, which have shown great promise both in generating contrast as well as, in some cases, providing therapeutic benefits [15, 22–27]. These elements have the potential to deliver heightened contrast and spectral information in comparison to iodine and extends enhanced accuracy in imaging both below and above the K-edge [28–30].

In the following section, we will cover the application of contrast agents in CT across various clinical practices, ranging from oncology to cardiology, highlighting the benefits of SPCCT in contrast agent utilization. Subsequently, we will review recent developments in the creation of new contrast agents for SPCCT. We will explore in detail the three categories that have garnered increased attention for developing novel contrast agents for SPCCT. Additionally, we will examine the safety and biocompatibility of these contrast agents. Finally, the chapter will conclude by addressing challenges and outlining future directions.

4.2 Clinical Relevance

The application of contrast agents in CT is highly significant in clinical practice. Their use across various medical specialties, from emergency medicine to oncology, emphasizes their indispensable role in enhancing diagnostic accuracy, improving patient outcomes, and effectively monitoring treatment progress [30, 31]. SPCCT, from a clinical perspective, represents a transformative advancement in diagnostic imaging that offers clinicians greater detail and specificity in patient assessment. It leads to improved diagnostic accuracy, reduction in contrast and radiation doses, and patient management and outcomes [31].

In the fast-paced environment of trauma and emergency medicine, the use of contrast agents is critically important. They play a vital role in rapidly assessing injuries, identifying internal bleeding, and guiding prompt and appropriate interventions [32, 33]. This can be crucial in life-threatening situations, where internal injuries and bleeding might not be immediately apparent but nevertheless require urgent medical attention [34, 35].

4.3 Infectious Disease

Beyond emergency presentations, contrast agents can also be employed in imaging infectious diseases [36]. They are particularly useful in identifying areas of infection, such as abscesses and inflammation. During the COVID-19 pandemic, contrast imaging played a role in assessing lung involvement and complications associated with the virus [37]. Its ability to delineate types of infection responses can help differentiate between infectious and inflammatory conditions. This aids clinicians in making precise diagnoses and formulating effective treatment plans for infectious diseases.

In cardiology, intravenous injection of contrast agents enables detailed visualization of the coronary arteries, cardiac chambers, and myocardial perfusion by opacifying the blood [38]. High-definition SPCCT images of coronary arteries, enables cardiologists to assess plaque composition and stenosis more accurately. This level of detail assists in risk stratification for coronary artery disease and guides therapeutic decision-making, potentially reducing the need for invasive procedures like cardiac catheterization [39, 40].

In oncology, contrast-enhanced imaging is indispensable for tumor detection, characterization, and treatment monitoring [41]. It provides critical information about tumor size, vascularity, and response to therapy, thereby playing an integral role in personalized cancer care and assessment of tumor progression or regression with greater accuracy. For example, dual-energy CT (DECT) aids in monitoring the response to therapy by providing detailed images of how tumors change over time [42, 43].

SPCCT and DECT's sensitivity allows for the use of lower doses of contrast media while still achieving high-quality images [14]. The diagnostic accuracy improves patient safety and can significantly influence patient outcomes across a wide range of medical specialties.

4.4 Recent Developments

SPCCT has witnessed significant strides in contrast agent development [44]. The introduction of specialized contrast agents tailored to the unique capabilities of SPCCT has opened new avenues for precise and targeted imaging for cardiovascular disorders, cancer, and neurological conditions by offering insights into tissue composition and function [14, 45]. Notable examples include the application of contrast agents for the assessment of vascular permeability in cancer imaging and the detection of early-stage atherosclerosis in cardiovascular studies [45]. These diverse applications underscore the versatility and adaptability of SPCCT contrast agents, highlighting their potential for improving diagnostic approaches across a spectrum of diseases. In the following sections, we will focus on studies that have reported on the use of contrast agents with SPCCT. Note that this chapter is not intended to be a comprehensive review of this topic. Therefore, additional studies can be found in the literature that may be of interest to the reader. We will discuss three categories that have been drawing particular attention: (1) using currently approved contrast agents and combinations thereof, (2) developing new contrast agents typically based on nanoparticles (NP), and (3) using novel agents to track cells and other materials such as hydrogels.

4.4.1 Dual Contrast

The use of two contrast agents, that may vary in delivery method or composition, for SPCCT can offer enhanced imaging capabilities by increasing the area reached by the contrast agent and leveraging the unique spectral characteristics of each material. When two contrast agents with distinct spectral signatures are used simultaneously, it becomes possible to differentiate and visualize multiple anatomical structures or physiological processes at the same time [46]. Since each contrast agent has a unique X-ray attenuation profile, material-specific images can be generated for each agent [47].

In terms of dual delivery methods, the combination of intravenous (IV) and intraperitoneal (IP) injections in SPCCT enhances imaging findings. IV injections facilitate rapid and widespread distribution of the contrast agent, which is advantageous for imaging vascular structures. Conversely, IP injections specifically target the peritoneal cavity, making them well-suited for detailed abdominal assessments [48].

Furthermore, Symons et al. explored the potential of photon-counting CT for in vivo simultaneous differentiation of three contrast agents in dogs. Bismuth subsalicylate was administered orally 24–72 h before imaging. Subsequently, CT scans were conducted during the intravenous injection of gadoterate meglumine followed by the intravenous injection of iopamidol 3.5 min later [49]. Their study demonstrated the feasibility of utilizing photon-counting CT with four energy thresholds (25, 50, 75, and 90 keV) to differentiate three K-edge contrast agents (iodine, gadolinium, and bismuth) in vivo. Beyond aiding in the differentiation of organs and tissues, these advancements contribute to reducing the radiation doses administered to patients since one scan can replace the several needed with conventional CT [49].

Take, for instance, the examination of gastrointestinal content. In traditional gastrointestinal imaging, separate studies are needed for polyp analysis and tagging bowel contents. However, advancements now allow the simultaneous use of two contrast agents, such as a single intravenous agent for improved polyp visualization and an oral agent for tagging bowel contents, streamlining the process [48]. Muenzel et al. conducted a study to explore the potential of SPCCT in differentiating between gadolinium- and nonionic iodine-based contrast materials within a colon phantom

[48]. Gadolinium-based contrast agents have been widely utilized in the field of medical imaging, particularly as contrast agents for magnetic resonance imaging (MRI) by binding to organic molecules and for X-ray computed tomography (CT) due to gadolinium's high atomic number (64). Gadolinium chelates, which are compounds binding gadolinium ions to organic molecules, represent the most commonly employed form of gadolinium contrast agents [50, 51]. The findings of Muenzel et al. demonstrated clear differentiation in the distribution of contrast materials, with gadolinium localized in the polyp and iodine in the colon's lumen. Moreover, SPCCT image-based analysis of contrast agent concentrations in the colon and polyp aligned well with the actual values (Fig. 4.2) [48]. This study demonstrated that SPCCT along with two contrast agents with different methods and delivery profiles would enhance diagnostic efficiency.

A developing application of SPCCT involves peritoneal imaging, which helps assess invasive tumors prone to spreading to the peritoneal cavity such as those originating from the ovaries, colorectal region, stomach, appendix, or primary peritoneal area [15, 52]. Si-Mohamed et al. investigated this area by utilizing an SPCCT scanner to distinguish between two contrast agents (a macrocyclic gadolinium chelate and an iodinated agent) in the peritoneal imaging of rats [53]. They simultaneously utilized the contrast agents within the vascular and peritoneal compartments, revealing superior diagnostic quality in SPCCT images with improved resolution compared to CT-like images. Excellent intraperitoneal diffusion was observed, resulting in comparable peritoneal opacification indices on SPCCT images, which was complemented by overlay contrast material maps. This advantageous feature allowed for a distinct separation of contrast agents, facilitating specific quantification of physiological enhancement in vital organs such as the liver, spleen, and kidney. Additionally, the maps enabled precise visualization of urinary clearance in the renal pelvis and bladder (Fig. 4.3) [53].

Liver imaging stands to significantly benefit from the implementation of SPCCT [54]. CT is frequently selected as the initial assessment for focal and diffuse liver disorders due to its good tolerability, minimal image artifacts, and the ability to capture images of the entire abdomen and pelvis quickly in a single breathhold. Previously, various protocols, such as unenhanced, single-phase, dual-phase, and triphasic contrast-enhanced protocols were employed for CT imaging of the liver [55, 56]. The standard CT liver protocol typically involves arterial and portal venous contrast-enhanced series, along with a native unenhanced scan if necessary [57]. However, this protocol, where as many as three scans are used, entails a significant radiation dose. SPCCT can come into play, since it can provide multiple sets of diagnostic information from a single, low radiation dose scan. In an in silico study, Muenzel et al. utilized hepatic SPCCT for simultaneous mapping of two contrast agents, i.e., gadolinium and iodine agents. Their findings suggested that dual-contrast SPCCT holds the potential to visualize characteristic arterial and portal venous enhancement within a single scan, presenting a notable advancement in liver diagnostics [57].

Subsequently, Si-Mohamed and colleagues conducted an animal study to illustrate the in vivo practicality of obtaining dual-contrast multi-phase hepatic K-edge images in a single acquisition using an SPCCT system [58]. Through the sequential

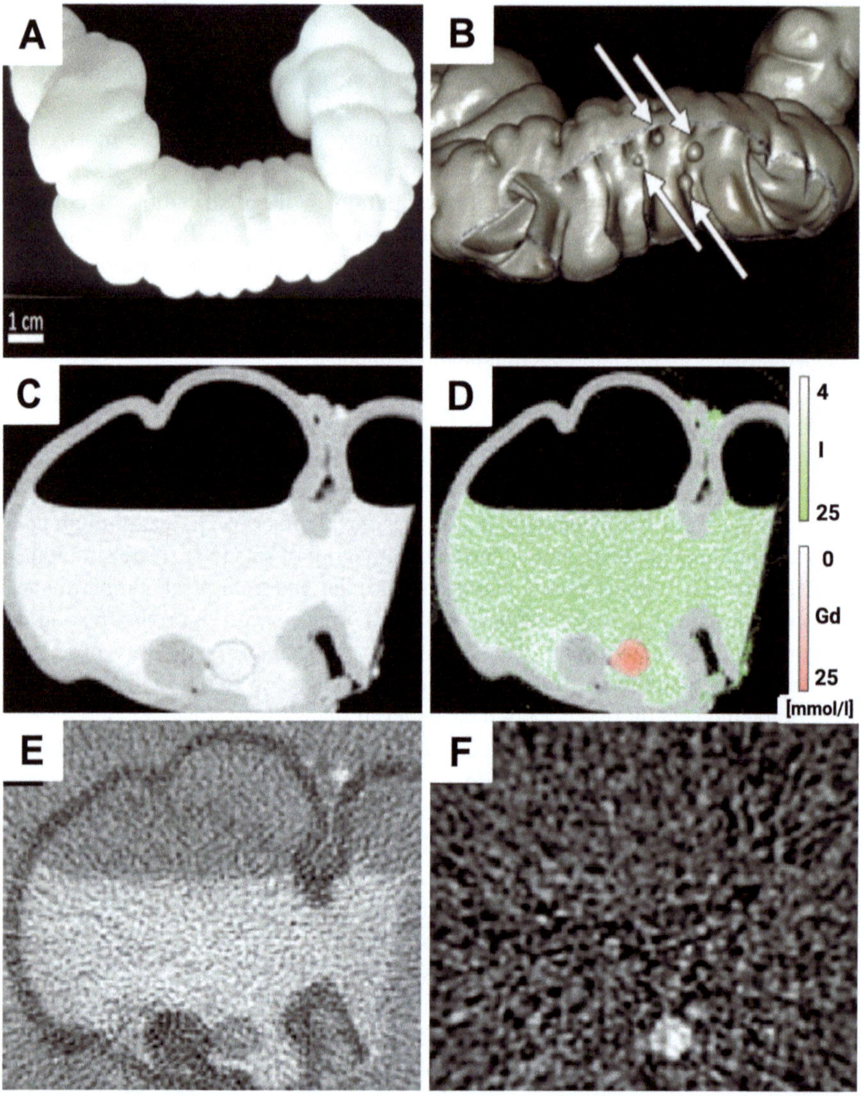

Fig. 4.2 (**a**) Image of the colon phantom. (**b**) 3D CT data of the phantom acquired with a conventional scanner. (**c**) Conventional CT scan of the colon phantom. (**d**) Conventional CT scan with an overlay of iodine (green) and gadolinium (red). (**e**) Iodine image. (**f**) Gadolinium image. (Figure reproduced with permission from reference [48])

administration of iodine and gadolinium-based contrast agents, a single SPCCT scan facilitated the visualization of portal venous and arterial phases within the livers of healthy rabbits. The distinct enhancement observed in the iodine and gadolinium maps, primarily concentrated in the portal vein and hepatic artery, respectively, which allows for clear differentiation [58]. The observed elevated

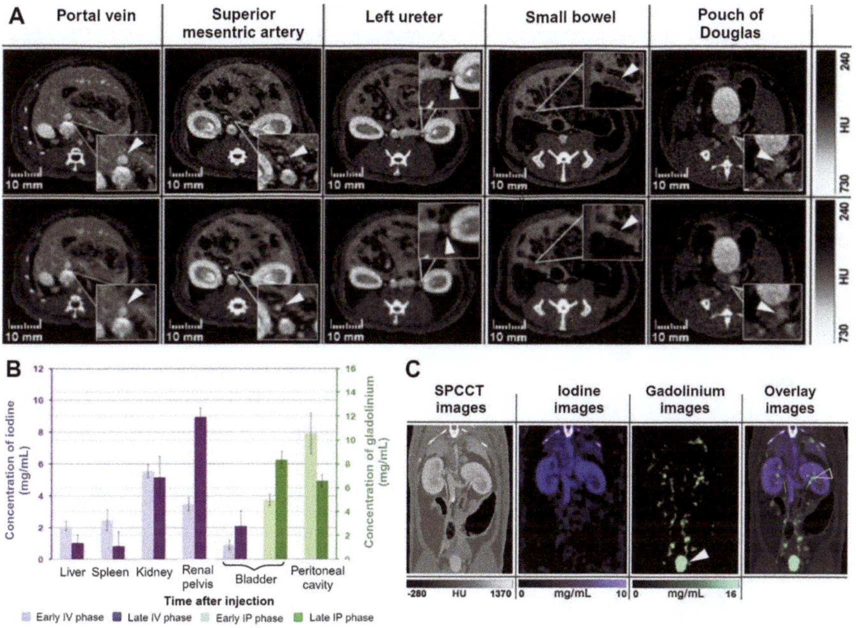

Fig. 4.3 (**a**) Comparison of SPCCT images at 250 μm (photon-counting CT) and 600 μm (single-energy CT-like) reveals anatomical structures, such as those noted here, with magnification boxes for specific features (arrowheads). (**b**) SPCCT imaging allows specific quantification of contrast agents, as exemplified in the peritoneal cavity and organ concentrations displayed here (iodine concentrations and gadolinium concentrations are displayed in purple and green, respectively). (**c**) SPCCT images and material contrast maps reveal the urinary tract excretion of the IP agent, which is distinctly visible only in the contrast maps. (Figure reproduced with permission from reference [53])

iodine concentration in the portal vein and minimal gadolinium enhancement align with anticipated vascular perfusion patterns, highlighting the potential of SPCCT in liver imaging and diagnosis (Fig. 4.4) [58].

Moreover, the capabilities of SPCCT, combined with a two-contrast agent injection protocol, can prove advantageous in cardiovascular imaging [40]. For instance, in a study conducted by Dangelmaier et al., SPCCT was employed with a two-contrast agent approach using iodine and gadolinium. The researchers successfully achieved accurate identification of endoleaks with a single CT acquisition [59]. In their study, they simulated enhanced blood in the stent-lined section of an abdominal aortic aneurysm phantom by filling it with a blend of iodine and gadolinium, while neighboring compartments were filled with either of the contrast agents or calcium chloride to imitate intra-aneurysmal calcifications and replicate endoleaks with varying flow rates. Following this, material-specific maps were employed to differentiate the distributions within the compartments containing iodine, gadolinium, or calcium [59]. The results demonstrated the ability of SPCCT to capture

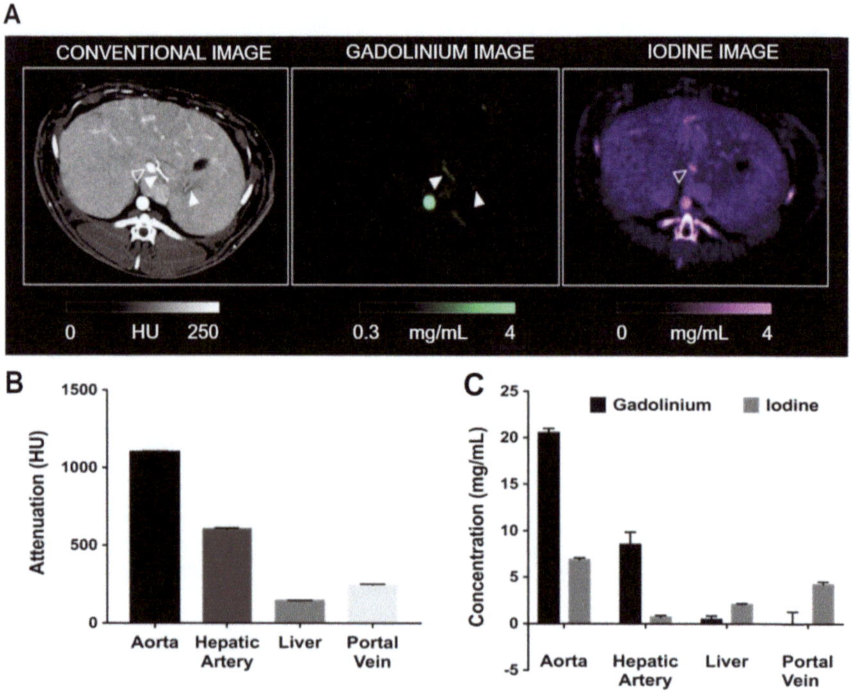

Fig. 4.4 (**a**) Three types of dual-phase SPCCT imaging of the liver: Conventional, gadolinium and iodine image. (**b**) Attenuation values (HU) and (**c**) concentrations of gadolinium and iodine in the structures noted. (Figure reproduced with permission from reference [58])

endoleak dynamics and reliably distinguish them from intra-aneurysmatic calcifications in a single scan. This innovative approach may enhance diagnostic precision, because it avoids motion issues between scans, and also allows for a substantial reduction in radiation exposure [59]. Since patients who have abdominal aortic aneurysms require frequent CT scanning to monitor their condition, the potential reduction in radiation would be very significant.

4.4.2 Nanoparticles

In the exploration of alternative options for creating innovative contrast agents, researchers have investigated a range of materials, placing particular emphasis on nanoparticles. Nanoparticles have garnered considerable attention in the field of CT contrast agents due to the enormous payloads that they can carry (as much as tens of millions of contrast-generating atoms per nanoparticle), which is particularly valuable for a low-sensitivity technique such as CT [60]. These high payloads make nanoparticles promising for imaging specific disease sites, offering the potential for the development of advanced contrast agents [61]. In addition, since iodinated agents are small molecules and SPCCT can image multiple agents simultaneously, there is a rationale to develop new agents with different structures to the iodinated agents to enable differential diagnoses. This is not to say that alternative small molecules, chelates, or cluster-based agents would not be of interest as novel SPCCT agents; however, there has been scant research on such agents, and the bulk of reports have focused on nanoparticles (Fig. 4.5) [62].

A study by Kim et al. aimed to identify elements suitable as candidates for novel SPCCT contrast agents [15]. The researchers applied specific criteria to exclude elements with high toxicity or radioactivity, such as thulium and radium. The study ultimately selected gadolinium, ytterbium, tantalum, tungsten, gold, and bismuth as the most suitable elements for contrast agent development. The experiments carried out demonstrated that gadolinium, ytterbium, and tantalum in particular exhibit elevated contrast-to-noise ratio in SPCCT [15]. These results suggest that these elements show promise as superior candidates for further refinement into novel contrast agents tailored for SPCCT imaging.

Gadolinium underwent additional scrutiny in a study by Halttunen et al., where they developed Nano-GdF_3 [63]. This nano-contrast agent exhibited an intense signal in the heart and liver of the animals post-injection, reaching a maximum at 30 min for the heart and 24 h for the liver. Regardless of gadolinium concentration, the strong contrast gradually faded after more than 2 h, highlighting the nanoparticles' prolonged presence in the blood pool. This result demonstrated a slow uptake by the reticuloendothelial system, affirming the stability in the blood. No mortality was noted in biodistribution studies with mice and rats. Conventional CT technology displayed a strong HU signal, facilitating the visualization of small vessels and arteries, while K-edge imaging provided a specific depiction of the contrast media within the animal vasculature. Combining both modalities enabled observation of

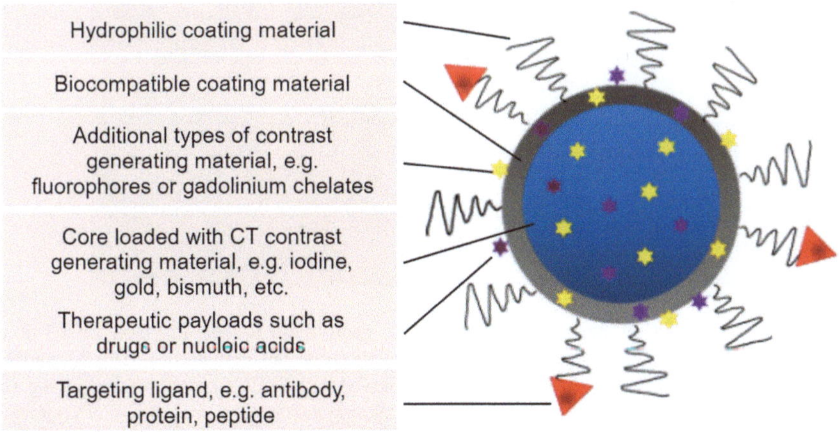

Fig. 4.5 Schematic representation illustrating a general concept of nanoparticle contrast agents in CT. (Figure reproduced with permission from reference [62])

abdominal vascularization and particle localization in the bloodstream, holding promise for studies on vascularization-associated pathologies like ischemia or cancer angiogenesis. Additionally, SPCCT quantified gadolinium concentration at 17.3 mg/mL, showcasing the technology's potential for accurate concentration determination in a given volume (Fig. 4.6) [63].

Although gold is a relatively expensive element, it is highly biocompatible and gold nanoparticles can be synthesized in almost any conceivable size or shape, which has encouraged an enormous number of studies on the biomedical applications of it [64]. Gold nanoparticles have shown considerable promise as SPCCT agents specifically, as evidenced by studies from Cormode et al. [15, 16] This research illustrated the practicality of using SPCCT to differentiate between gold nanoparticles (AuNP) and an iodinated contrast agent. Phantoms containing calcium phosphate and contrast agents underwent imaging, resulting in conventional CT, gold K-edge, iodine, and water images. These images accurately distinguished and measured gold and iodine concentrations in the mixtures within the phantom [16].

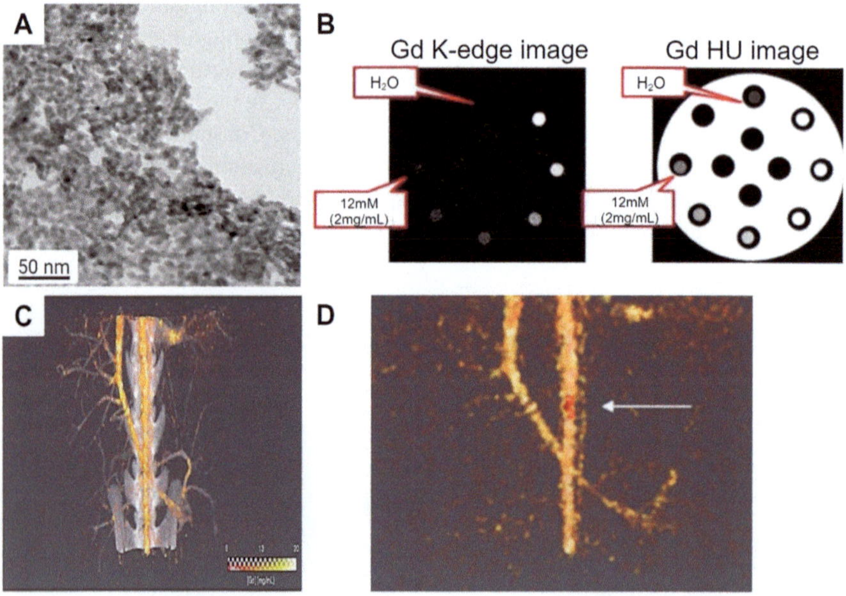

Fig. 4.6 (**a**) TEM of GdF_3 nanoparticles. (**b**) SPCCT gadolinium K-edge image on the left illustrates a phantom containing GdF3 suspensions with concentrations ranging from 2 to 20 mg/mL, while the right image presents the same phantom observed with SPCCT conventional imaging. (**c**) Conventional HU image obtained through SPCCT depicting the rat's abdomen after GdF_3 injection at 5 min. (**d**) A highlighted box (in red) delineates the region for Gd^{3+} quantification in the aorta. (Figure reproduced with permission from reference [63])

In vivo experiments were carried out on rabbits at various time points following the administration of AuNP and iodinated contrast agents. The contrast material maps effectively separated gold and iodine distributions in tissues, enabling the quantification of contrast agent concentrations consistent with expected pharmacokinetics. Additionally, rapid and repetitive scanning facilitated high-temporal-resolution measurements of the contrast agent kinetics (Fig. 4.7) [16].

To summarize, the study demonstrated that a clinical-scale, high count rate SPCCT system has the ability to differentiate between gold and iodine contrast media within various organs in vivo. This finding not only suggests potential applications in preclinical and clinical settings for investigating various biological processes and diseases but also underscores the importance of developing contrast agents specifically designed for SPCCT. Such agents could play a crucial role in expanding CT-based molecular imaging and enabling multi-phase imaging in a single scan [16].

Beyond gold's application for dual contrast agent SPCCT, studies including those conducted by Is-Mohamed et al. also examined the specific identification and quantification of macrophage burden in atherosclerotic New Zealand white rabbits. This was achieved by conducting imaging before and at various intervals up to 2 days following the intravenous injection of 12.5 nm, PEG-coated gold nanoparticles. Gold K-edge images, specifically at the 2-day mark post-injection, displayed the absorption of gold nanoparticles by macrophage foam cells within the aorta wall. Gold nanoparticle uptake by these cells was further corroborated by histologic sections, transmission electron microscopy, and quantification via inductively coupled plasma optical emission spectrometry (ICP) in ex vivo aortas. Hence, SPCCT

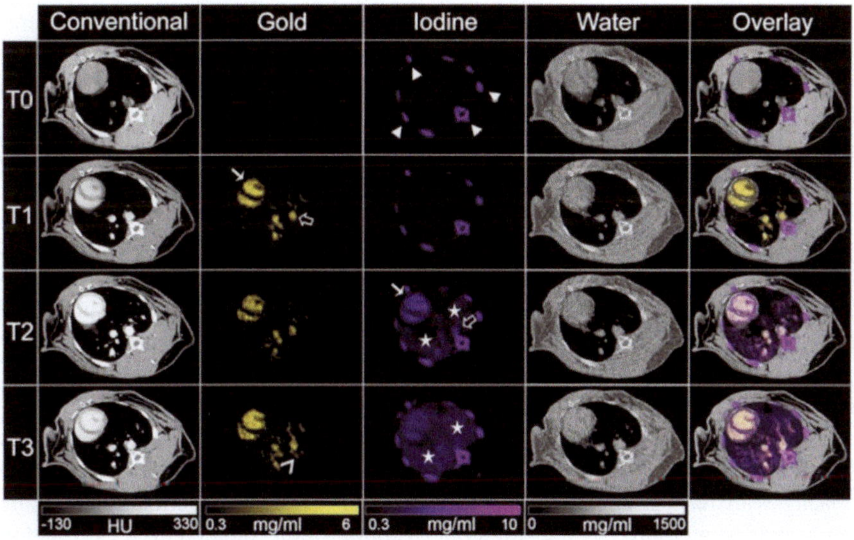

Fig. 4.7 Photon-counting CT images of the chest of a rabbit featuring conventional, gold, iodine, water, and overlay images across four distinct time points. Bones are marked with arrowheads, the heart with white arrows, the aorta with black arrows, and the lungs with stars. (Figure reproduced with permission from reference [16])

with gold nanoparticles facilitated the precise identification and quantification of macrophage burden in atherosclerotic plaques (Fig. 4.8) [65].

In addition to gold, hafnium has also emerged as a significant element used in SPCCT, going beyond the elements identified by Kim et al. [15] In the field of nanomedicine, hafnium nanoparticles (Hf NPs) serve various roles, acting as radiosensitizers in cancer radiotherapy, contrast agents for diagnostic X-ray imaging, biosensors, and radioluminescent scintillators [15, 66–68]. Notably, hafnium-based agents offer superior contrast compared to iodine [69]. The advantage of hafnium is that it is the element before tantalum in the periodic table and therefore has a similar excellent contrast-generating performance in SPCCT [67]. These nanoparticles find applications in various contexts, including atherothrombosis. Sebti et al. investigated the potential application of uncoated hafnium oxide nanoparticles (HfO_2 NPs), synthesized through a reaction of hafnium(IV) chloride and subsequently coated with fucoidan (a sulfated polysaccharide known for its high affinity for P-selectin, a glycoprotein biomarker in atherothrombosis), and citrate as a control [70]. Both coatings exhibited good colloidal stability in water and glucose. The HfO_2 NPs demonstrated good biocompatibility in vitro, even at high extracellular concentrations (42 mM). Efficient X-ray attenuation in SPCCT Spectral photon-counting CT (SPCCT) was demonstrated via phantom imaging. Based on the data they concluded that these HfO_2@fucoidan NPs emerged as a promising candidate for X-ray imaging, offering molecular-scale diagnostic capabilities for atherothrombosis [70].

Additionally, ostadhossein demonstrated the effectiveness of Hf NPs in visualizing the gastrointestinal tract [71]. In their case, they produced the Hf NPs via the reduction of Hf nitrate at room temperature, resulting in ~2 nm HfO_2 nanodots coated with polyvinylpyrrolidone (PVP) (P-HfO_2 NP). After incubation with probiotic bacteria, the lyophilized bacteria were loaded into capsules for oral administration, serving as radiopaque contrast agents. The P-HfO_2 NP showed cytocompatibility at concentrations as high as 10 mg/ml. K-edge imaging of P-HfO_2 NP in Sprague Dawley rats allowed the delineation of the esophageal lining, with distinct contrast observed for P-HfO_2 NPs post 30 min of administration (Fig. 4.9) [70, 71].

Another interesting development in the study of nanoparticles as contrast agents for SPCCT was made by Balegamire et al., where they developed iodinated polymer nanoparticles with a size of approximately 150 nm in diameter as a contrast agent for SPCCT [72]. This was achieved through precipitation of poly(vinyl alcohol) that had triiodobenzoyl groups covalently attached. PCL-b-PEG was used to provide a PEGylated shell, which stabilized the resulting nanoparticles for 8 months in water and human serum. In vitro experiments demonstrated 100% cell viability after 5 h of exposure to concentrations ranging from 0 to 10 mg/mL of iodinated nanoparticles. Radiopacity was tested in a phantom using iodinated polymer nanoparticles suspended in water at varying concentrations, revealing a linear increase in radiopacity with respect to iodine concentration and a good correlation between the concentration measured by the "Material Decomposition" method in SPCCT and in the ICP. Intravenous injections of these nanoparticles to rats, together with conventional CT and μCT imaging, allowed for specific monitoring of the blood vessels over several hours and resulted in substantial radiopacity (200–250 HU). This radiopacity remained above 150 HU for at least 5 h and returned to the initial value (50

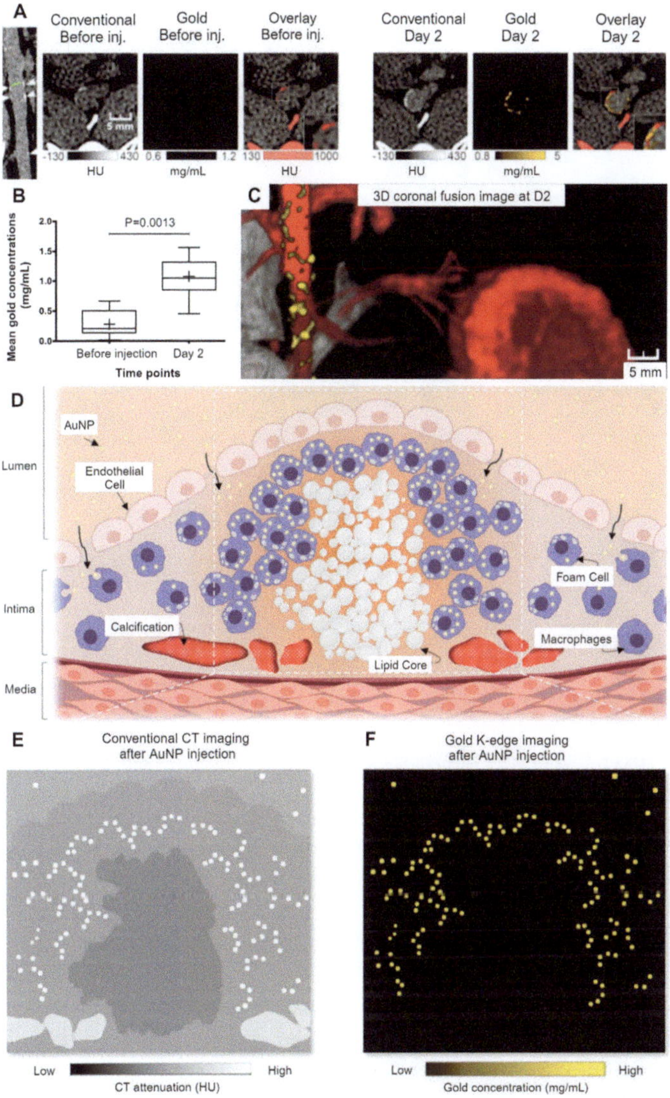

Fig. 4.8 In vivo molecular imaging of macrophage burden in an atherosclerotic rabbit model using color K-edge imaging in combination with gold nanoparticles for inflammation imaging and iodinated contrast media for lumen imaging. (**a**) Cross-sections of a calcified atherosclerotic plaque before injection, with gold uptake on color K-edge imaging at day 2 with a mean gold nanoparticle concentration of 2.74 mg/mL in the rabbit abdominal aortic wall. (**b**) Quantitative analysis reveals changes in gold concentration before and after injection. (**c**) Simultaneous bi color imaging of lumen via use of iodinated contrast agent and inflammation via use of gold nanoparticles. (**d**) Schematic of molecular K-edge imaging detects and quantifies macrophages in calcified plaque with gold nanoparticles. (**e**) A schematic highlights challenges in conventional CT for macrophage burden analysis. (**f**) Schematic representation of PCCT gold K-edge imaging offering specific, non-invasive macrophage burden quantification. (Figure reproduced with permission from reference [65])

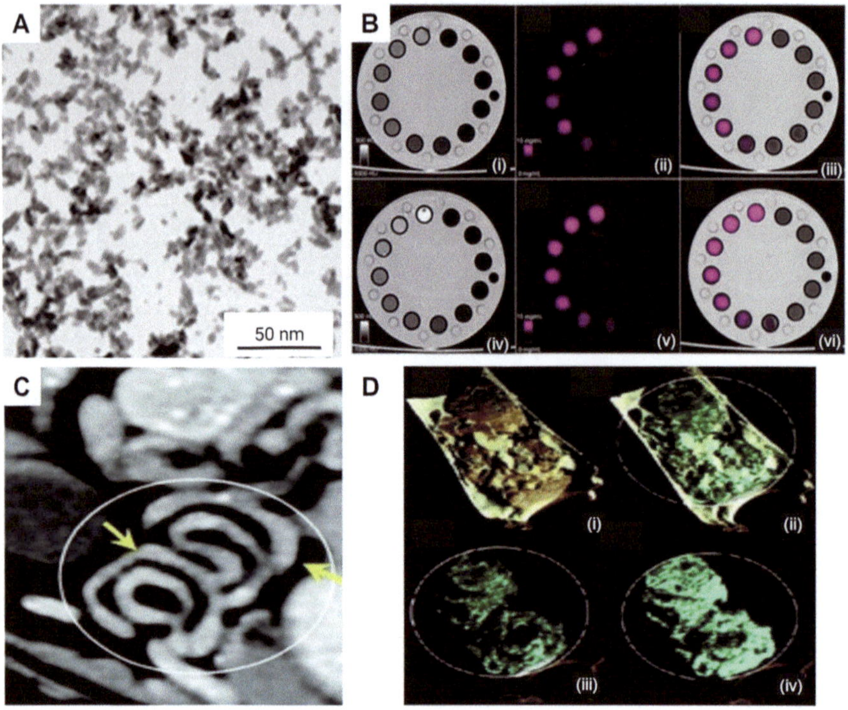

Fig. 4.9 (**a**) TEM image of HfO_2@fucoidan NPs. (**b**) SPCCT images include conventional images (i and iv), Hafnium K-edge images (purple, ii and v), and overlay of K-edge with conventional images (iii and vi) for HfO_2@citrate NPs (top row) and HfO_2@fucoidan (bottom row). (**c**) Photon counting CT image of the GI tract with rats fed before probiotic-Hf capsule gavage. (**d**) Corresponding material-weighted photon counting CT images (i) depicting lipid (yellow) and (ii) HfO_2 NPs (cyan); (iii) HfO_2 NPs (cyan) and HA (gray), and (iv) HfO_2 NPs (cyan). (Figure reproduced with permission from references [70, 71])

HU) after 24 h. In addition, it was observed that the iodinated polymer nanoparticles gradually accumulated in the liver and spleen after 3 h and up to 5 days post-injection. Therefore, this substance serves as a "blood pool" contrast agent with an extended residence time in the bloodstream, which can be beneficial for angiography over long durations without multiple injections (Fig. 4.10) [72].

4.4.3 Cell Tracking and Hydrogel Implants

Amid the diverse applications of SPCCT, the realm of cell tracking has emerged as an area of increasing interest. Cell tracking can provide non-invasive and real-time insights into the dynamics of cell therapies, regenerative medicine, and immune responses. Examples of such topics include stem cell-based therapies and CAR T-cell cancer therapeutics [73]. Cell tracking with SPCCT provides an

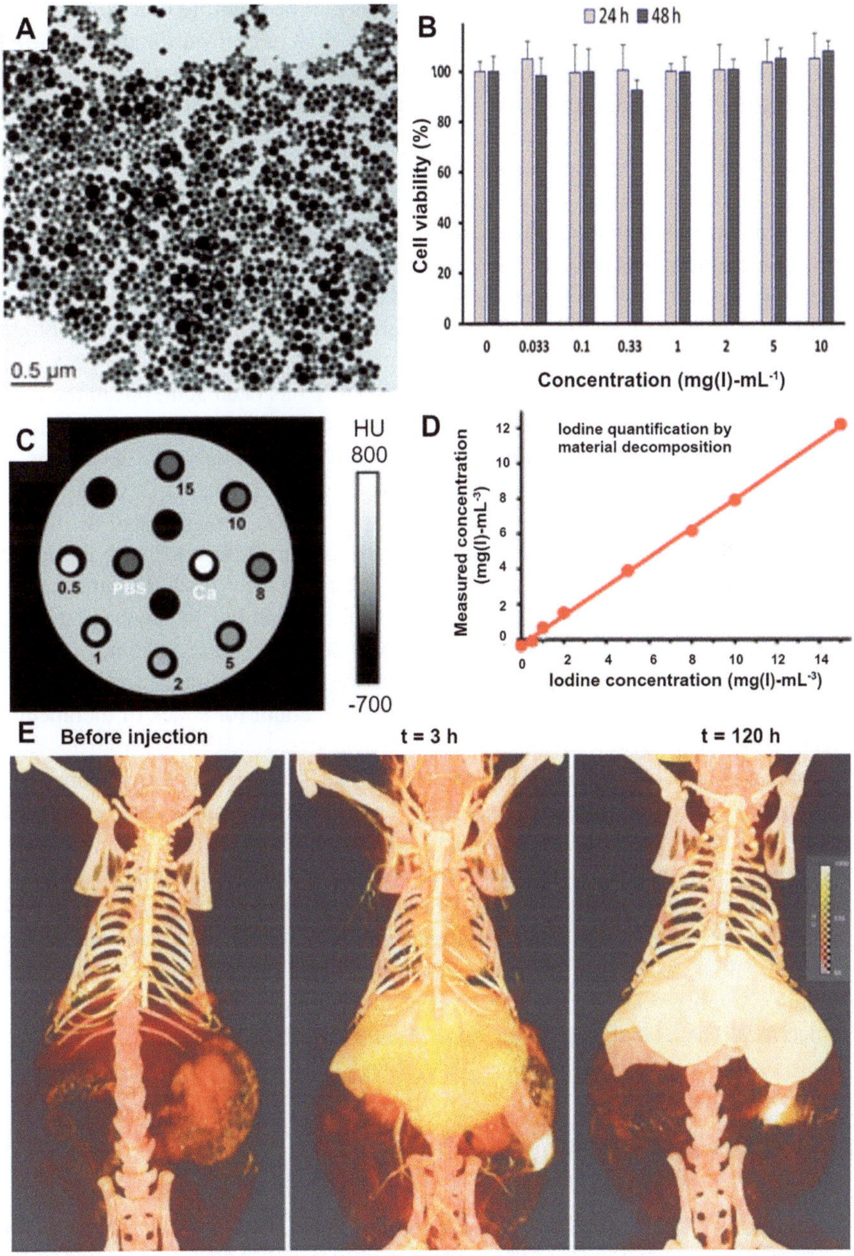

Fig. 4.10 (**a**) TEM of TIB-PVAL nanoparticles (**b**) Cell viability, after 5 h of exposure to varying concentrations of iodinated nanoparticles. (**c**) A phantom with tubes filled with iodinated polymer nanoparticles in water, imaged using an SPCCT system. *Ca* calcium, *PBS* phosphate-buffered saline; numerical values indicate iodine concentrations in mg(I) mL^{-1}. (**d**) Quantitative iodine measurement through material decomposition. (**e**) CT images of rats before, 3 h, and 120 h after a 1.5 mL intravenous injection of iodinated polymer nanoparticles. (Figure reproduced with permission from reference [72])

ability to detect specific cells since labeled cells can be identified on post-administration K-edge maps, whereas other techniques such as conventional CT or MRI require pre- and post-administration comparisons, these conventional methods are prone to errors and allow areas of cell accumulation to go unnoticed [74]. In this section, we will explore exciting research that involves SPCCT in cell tracking. Moreover, we will also discuss the application of SPCCT to the tracking of hydrogel implants. Several hydrogels are FDA-approved for purposes such as drug administrations or tissue protection during radiotherapy. Labeling a hydrogel with radiopaque materials can allow it to be imaged with SPCCT, which is an appealing method due to the specificity of detection and the ability to simultaneously image the hydrogel as well as measure other parameters including blood flow or additional payloads.

Particular progress in tracking cells using SPCCT has been achieved in the field of brain damage [45]. Cuccione et al. reported that SPCCT imaging enabled the monitoring of bone marrow-derived M2-polarized macrophage cells labeled with 11-MUDA gold nanoparticles in brain-damaged rats for up to 2 weeks post-injection, providing specificity and quantification. SPCCT imaging also facilitated the differentiation of labeled cells with a supporting scaffold labeled with iodinated nanoparticles, with a detection limit as low as 5000 cells per voxel (250 μm in each dimension). Furthermore, SPCCT imaging has the capability to illustrate the 3D distribution of cells throughout the entire brain and track their outcomes over an extended duration. This is important for evaluating cell engraftment and determining whether the early disappearance of cells could account for a lack of therapeutic effectiveness [45].

Dong et al. also highlighted the potential of SPCCT combined with nanoparticle labeling to effectively monitor hydrogels [11]. In their study, they conducted research on the synthesis of ultrasmall ytterbium nanoparticles (YbNP). The researchers assessed the cytotoxicity and contrast-generating properties of these nanoparticles in vitro across various CT imaging systems including SPCCT. In particular, YbNP were investigated for their suitability in labeling and monitoring hydrogels, showcasing their presence within hydrogels for in vivo localization and tracking. Notably, the in vivo imaging results demonstrated that YbNP outperformed AuNP as a label, generating greater contrast (Fig. 4.11) [11].

4.5 Safety and Biocompatibility

The context of SPCCT, similar to that for conventional CT or for a contrast agent for any modality, agent safety is of the highest importance [75]. The safety of contrast agents is a multifactorial concept in medical imaging. In general, contrast agents should be non-toxic, non-allergenic, and devoid of significant side effects. Additionally, they should be biocompatible, which refers to the ability of the contrast agents to exist in harmony with the body's biological systems without eliciting immune responses and inflammation [20]. In the case of SPCCT, where contrast

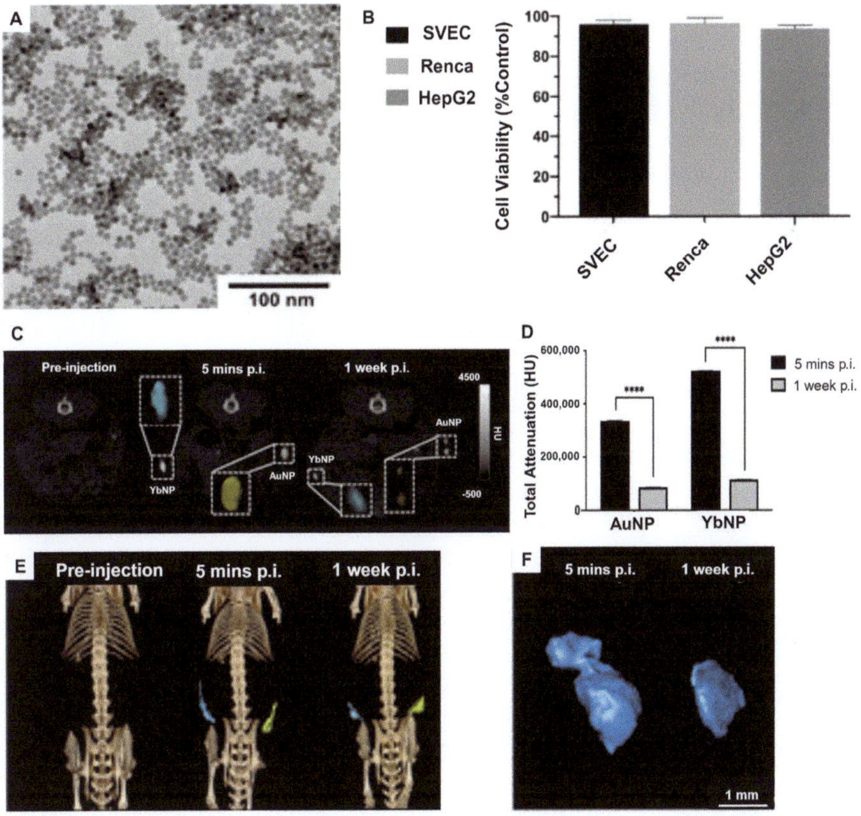

Fig. 4.11 (**a**) Transmission electron micrograph of YbNP. (**b**) Assessment of cell viability after 8 h of incubation with YbNP-loaded hydrogels. (**c**) In vivo assessment of contrast generation from YbNP and AuNP-loaded hydrogels. (**d**) Quantification of total CT attenuation attributed to hydrogels. (**e**) Representative 3D images derived from CT scans of a mouse injected with YbNP and AuNP-loaded hydrogels at various time points. (**f**) 3D reconstructions of a YbNP-loaded hydrogel derived from such CT images. (Figure reproduced with permission from reference [11])

agents may be designed to target specific tissues and molecular pathways, their biocompatibility ensures that they do not induce any unintended biological response.

Traditional CT contrast agents are typically small-molecule iodinated compounds [75]. While effective, they have limitations such as a relatively short circulation time in the bloodstream and potential toxicity at high doses [62]. Nanoparticles based on heavy elements, like gold (Au), are more effective in absorbing X-rays in the diagnostic range, which is fundamental for contrast enhancement in CT imaging. Depending on the design, they have an extended circulation time, allowing for improved imaging of blood vessels and blood flow [62]. Gold nanoparticles are known for their biocompatibility, meaning they are less likely to cause adverse reactions within the body, even at high concentrations. This makes them safer for use, especially in cases where repeated imaging is required [22]. However, it should be noted that if nanoparticles accumulate in organs of the RES, they can potentially obscure or distort imaging results, leading to inaccurate diagnosis [76]. More importantly, prolonged retention of an agent in this manner would likely concern regulatory agencies due to the potential for unknown long-term side effects and preclude their approval for clinical use. Solutions to this issue include designing agents that can degrade into excretable components in vivo or by use of sub-5 nm nanoparticles that can be excreted via the kidneys, thereby reducing potential safety concerns [11].

4.6 Challenges and Future Directions

The challenges and future directions of contrast agents in SPCCT present a multifaceted landscape. One prominent challenge lies in optimizing the parameters for CT examinations, especially with the added complexities introduced by SPCCT capabilities. Striking a delicate balance between high sensitivity and specificity while minimizing contrast agent usage poses a significant hurdle. Furthermore, the evolution of injection protocols, necessitated by altered pharmacokinetics and potential interactions with new contrast agents, requires thorough reevaluation [12].

Looking ahead, the exploration of novel contrast agents for SPCCT opens exciting possibilities. The development of agents tailored to specific imaging needs, from clinical cell tracking to peritoneal imaging, highlights the potential for innovation. As researchers further study nanoparticle applications, such as gold and hafnium-based agents, and explore the use of iodinated polymer nanoparticles, the potential of SPCCT in changing contrast-enhanced imaging becomes evident.

In the future, advancements in SPCCT technology and the development of next-generation multicolor clinical scanners are anticipated. Researchers aim to refine contrast agent combinations, focusing on contrast volume, concentration, injection protocols, scan delays, and image reconstruction parameters. This quest for optimization will likely involve extensive research and clinical trials over the years.

As alluded to above, regulatory considerations are crucial for the approval of contrast agents. Agencies like the U.S. Food and Drug Administration (FDA) and the European Medicines Agency (EMA) have established guidelines that mandate laboratory testing and clinical trials to develop efficacy and safety [77]. The approval process involves assessing the contrast agents' interaction with the body, their metabolism and excretion, and potential long-term effects. Despite the considerable clinical benefits of SPCCT and the expected benefits of novel agents, the path from discovery to clinical approval is a lengthy process that involves a considerable sum of money and the collaboration of highly skilled teams [78].

Despite the challenges noted above, the clinical uptake of SPCCT in the last couple of years has been truly remarkable and it is expected that SPCCT scanners will represent a substantial part of the medical imaging done within the next decade [79]. The emergence of this imaging market will create an impetus for more research into novel contrast agents for SPCCT both in terms of innovation and exploitation of existing agents as well as the clinical translation of novel agents. Overall, the outlook for SPCCT agents in the coming years is very encouraging.

Acknowledgments This work was supported in part by NIH grants R21-EB029158 (DPC) and F31-EB034165 (DNR).

References

1. Power SP, Moloney F, Twomey M, James K, O'Connor OJ, Maher MM. Computed tomography and patient risk: facts, perceptions and uncertainties. World J Radiol. 2016;8:902.
2. Flohr T, Petersilka M, Henning A, Ulzheimer S, Ferda J, Schmidt B. Photon-counting CT review. Phys Med. 2020;79:126.
3. Mak GS, Truong QA. Cardiac CT: imaging of and through cardiac devices. Curr Cardiovasc Imaging Rep. 2012;5:328.
4. Dodd JD, Leipsic J. Cardiovascular CT and MRI in 2019: review of key articles. Radiology. 2020;297:17.
5. Wells ML, Hansel SL, Bruining DH, Fletcher JG, Froemming AT, Barlow JM, et al. CT for evaluation of acute gastrointestinal bleeding. Radiographics. 2018;38:1089.
6. Furukawa A, Sakoda M, Yamasaki M, Kono N, Tanaka T, Nitta N, et al. Gastrointestinal tract perforation: CT diagnosis of presence, site, and cause. Abdominal Imaging. 2005;30:524.
7. Hounsfield GN. Computerized transverse axial scanning (tomography): I. Description of system. Br J Radiol. 1973;46:1016–22.
8. Willemink MJ, Leiner T, de Jong PA, de Heer LM, Nievelstein RA, Schilham AM, et al. Iterative reconstruction techniques for computed tomography part 2: initial results in dose reduction and image quality. Eur Radiol. 2013;23:1632–42.
9. Willemink MJ, de Jong PA, Leiner T, de Heer LM, Nievelstein RAJ, Budde RPJ, et al. Iterative reconstruction techniques for computed tomography part 1: technical principles. Eur Radiol. 2013;23:1623–31.
10. Goldman LW. Principles of CT and CT technology. J Nucl Med Technol. 2007;35:115–28.
11. Dong YC, Kumar A, Rosario-Berríos DN, Si-Mohamed S, Hsu JC, Nieves LM, et al. Ytterbium nanoparticle contrast agents for conventional and spectral photon-counting CT and their applications for hydrogel imaging. ACS Appl Mater Interfaces. 2022;14:39274–84.
12. Tortora M, Gemini L, D'Iglio I, Ugga L, Spadarella G, Cuocolo R. Spectral photon-counting computed tomography: a review on technical principles and clinical applications. J Imaging. 2022;8:112.
13. Kim J. Development of nanoparticle-based contrast agents for applications with conventional and photon-counting CT imaging. Philadelphia: University of Pennsylvania; 2021.
14. Sandfort V, et al. Spectral photon-counting CT in cardiovascular imaging. J Cardiovasc Comput Tomogr. 2021;15:218.
15. Kim J, Persson M, Pourmorteza A, Noël PB, Fleischmann D, Willemink MJ. Assessment of candidate elements for development of spectral photon-counting CT specific contrast agents. Sci Rep. 2018;8:12119.
16. Cormode DP, Si-Mohamed S, Bar-Ness D, Sigovan M, Naha PC, Balegamire J, et al. Multicolor spectral photon-counting computed tomography: in vivo dual contrast imaging with a high count rate scanner. Sci Rep. 2017;7:4784.
17. Si-Mohamed SA, Boccalini S, Villien M, Yagil Y, Erhard K, Boussel L, Douek PC. First experience with a whole-body spectral photon-counting CT clinical prototype. Invest Radiol. 2023;58:459.

18. Greffier J, Si-Mohamed SA, Lacombe H, Labour J, Djabli D, Boccalini S, et al. Virtual monochromatic images for coronary artery imaging with a spectral photon-counting CT in comparison to dual-layer CT systems: a phantom and a preliminary human study. Eur Radiol. 2023;33:5476.
19. Hrvoje L, Greenstaff MW. X-ray computed tomography contrast agents. Chem Rev. 2014; https://doi.org/10.1021/cr200358s.
20. Hasebroock KM, Serkova NJ. Toxicity of MRI and CT contrast agents. Expert Opin Drug Metab Toxicol. 2009;5:403.
21. Schirra CO, Brendel B, Anastasio MA, Roessl E. Spectral CT: a technology primer for contrast agent development. Contrast Media Mol Imaging. 2014;9:62.
22. Si-Mohamed S, Cormode DP, Bar-Ness D, Sigovan M, Naha PC, Langlois J-B, et al. Evaluation of spectral photon counting computed tomography K-edge imaging for determination of gold nanoparticle biodistribution: in vivo. Nanoscale. 2017;9:18246.
23. Dong YC, Hajfathalian M, Maidment PSN, Hsu JC, Naha PC, Si-Mohamed S, et al. Effect of gold nanoparticle size on their properties as contrast agents for computed tomography. Sci Rep. 2019;9:14912.
24. Bonitatibus PJ, Torres AS, Goddard GD, Fitzgerald PF, Kulkarni AM. Synthesis, characterization, and computed tomography imaging of a tantalum oxide nanoparticle imaging agent. Chem Commun. 2010;46:8956.
25. Rabin O, Perez JM, Grimm J, Wojtkiewicz G, Weissleder R. An X-ray computed tomography imaging agent based on long-circulating bismuth sulphide nanoparticles. Nat Mater. 2006;5:118.
26. Zhou ZB, Kong CY, Shi X, Wang M, Liu W, et al. Tungsten oxide nanorods: an efficient nanoplatform for tumor CT imaging and photothermal therapy. Sci Rep. 2014;4:3653.
27. Alric C, Taleb J, Le Duc G, Mandon C, Billotey C, Le Meur-Herland A, et al. Gadolinium chelate coated gold nanoparticles as contrast agents for both X-ray computed tomography and magnetic resonance imaging. J Am Chem Soc. 2008;130:5908.
28. Amato C, Klein L, Wehrse E, Rotkopf LT, Sawall S, Maier J, et al. Potential of contrast agents based on high-Z elements for contrast-enhanced photon-counting computed tomography. Med. Phys. 2020;47:6179.
29. Nowak T, Hupfer M, Brauweiler R, Eisa F, Kalender WA. Potential of high-Z contrast agents in clinical contrast-enhanced computed tomography. Med Phys. 2011;38:6469.
30. Jost G, McDermott M, Gutjahr R, Nowak T, Schmidt B, Pietsch H. New contrast media for K-edge imaging with photon-counting detector CT. Invest Radiol. 2023;58:515.
31. Widmark JM. Imaging-related medications: a class overview. Proc Baylor Univ Med Cent. 2007;20:408.
32. Hassan R, Aziz AA. Computed tomography (CT) imaging of injuries from blunt abdominal trauma: a pictorial essay. Malays J Med Sci. 2010;17(2):29–39.
33. Treskes KT, Saltzherr P, Edwards MJR, Beuker BJA, Van Lieshout EMM, Hohmann J, et al. Refining the criteria for immediate total-body CT after severe trauma. Eur. Radiol. 2020;30:2955.
34. Ahmed N, Kassavin D, Kuo YH, Biswal R. Sensitivity and specificity of CT scan and angiogram for ongoing internal bleeding following torso trauma. Emerg Med J. 2013;30:e14.
35. Rossaint R, Afshari A, Bouillon B, Cerny V, Cimpoesu D, Curry N, et al. Management of bleeding following major trauma: an updated European guideline. Crit Care. 2010;14:R52.
36. Si-Mohamed SA, Miailhes J, Rodesch P-A, Boccalini S, Lacombe H, Leitman V, et al. Spectral photon-counting ct technology in chest imaging. J Clin Med. 2021;10:5757.
37. Tee A, Wong A, Yusuf GT, Rao D, Sidhu PS. Contrast-enhanced ultrasound (CEUS) of the lung reveals multiple areas of microthrombi in a COVID-19 patient. Intensive Care Med. 2020;46:1660.
38. Rotaru L, Nanea T. Assessment of myocardial perfusion using contrast echocardiography - case report. J Med Life. 2015;8(4):471–5.

39. Albrecht MH, De Cecco CN, Schoepf UJ, Spandorfer A, Eid M, De Santis D, et al. Dual-energy CT of the heart current and future status. Eur J Radiol. 2018;105:110–8.
40. Rotzinger DC, Racine D, Becce F, Lahoud E, Erhard K, Si-Mohamed SA, et al. Performance of spectral photon-counting coronary CT angiography and comparison with energy-integrating-detector CT: objective assessment with model observer. Diagnostics. 2021;11:2376.
41. Zhou Y, Zhu Y, Li J. Advantages of CT nano-contrast agent in tumor diagnosis a retrospective study. Medicine. 2021;100:e27044.
42. Goo HW, Goo JM. Dual-energy CT: new horizon in medical imaging. Korean J Radiol. 2017;18:555.
43. Odisio EG, Truong MT, Duran C, de Groot PM, Godoy MC. Role of dual-energy computed tomography in thoracic oncology. Radiol Clin North Am. 2018;56:535–48.
44. Thivolet A, Si-Mohamed S, Bonnot P-E, Blanchet C, Képénékian V, Boussel L, et al. Spectral photon-counting CT imaging of colorectal peritoneal metastases: initial experience in rats. Sci Rep. 2020;10:1–13.
45. Cuccione E, Chhour P, Si-Mohamed S, Dumot C, Kim J, Hubert V, et al. Multicolor spectral photon counting CT monitors and quantifies therapeutic cells and their encapsulating scaffold in a model of brain damage. Nanotheranostics. 2020;4:129.
46. Parakh A, Lennartz S, An C, Rajiah P, Yeh BM, Simeone FJ, et al. Dual-energy CT images: pearls and pitfalls. Radiographics. 2021;41:98.
47. McCollough CH, Leng S, Yu L, Fletcher JG. Dual- and multi-energy CT: principles, technical approaches, and clinical applications. Radiology. 2015;276:637.
48. Muenzel D, Bar-Ness D, Roessl E, Blevis I, Bartels M, Fingerle AA, et al. Spectral photon-counting cT: initial experience with dual-contrast agent K-edge Colonography. Radiology. 2017;283:723.
49. Symons R, Krauss B, Sahbaee P, Cork TE, Lakshmanan MN, Bluemke DA, et al. Photon-counting CT for simultaneous imaging of multiple contrast agents in the abdomen: an in vivo study. Med Phys. 2017;44:5120.
50. Ramalho J, Semelka RC, Ramalho M, Nunes RH, AlObaidy M, Castillo M. Gadolinium-based contrast agent accumulation and toxicity: an update. AJNR Am J Neuroradiol. 2016;37:1192.
51. Perry HL, Botnar RM, Wilton-Ely JDET. Gold nanomaterials functionalised with gadolinium chelates and their application in multimodal imaging and therapy. Chem Commun. 2020;56:4037.
52. Patel CM, Sahdev A, Reznek RH. CT, MRI and PET imaging in peritoneal malignancy. Cancer Imaging. 2011;11:123.
53. Si-Mohamed STA, Bonnot PE, Bar-Ness D, Képénékian V, Cormode DP, et al. Improved peritoneal cavity and abdominal organ imaging using a biphasic contrast agent protocol and spectral photon counting computed tomography K-edge imaging. Invest Radiol. 2018;53:629.
54. Baron RL. Understanding and optimizing use of contrast material for CT of the liver. Am J Roentgenol. 1994;163:323.
55. Kulkarni NM, Fung A, Kambadakone AR, Yeh BM. Computed tomography techniques, protocols, advancements, and future directions in liver diseases. Magn Reson Imaging Clin North Am. 2021;29:305.
56. Kartalis N, Brehmer K, Loizou L. Multi-detector CT: liver protocol and recent developments. Eur J Radiol. 2017;97:101.
57. Muenzel D, et al. Photon counting CT of the liver with dual-contrast enhancement. In: Medical Imaging 2016: Physics of Medical Imaging; 2016.
58. Si-Mohamed S, Tatard-Leitman V, Laugerette A, Sigovan M, Pfeiffer D, Rummeny EJ, et al. Spectral photon counting computed tomography (SPCCT): in-vivo single-acquisition multi-phase liver imaging with a dual contrast agent protocol. Sci Rep. 2019;9:8458.
59. Dangelmaier J, Bar-Ness D, Daerr H, Muenzel D, Si-Mohamed S, Ehn S, et al. Experimental feasibility of spectral photon-counting computed tomography with two contrast agents for the detection of endoleaks following endovascular aortic repair. Eur Radiol. 2018;28:3318.

60. Jakhmola A, Anton N, Vandamme TF. Inorganic nanoparticles based contrast agents for X-ray computed tomography. Adv Healthc Mater. 2012;1:413.
61. Kim D, Shin K, Kwon SG, Hyeon T. Synthesis and biomedical applications of multifunctional nanoparticles. Adv Mater. 2018;30:e1802309.
62. Cormode DP, Naha PC, Fayad ZA. Nanoparticle contrast agents for computed tomography: a focus on micelles, vol. 9. Imaging: Contrast Media Mol; 2014. p. 37.
63. Halttunen N, Lerouge F, Chaput F, Vandamme M, Karpati S, Si-Mohamed S, et al. Hybrid Nano-GdF3 contrast media allows pre-clinical in vivo element-specific K-edge imaging and quantification. Sci Rep. 2019;9:12090.
64. Mieszawska AJ, Mulder WJM, Fayad ZA, Cormode DP. Multifunctional gold nanoparticles for diagnosis and therapy of disease. Mol Pharm. 2013;10:831.
65. Si-Mohamed SA, Sigovan M, Hsu JC, Tatard-Leitman V, Chalabreysse L, Naha PC, et al. In vivo molecular K-edge imaging of atherosclerotic plaque using photon-counting CT. Radiology. 2021;300:98.
66. Kiran I, Shad NA, Sajid MM, Mahmood HZ, Javed Y, Hanif M, et al. Emerging nanomaterials as radio-sensitizer in Radiotherapy. Cham: Springer; 2022.
67. Berger M, Bauser M, Frenzel T, Hilger CS, Jost G, Lauria S, et al. Hafnium-based contrast agents for X-ray comsputed tomography. Inorg Chem. 2017;56:5757.
68. Frenzel T, Bauser M, Berger M, Hilger CS, Hegele-Hartung C, Jost G, et al. Characterization of a novel hafnium-based x-ray contrast agent. Invest Radiol. 2016;51(12):776–85.
69. Ding S, Chen L, Liao J, Huo Q, Wang Q, Tian G, et al. Harnessing hafnium-based nanomaterials for cancer diagnosis and therapy. Small. 2023;19:e2300341.
70. Sebti Y, Si-Mohamed S, Aid R, Geinguenaud F, Chalal M, Lalatonne Y, et al. Optical and X-ray attenuation properties of hafnium oxide nanoparticles surface functionalized with fucoidan: toward the early diagnosis of atherothrombotic diseases. Mater Adv. 2023;4:1011.
71. Ostadhossein F, Moitra P, Gunaseelan N, Nelappana M, Lowe C, Moghiseh M, et al. Hitchhiking probiotic vectors to deliver ultra-small hafnia nanoparticles for 'color' gastrointestinal tract photon counting X-ray imaging. Nanoscale Horizons. 2022;7:533.
72. Balegamire J, Vandamme M, Chereul E, Si-Mohamed S, Maache SA, Almouazen E, et al. Iodinated polymer nanoparticles as contrast agent for spectral photon counting computed tomography. Biomater Sci. 2020;8:5715.
73. Tavakoli C, Cuccione E, Dumot C, Balegamire J, Si-Mohamed SA, Kim J, et al. High-resolution synchrotron K-edge subtraction CT allows tracking and quantifying therapeutic cells and their scaffold in a rat model of focal cerebral injury and can serve as a reference for spectral photon counting CT. Nanotheranostics. 2023;7:176.
74. Wiart M, Tavakoli C, Hubert V, Hristovska I, Dumot C, Parola S, et al. Use of metal-based contrast agents for in vivo MR and CT imaging of phagocytic cells in neurological pathologies. J Neurosci Methods. 2023;383:109729.
75. Lusic H, Grinstaff MW. X-ray-computed tomography contrast agents. Chem Rev. 2013;113:1641.
76. Lankveld DPK, Oomen AG, Krystek P, Neigh A, Troost-de Jong A, Noorlander CW, et al. The kinetics of the tissue distribution of silver nanoparticles of different sizes. Biomaterials. 2010;31:8350.
77. Muskula PR, Main ML. Safety with echocardiographic contrast agents. Circ Cardiovasc Imaging. 2017:10. https://doi.org/10.1161/CIRCIMAGING.116.005459.
78. Baldrick P. Nonclinical safety testing of imaging agents, contrast agents and radiopharmaceuticals. J Appl Toxicol. 2021;41:95–104.
79. Michelfeit F. Siemens Healthineers aims to expand photon-counting scanning technology to 1 billion patients over next decade; 2023.

Chapter 5
Clinical Potential of CZT-Based Photon Counting Detector CT for Lung Cancer Imaging

Tomoaki Sasaki, Hirofumi Kuno, Xiaohui Zhan, Ruoqiao Zhang, Zhou Yu, Hiroki Taguchi, and Tatsushi Kobayashi

Contents

5.1 Introduction

Photon-counting detector computed tomography (PCD-CT) is an emerging technique with the potential to significantly influence clinical practice. Initial studies suggest that PCD-CT may offer several advantages over energy-integrated detector CT (EID-CT), including improved spatial resolution, reduced image noise, enhanced iodine quantification, multi-energy spectral imaging capabilities, and lower

T. Sasaki · H. Kuno · T. Kobayashi
National Cancer Center Hospital East, Kashiwa, Japan

X. Zhan · R. Zhang (✉) · Z. Yu
Canon Medical Research USA, Inc., Vernon Hills, IL, USA
e-mail: rzhang@MRU.MEDICAL.CANON

H. Taguchi
Canon Medical System Corporation, Otawara, Tochigi, Japan

S. A. Si-Mohamed, K. K. Iniewski (eds.), *Clinical Applications of Spectral Photon Counting Computed Tomography Technology*,
https://doi.org/10.1007/978-3-031-85078-3_5

radiation dose [1–14]. However, further research is needed to fully validate these potential benefits. Notably, PCD-CT can provide both super high-resolution and spectral imaging simultaneously.

For chest imaging, PCD-CT's high spatial resolution and spectral capabilities have been reported to improve the visibility of normal lung structure [10, 15, 16], interstitial pneumonia [17, 18], pulmonary emboli [19, 20], and lung cancer screening [21]. To our knowledge, there are few reports specifically addressing the advantages of PCD-CT for lung cancer diagnosis, despite the many roles that CT is expected to play in lung cancer practice [22]. Since lung cancer is the leading cause of cancer-related death worldwide, improvements in CT imaging quality will impact lung cancer management and ultimately reduce mortality. Therefore, this chapter aims to discuss the potential of PCD-CT, particularly Cadmium Zinc Telluride–based PCD-CT, for lung cancer imaging.

5.2 Roles of Computed Tomography (CT) Imaging in Lung Cancer

Chest CT is now an indispensable tool in lung cancer management [22]. There are four major roles of CT for lung cancer: (1) lung cancer screening using low-dose CT for healthy candidates, (2) qualitative evaluation of a lung nodule, (3) staging of lung cancer, and (4) screening for recurrence and metastasis after treatment.

First, lung cancer screening using low-dose CT is one of the most important roles that CT fulfills. Two large clinical trials in Europe [23] and the United States [24] have demonstrated the effectiveness of low-dose CT in reducing lung cancer mortality. Low-dose CT enables the detection of early-stage lung cancer that could be potentially resected. However, many of the pulmonary nodules detected were false positives, leading to the development of guidelines for managing pulmonary nodules [25, 26].

Second, CT is the best modality for the qualitative evaluation of pulmonary nodules [22]. The qualitative evaluation considers several factors, including nodule size [27, 28], shape [29–31], density (whether the nodule is pure ground-glass, part-solid, or solid) [28], and accompanying findings such as cavity [29], cyst [29], calcification [31] or fat [31]. Solid lung nodules larger than 8 mm have an increased likelihood of malignancy [26], making accurate size measurement on CT crucial. Lobulated and spiculated nodules are more likely to be malignant than round, oval, or polygonal nodules [31]. Nodule density can be well correlated with histopathology. Pure ground-glass nodules are often associated with lepidic growth patterns with good prognosis [32]. Part-solid nodules are highly indicative of malignancy, especially adenocarcinoma, compared to solid nodules [31]. It is also important to assess the nodule's relationship with surrounding structures, such as the airway, vessels, pleura [30], chest wall, and coexisting lung conditions like emphysema or interstitial lung disease, which can increase lung cancer risk. Furthermore, the growth rate of tumor doubling time calculated from two interval CTs is also useful for estimating the possibility of malignancy [33]. A mean diameter increase of more than 1.5 mm over 12 months is considered an indicator of malignant potential [26].

Third, CT plays a major role in staging lung cancer according to the TNM classification [34], particularly in evaluating the T-category for tumor size and local invasion, such as chest wall or mediastinal involvement [27, 35]. Meanwhile, the four-dimensional respiratory-gated CT technique is also being explored to assess chest wall invasion in lung cancer [36]. CT is also crucial for evaluating the N and M categories, as it can detect intrathoracic metastases, including intrapulmonary metastasis, pleural dissemination, and pericardial involvement [27]. However, mediastinal nodal evaluation via CT remains challenging due to its relatively lower sensitivity of 57–68% and specificity of 76–82% compared to 18F-fluorodeoxyglucose (FDG) positron emission tomography/computed tomography (PET/CT), which has a sensitivity of 79–85% and specificity of 87–92% [37]. Moreover, CT's role in the evaluation of extrathoracic metastasis, especially those in distant structures such as the adrenal gland or bones, is still limited compared to PET/CT or magnetic resonance imaging due to low diagnostic accuracy [38].

Fourth, following curative treatment for lung cancer, periodic CT scans are recommended as the optimal imaging technique for evaluating recurrence every 6 months for 2 years, as well as for screening of new primary lung cancers [39].

5.3 Key Considerations for Chest CT Imaging

To obtain high-quality PCD-CT images of the lungs, it is crucial to understand how the physical and anatomical characteristics of the lungs affect CT imaging, allowing for the selection of optimal CT parameters. Lungs are primarily composed of air, allowing the appearance of a normal lung with high contrast compared to lung cancer on imaging. The high contrast on CT generally allows for a reduction in tube current while preserving image quality. With a lower tube current, a smaller focal spot can be used, which enhances spatial resolution by providing sharper CT images [40]. However, there are two structures that typically degrade the image quality of CT in the thorax: (1) the surrounding bony structures and (2) the heart. The scapula or humerus can worsen the image quality due to the presence of streaking or cupping artifacts in the lung apex. PCD-CT has the potential to reduce beam hardening artifacts from the bony thorax due to its efficient use of photons in the detector (Fig. 5.1). Additionally, chest CT, which is usually non-cardiac gated, is susceptible to image degradation originating from cardiac motion, especially in the lingula, appearing as motion artifacts. The motion artifacts can be reduced by improving the temporal resolution through strategies such as increasing the helical pitch or shortening the rotation time (Fig. 5.2).

a) b)

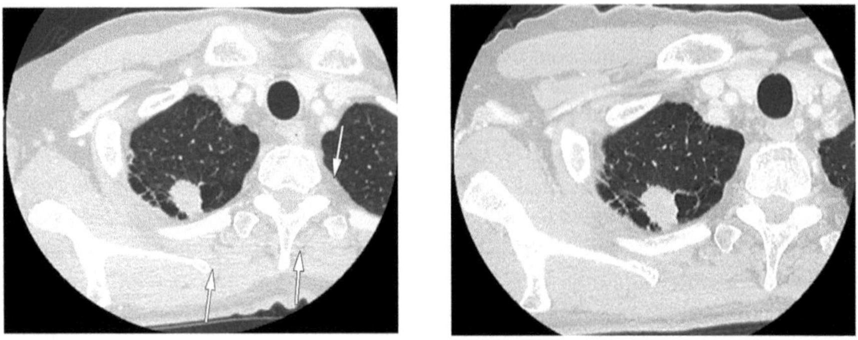

Fig. 5.1 (**a**) Energy-integrated detector computed tomography (EID-CT) was performed to examine the right lung nodule on follow-up, showing several streak artifacts (arrows). (**b**) Photon-counting detector computed tomography (PCD-CT) was performed 3 weeks later, showing fewer streak artifacts compared to EID-CT

a) b)

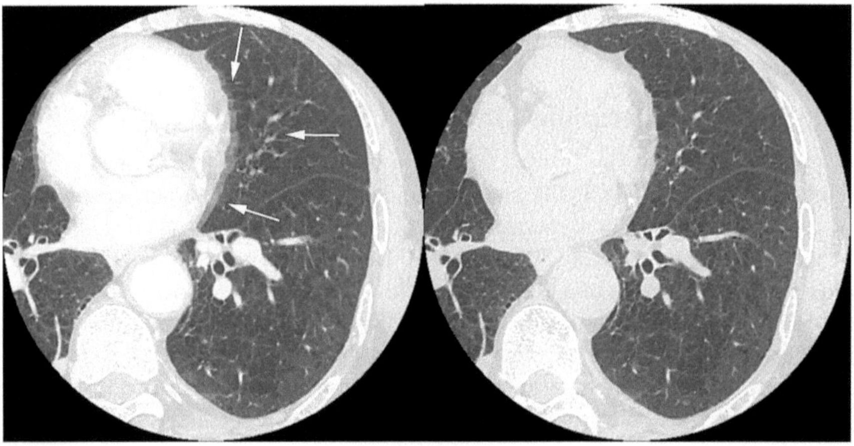

Fig. 5.2 PCD-CT images scanned at tube voltage 120 kVp, tube current 250 mA, rotation time 500 ms, and spiral helical pitch factor (**a**) 0.828 and (**b**) 1.203. (**a**) The image quality of the vessels in the lingula and the cardiac border (arrows) was degraded at the slower pitch factor. (**b**) At a faster pitch factor, there are fewer motion artifacts

5.4 Imaging Benefits of PCD-CT

PCD-CT marks a significant advancement in medical imaging physics, primarily due to its advanced photon detection and spectral characterization capabilities enabled by a new detector technology. Unlike conventional EID-CT, which aggregates the energy from multiple X-ray photons, PCD-CT employs semiconductor materials—such as cadmium telluride (CdTe), cadmium zinc telluride (CZT), or silicon—that allow direct photon-to-electric conversion without intermediate scintillation. Each X-ray photon absorbed in the semiconductor generates electron-hole pairs proportional to the photon's energy. An applied electric field then directs these charges to small pixelated anodes to produce a current pulse, which is then converted by a pulse-shaping circuit into a voltage pulse whose amplitude is proportional to the photon's energy. The voltage pulse is evaluated against a predefined threshold in a pulse height comparator circuit and then recorded by the readout electronics. For spectral data acquisition, multiple counters set at different energy thresholds are simultaneously read. Each threshold produces a distinct data stream, which captures only those X-ray photons with energies above the respective threshold, enabling spectral characterization of the measured signal.

In this chapter, we explore the imaging advantages of CZT-based PCD-CT using a clinical research CZT-based Photon Counting Detector CT (TSX-501R, Canon Medical Systems, Otawara, Japan). CZT (and CdTe) have high effective atomic numbers and therefore allow for strong X-ray absorption even with a relatively thin layer. Compared to CdTe, CZT typically exhibits lower dark current and is less susceptible to polarization effects [41]. Canon's clinical research CZT-based PCD-CT is based on the Canon Aquilion Precision CT gantry. The CZT-based photon-counting detector array spans the entire 500 mm field of view (FOV) in the fan angle and covers up to 39.6 mm in the z-axis at the isocenter. The scanner has an X-ray source with a 0.4 × 0.5 mm focal spot and operates in two modes: normal-resolution (NR) mode with 64 × 0.62 mm collimation and super-high-resolution (SHR) mode with 192 × 0.21 mm collimation. In SHR mode, each detector pixel readout is processed individually for image reconstruction. In NR mode, a 3 × 3 grid of micro pixels is summed for image reconstruction, yielding the same in-plane detection pitch as Canon's conventional EID-CT (Fig. 5.3). Each micro detector pixel can output up to six energy bins of measurements, starting from 20 keV with flexible threshold setting. The PCD-CT system supports both counting and spectral modes. Counting mode generates images based on photon events

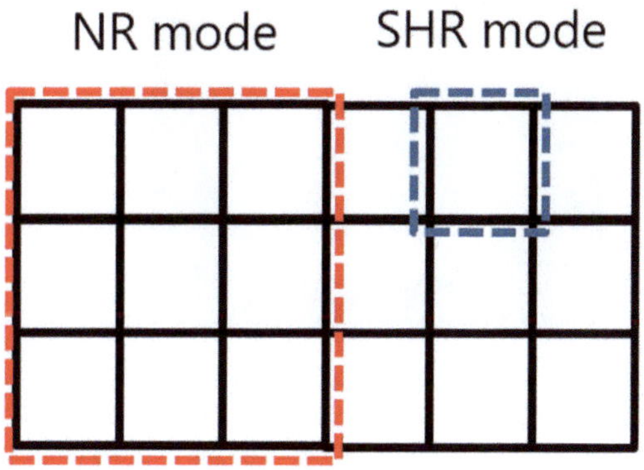

Fig. 5.3 An illustration of the clinical research CZT-based PCD-CT detector pixel readout schematic: the normal resolution (NR) mode combines the readout of a 3 × 3 micro-pixels, and the SHR mode utilizes the readout of individual micro pixels for processing and image reconstruction

above 30 keV, producing images comparable to those obtained from a conventional EID-CT system at the same X-ray tube voltage. Spectral mode produces images with five closed energy bins with threshold settings of 30/45/55/65/80 keV, respectively. These energy bin measurements are used to create two basis material sinograms through projection-domain material decomposition, from which basis material images are reconstructed and virtual monoenergetic images are then synthesized [42].

One of the most compelling benefits of PCD-CT is its potential for significant radiation dose reduction. In PCDs, the minimum threshold energy for X-ray photon counting is typically around 20 keV, which is set well above the electronic noise floor, ensuring that electronic noise does not impact PCD count rates. As a result, even under low X-ray flux, the signal reflects only the statistical quantum noise of X-ray photons [4, 8]. Conversely, conventional EIDs are heavily influenced by electronic noise at low X-ray flux, leading to a sharp increase in image noise, noise streaks, and reduced CT value stability. PCDs, therefore, deliver significantly clearer, lower-noise images, allowing for the potential to further reduce radiation dose. Studies have shown that photon-counting detectors improve dose efficiency while maintaining high image quality, even in low-dose scenarios. For instance, photon-counting CT has demonstrated a 30–60% reduction in radiation dose while preserving or enhancing image resolution, which is particularly advantageous in pediatric and vulnerable populations [14]. The removal of electronic noise also reduces CT number bias and image artifacts from photon starvation, which potentially allows for lower doses to achieve the same image quality as EID-CT in clinical settings. Figure 5.4 illustrates a 40 cm diameter water phantom scanned at 50 mAs with EID-CT and PCD-CT. PCD-CT NR counting image shows improved

a) b)

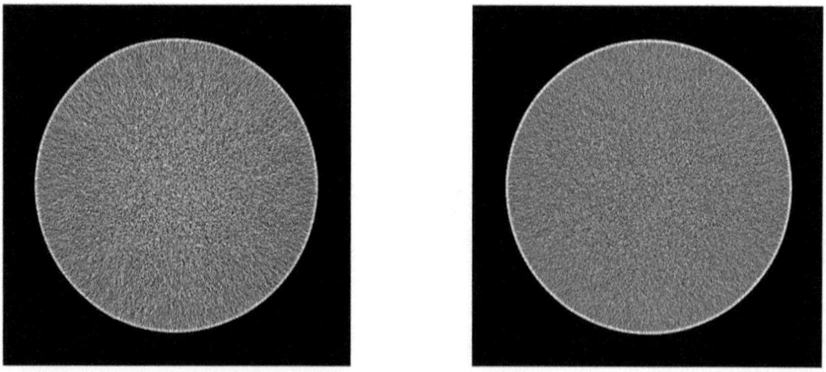

Fig. 5.4 Images of a 40 cm water phantom acquired at 50 mAs from (**a**) EID-CT single-energy and (**b**) PCD-CT NR counting mode. Display window WW/WL: 200/0 HU. PCD-CT NR counting image shows fewer noisy streaks and finer noise texture compared to EID-CT. PCD-CT image also has improved CT number uniformity over the entire phantom while the EID-CT has noticeable CT number bias toward the center (~14 HU)

CT number uniformity across the entire phantom, whereas the EID-CT image has a noticeable CT number bias towards the center (~14 HU). The PCD-CT image also displays reduced noise streaks and finer noise texture compared to EID-CT.

Another major advantage of PCD-CT is its capacity for high-resolution imaging, enabled by its smaller detector pixels compared to EID-CT [9, 14]. In conventional EID-CTs, opaque interlayers are required between detector pixels to prevent optical crosstalk, which reduces the active area on the detector plane as detector pixel size decreases. PCD-CT, however, achieves direct photon-to-electricity conversion, eliminating the need for such intermediate layers and allowing for significantly smaller detector pixels. This reduction in pixel size minimizes partial volume effects and enhances the clarity of fine anatomical details. When combined with a small focal spot on the X-ray tube and a sharp reconstruction kernel, PCD-CT operating in SHR mode can produce highly precise depictions of fine structures. Figure 5.5 demonstrates that Canon's PCD-CT system in functioning SHR mode with a small focal spot and a sharp lung reconstruction kernel achieves much greater spatial resolution than that of an NR mode. Figure 5.6 presents an example where a sharp lung kernel leads to undershoot/overshoot artifacts at high-contrast edges in NR images. In contrast, the same kernel produces sharp, crisp high-contrast boundaries in SHR images without causing artifacts. This is because PCD-CT SHR mode has a much higher intrinsic resolution than NR mode thanks to smaller pixel size and higher

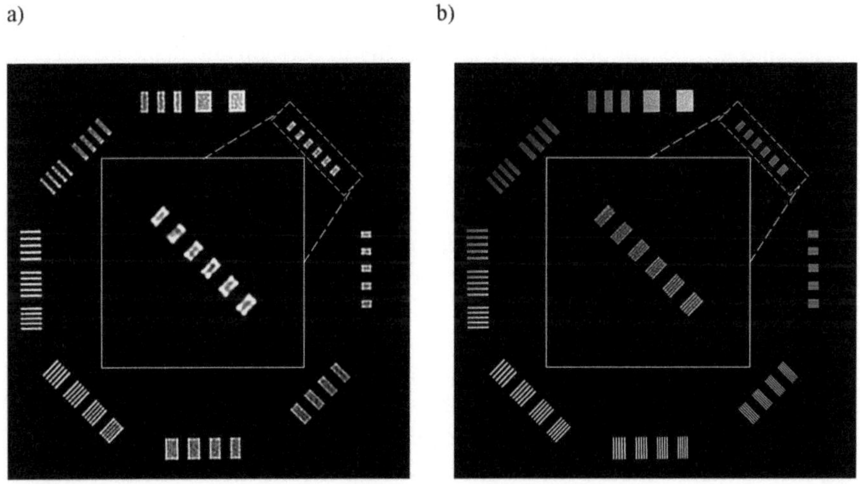

Fig. 5.5 Reconstructed images of Catphan CTP 714 module (spatial resolution up to 30 lp/cm) scanned with a clinical research CZT-based PCD-CT at tube voltage 120 kVp, tube current 250 mA, rotation time 1 s, focal spot size 0.4 mm × 0.5 mm. Images were reconstructed into 1024 × 1024 matrices with 2.5 mm thick slices, with a reconstructed FOV of 120 mm with the FBP method using a sharp lung kernel, into (**a**) normal-resolution mode, (**b**) super-high-resolution mode. The normal-resolution mode has a comparable spatial resolution as EID-CT system. The result shows the PCD-CT has the capability to resolve much higher spatial resolution under the super-high-resolution mode as compared with that of the normal-resolution mode in either PCD-CT or EID-CT systems

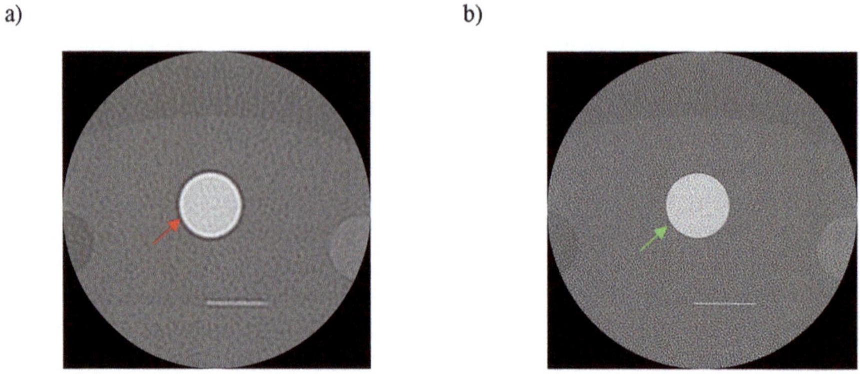

Fig. 5.6 Zoomed reconstruction of the Teflon pin on Catphan CTP 682 module. Data was acquired with tube voltage 120 kVp, tube current 250 mA, rotation time 1 s, and focal spot size 0.4 mm × 0.5 mm. Images were reconstructed into 512 × 512 matrices with 5 mm thick slices, with a reconstructed FOV of 60 mm with FBP method using a sharp lung kernel, into (**a**) NR mode, (**b**) SHR mode. The NR image shows a strong undershoot/overshoot artifact at the high-contrast edge of the Teflon pin, while the SHR image presents a sharp edge without such artifact

sampling rate. Figure 5.7 compares the NR and SHR images of an anthropomorphic chest phantom at a matched noise level, achieved by adjusting hybrid-IR parameters. Even at matched noise level, SHR images exhibit superior in-plane and z-resolution compared to NR images.

The third key benefit of PCD-CT is its ability to perform simultaneous spectral imaging without the added hardware complexities in workflow or the spatial-temporal data inconsistencies often encountered with conventional dual-energy EID-CTs that use dual-source or rapid-kVp-switching techniques [43, 44]. Simultaneous spectral imaging in PCD-CT ensures both spatial and temporal consistency in spectral data acquisition, while the use of multiple energy bins improves accuracy and reduces noise in material decomposition [9]. This enables clearer differentiation between tissues with similar attenuation properties, such as distinguishing iodine contrast from calcifications. Additionally, Canon's PCD-CT performs material decomposition in the projection domain, which may be less prone to beam hardening effects and errors in detector response modeling compared to image-domain decomposition methods [45]. The synergy of high resolution and spectral imaging in PCD-CT also facilitates material differentiation in small objects, potentially improving diagnostic accuracy for small lesions. Figure 5.8 illustrates a tube phantom simulating various types of coronary stenosis and plaques with the presence of a coronary stent, scanned using a clinical research CZT-based PCD-CT (Canon Medical Systems, Tochigi, Japan). These SHR spectral images not only enhance visualization of small stenoses and plaques compared to NR images but also distinctly differentiate calcium plaques from the iodinated background through multi-energy information.

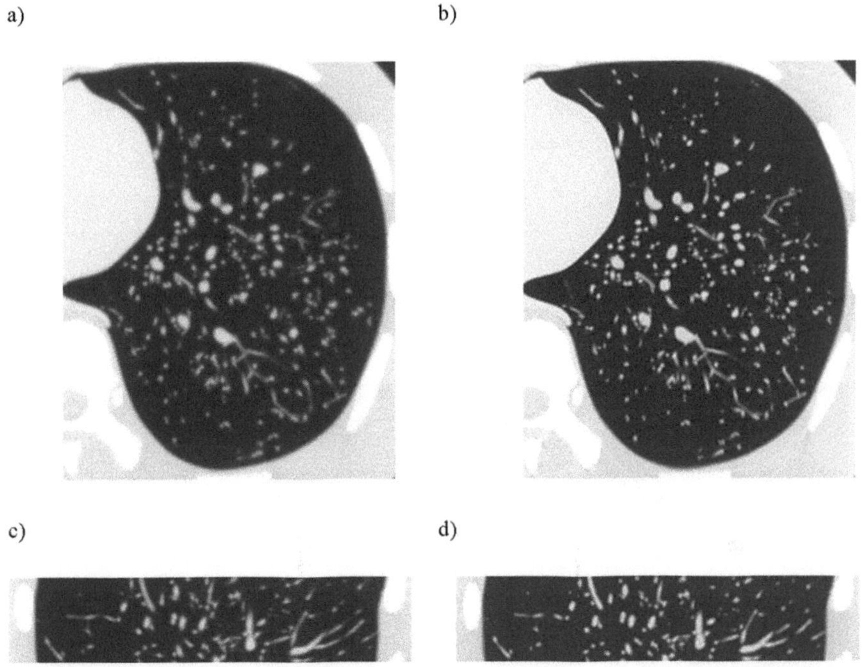

Fig. 5.7 Reconstructions of an anthropomorphic chest phantom (Kyoto Kagaku Co. Ltd., Japan) in NR mode (**a**, **c**) and SHR mode (**b**, **d**), where (**a**, **b**) show the axial plane and (**c**, **d**) show the sagittal view. Display window [-1350 150] HU. Both NR and SHR images were reconstructed into 1024 × 1024 matrices using off-line software. The NR image has 0.62 mm slice thickness, while the SHR has 0.21 mm thickness. Both NR and SHR images were reconstructed using hybrid-IR with in-house parameter settings, where the hybrid-IR strength for SHR was adjusted so that the noise of the SHR image matched that of the NR image. Even at a matched noise level, the SHR image still presents higher in-plane resolution and z-resolution than the NR image

5.5 Potential Advantages of PCD-CT for Lung Cancer Imaging

Since the radiation dose of PCD-CT is lower than that of EID-CT, lung cancer screening using low-dose radiation is one of the CT examinations that can benefit most from PCD-CT. Inoue et al. reported that PCD-CT for lung cancer screening significantly reduced radiation exposure and improved the image quality of pulmonary emphysema and lung nodule borders due to reduced image noise compared to EID-CT [21].

For the qualitative assessment of pulmonary nodules, PCD-CT may potentially offer better diagnostic accuracy for lung cancer. This potential improvement can be attributed to improved image quality and increased diagnostic confidence using high-resolution CT imaging reconstructed with smaller voxel sizes compared to EID-CT. Baller et al. reported significantly improved visibility of normal lung

(a)

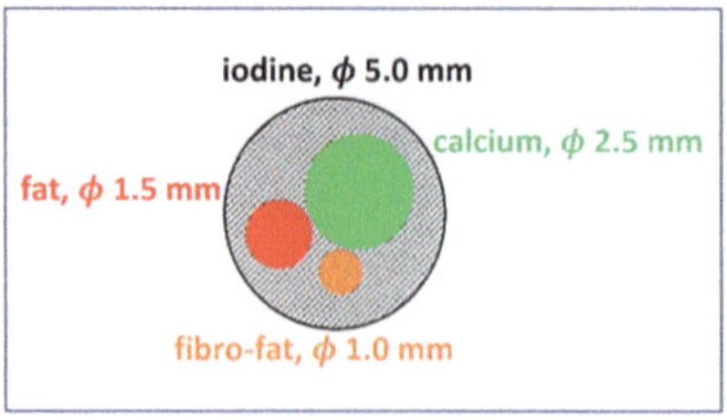

(b)

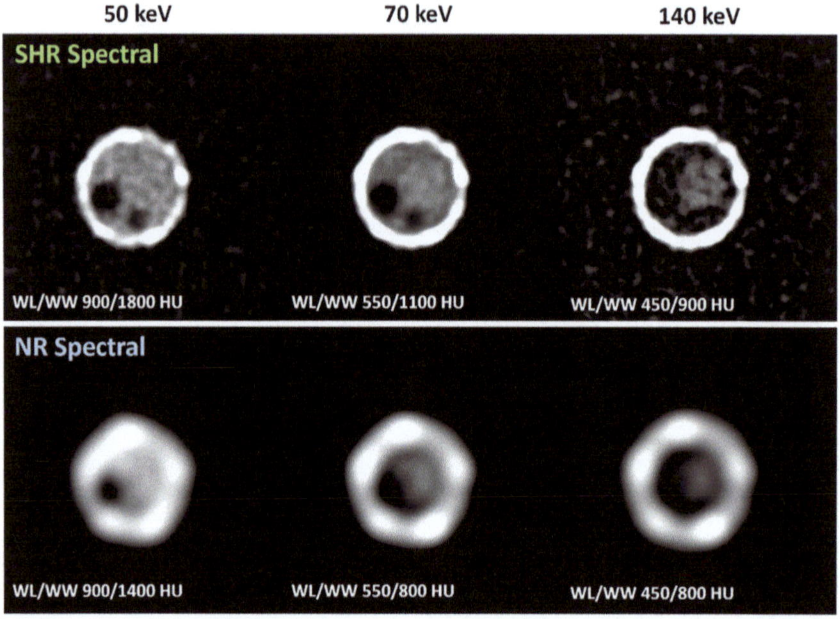

Fig. 5.8 (**a**) Illustration of a tube phantom mimicking different types of coronary stenosis and plaques. In addition, a stent was placed outside of this tube phantom during imaging. (**b**) Zoomed reconstructions from NR Spectral and SHR Spectral at various photon energy levels. Data was acquired with a clinical research CZT-based PCD-CT (Canon Medical Systems, Tochigi, Japan) at tube voltage 120 kVp, tube current 200 mA, rotation time 1 s, focal spot size 0.4 mm × 0.5 mm. Images were reconstructed into 1024 × 1024 matrices with 2.5 mm thick slices with a reconstructed FOV of 18 mm with the FBP method using a standard body kernel. Results show SHR VMIs (top) produce much better visualization of the small stenosis and plaque objects than NR VMIs (bottom). In addition, SHR 140 keV VMI (top right) provides clear discrimination of calcium plaque from the iodine background, which are otherwise undisguisable in SHR 70 keV VMI (top middle)

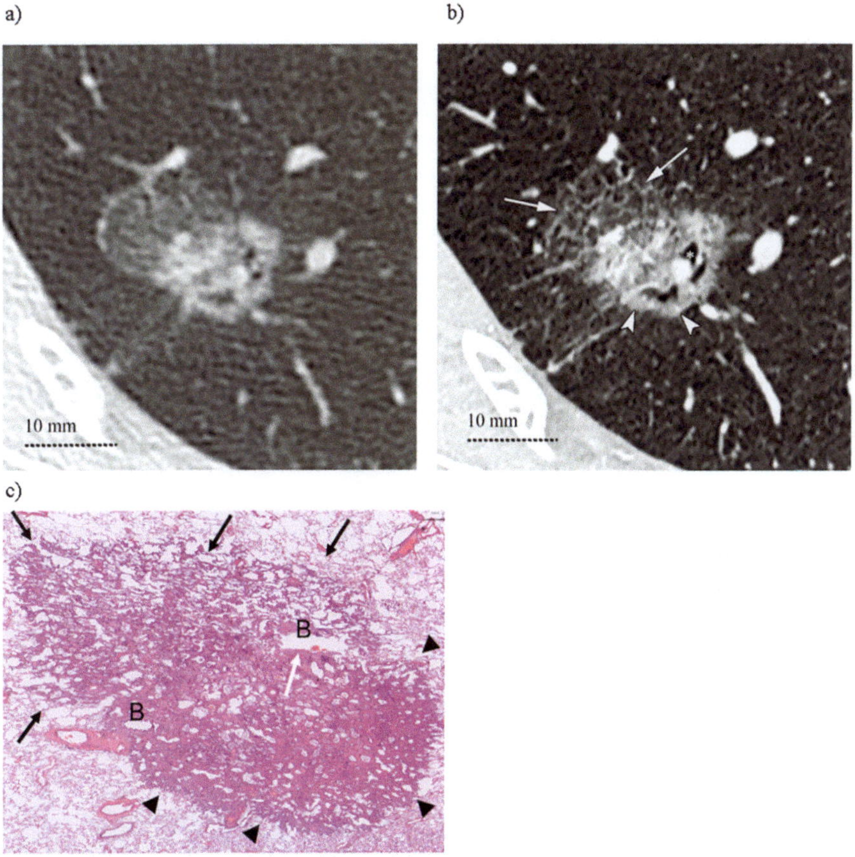

Fig. 5.9 CT images of a 76-year-old woman showing a part-solid nodule in the right lower lobe: (**a**) EID-CT (512 × 512 matrix, 1.0 mm-thickness), and (**b**) PCD-CT (1024 × 1024 matrix, 0.2 mm-thickness, scanned 3 weeks later) images. The image noise appears reduced in the (**b**) PCD-CD image and the structures (vessels and bronchiole (*)) are more sharply delineated. Although the ground glass opacity in the nodule (left upper side of the tumor) appears homogenous in (**a**) the EID-CT image, reticulation opacity (arrows) within the ground glass opacity is clearer in (**c**) the PCD-CT image. (**c**) Histopathological examination (overview) reveals that the tumor consists of a 50% lepidic growth pattern (black arrows) and 50% acinar pattern (arrowheads) with bronchioles (white arrow, **B**), which are well correlated with the (**b**) PCD-CT image

structures, such as central and peripheral airways, lung vessels, fissures, small lesions (ground-glass opacities (Fig. 5.9), macro- and micro-nodules (Fig. 5.10), inter- and intralobular lines, paraseptal and centrilobular emphysema, bullae/cysts, and air trapping) with PCD-CT [10]. However, these findings require further validation through larger, more comprehensive studies. Using small voxel sizes for high-resolution CT, Yanagawa et al. reported that detailed evaluation of the irregularity of air bronchiologram within lung adenocarcinomas was useful for diagnosing invasive lung cancer [46] (Fig. 5.11). Furthermore, such high-spatial-resolution CT

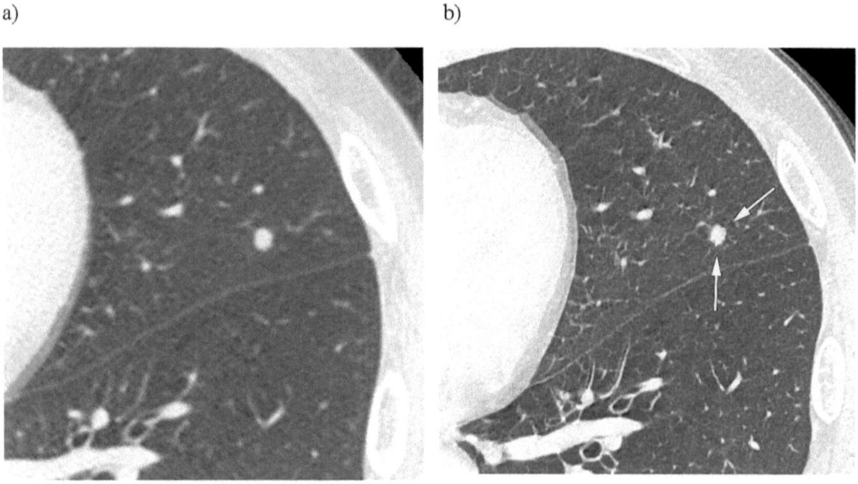

Fig. 5.10 CT images of a 78-year-old man with a history of colon cancer showing a small nodule in the left upper lobe: (**a**) EID-CT (512 × 512 matrix, 1.0 mm thickness), and (**b**) PCD-CT (1024 × 1024 matrix, 0.2 mm thickness, scanned 3 weeks later) images. (**a**) EID-CT image shows that the small nodule appears round or oval with a slightly irregular margin. However, (**b**) PCD-CT more clearly demonstrates the irregularity of the nodule (arrows), which is suggestive of malignancy. The nodule was confirmed to be a metastatic tumor from colon cancer after the surgical resection

image may be capable of rendering images close to histopathological images (Fig. 5.12), leading to improved quality of texture analysis in the future. Texture analysis is one of the techniques for differentiating lung adenocarcinoma from squamous cell carcinoma [47, 48].

Preliminary research suggests that spectral imaging holds promise for the qualitative diagnosis of pulmonary nodules, although most current reports on chest spectral imaging primarily focus on pulmonary perfusion imaging [16]. Several studies indicate that for assessing solitary lung nodules, iodine content imaging may negatively correlate with the grading of primary lung cancer [49–51] (Fig. 5.13). Tumors with higher grades tend to show lower iodine content, possibly reflecting necrosis within high-grade tumors [49, 50]. These preliminary findings require further validation through additional studies, including research utilizing PCD-CT and other advanced imaging techniques. Initial studies suggest that spectral imaging may help differentiate metastatic tumors from primary lung cancer. Sato et al. reported that ring-like peripheral high iodine concentration on spectral imaging could differentiate metastatic lung tumors from primary lung cancer, with a sensitivity of 52% and a specificity of 81% [52]. Moreover, spectral imaging also shows promise for the

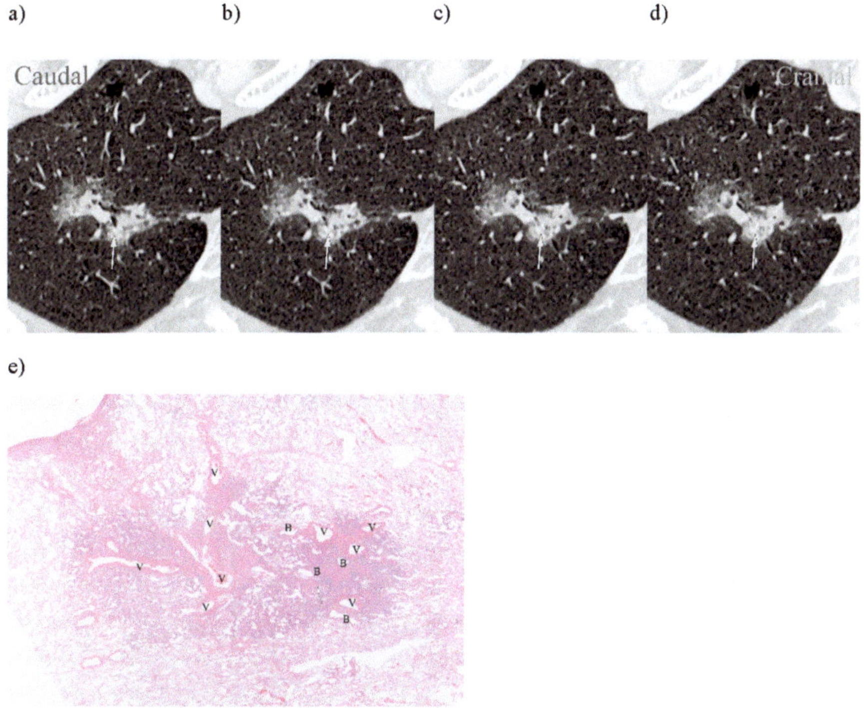

Fig. 5.11 CT images (1024 × 1024 matrix, 0.2 mm thickness) of a 74-year-old man showing a part-solid nodule in the right upper lobe: Continuous PCD-CT images in sequential order from caudal to cranial (**a-d**) show that the irregularity of the air bronchiologram in the solid component becomes progressively clearer towards the cranial image (arrows). Each image represents a consecutive 0.2 mm slice moving cranially through the nodule. (**e**) Histopathological examination (overview) reveals that the irregularity of the bronchiole surrounded by fibrosis is noted in the invasive adenocarcinoma tumor component (arrow). *B* bronchiole, *V* vessel

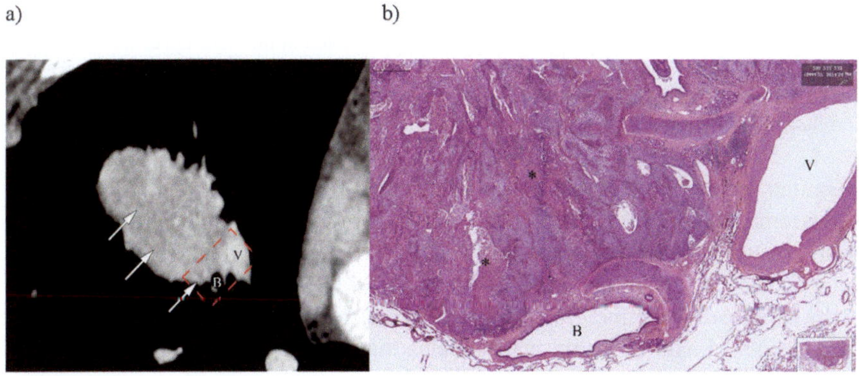

Fig. 5.12 (**a**) Chest PCD-CT image (1024 × 1024 matrix, 0.2 mm thickness) shows a lobulated heterogenous nodule in the right upper lobe of a 78-year-old man. The tumor includes multiple unenhanced spots (arrows). (**b**) Histopathological examination of the area correlated with the dotted red rectangle reveals multiple necrotic spots (*) and keratinization within the tumor nests, which are suggestive of squamous cell carcinoma. *B* bronchus, *V* vessel

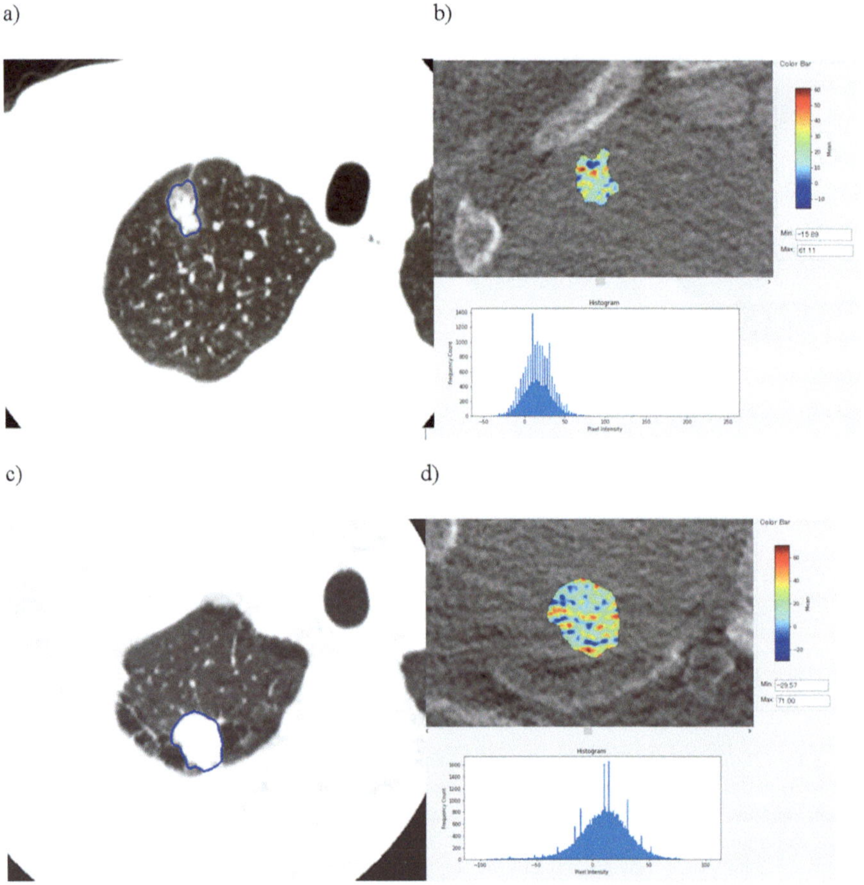

Fig. 5.13 (**a**) Minimally invasive adenocarcinoma in a 79-year-old man. A part-solid nodule is noted in the right upper lobe at 70 keV image. (**b**) Histogram analysis of the iodine content image calculates the mean iodine content in the whole tumor is 1.70 mg/mL. (**c**) Poorly differentiated adenocarcinoma in a 62-year-old man. A solid nodule is noted in the right upper lobe at 70 keV images. (**d**) Histogram analysis of the iodine content image calculates the mean iodine concentration in the whole tumor is 0.93 mg/mL. Iodine concentration in the lung tumor appears negatively correlated with the tumor grade

quantitative evaluation of fat with high accuracy [53, 54] (Fig. 5.14). Pulmonary hamartoma, the most frequent benign lung tumor, contains fat components in approximately 50–60% of cases, along with cartilage or calcification [31]. While fat components are rare in lung cancers, lung cancers can have negative CT values due to the partial volume effect from surrounding air. Accurate fat quantification derived from spectral imaging within lung nodules may present a promising strategy for distinguishing pulmonary hamartoma from lung cancer.

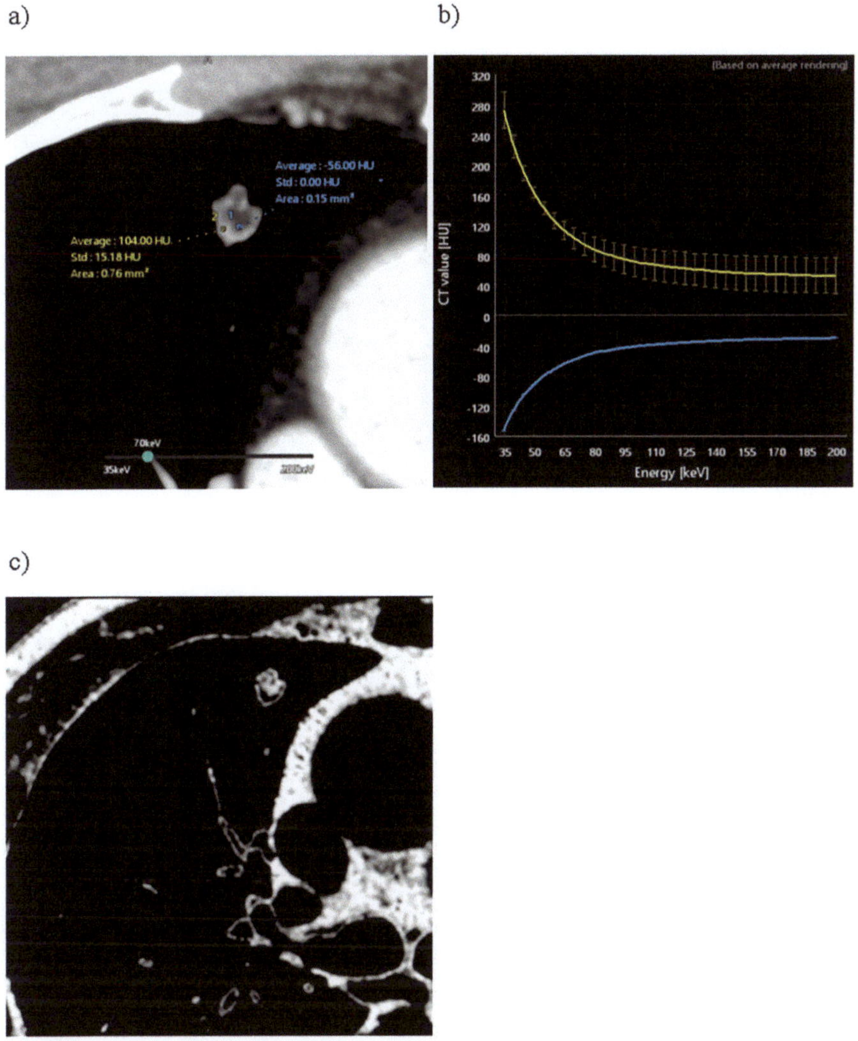

Fig. 5.14 CT images of a 72-year-old man diagnosed with pulmonary hamartoma. (**a**) Contrast-enhanced PCD-CT image reconstructed at 70 keV shows a lobulated nodule with a low-density area in the right upper lobe. (**b**) The blue line [mean CT value of the blue region-of-interest (ROI) of (**a**)] appears lower at low kilo-electron volt compared to the yellow line, which is suggestive of lipid component in the nodule. (**c**) Fat-enhanced CT image generated by PCD-CT shows the nodule appears as high density as subcutaneous adipose tissue, which is consistent with pulmonary hamartoma

For lung cancer staging, both high-resolution and spectral imaging with PCD-CT are promising approaches for evaluating the extent of the disease. For locoregional evaluation of advanced lung cancer in staging, iodine content imaging can help assess the extent of chest wall invasion (Fig. 5.15), with increased iodine content indicating true tumor invasion [55]. Additionally, high-resolution imaging may

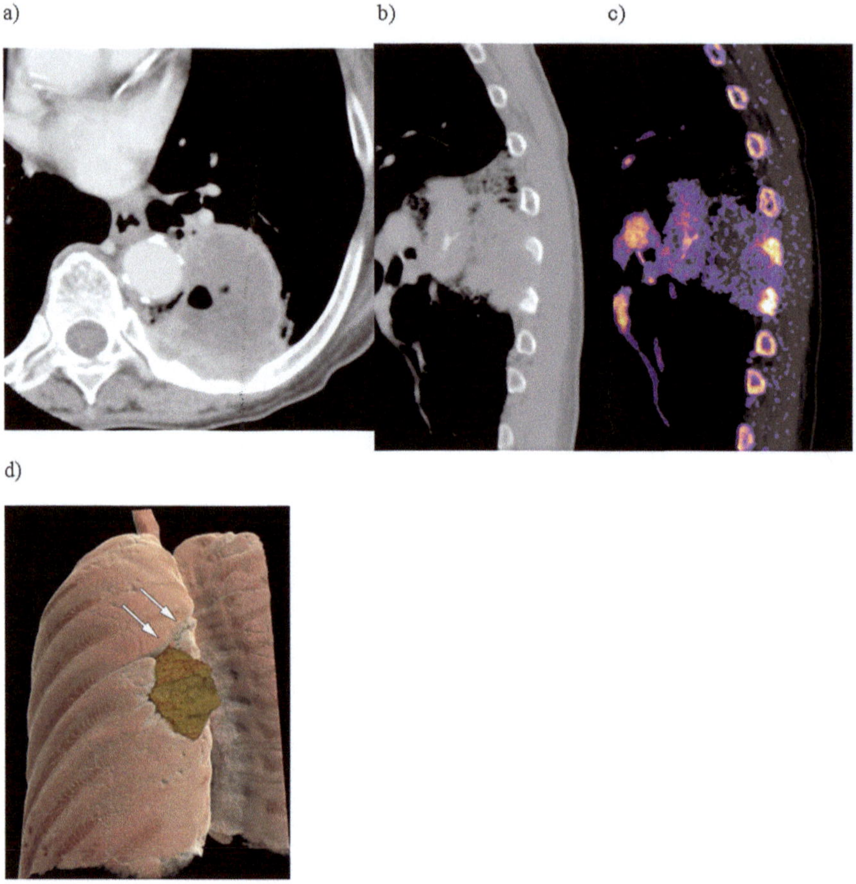

Fig. 5.15 CT images of a 78-year-old woman diagnosed with pulmonary adenosquamous carcinoma. (**a**) Contrast-enhanced PCD-CT image reconstructed at 70 keV shows a lobulated necrotic mass invading the chest wall with the rib destruction. (**b**) Sagittal PCD-CT image at 70 keV with bone window shows partially destructed ribs appearing as slightly increased density, which may be compatible with invasion. (**c**) Sagittal PCD-CT iodine overlay map image clearly reveals the destructed ribs, which can be easily reviewed for tumor invasion. (**d**) Simultaneously, a 3D volume-rendered image based on a super high-resolution image also shows the left major fissure clearly (arrows)

reveal clearer borders of intrathoracic structures like fissures. For the assessment of intrathoracic metastasis, small intrapulmonary metastasis and pleural dissemination may be more easily recognized with high-resolution imaging (Fig. 5.16).

Spectral imaging, rather than high-resolution imaging, may play a more significant role in assessing extrathoracic metastasis. While FDG-PET/CT shows high diagnostic performance for adrenal [56] and bone metastasis [22, 38], PCD-CT is expected to improve specificity. Spectral imaging for evaluating lipid content in a region of interest may help differentiate the adrenal gland, which can contain lipid

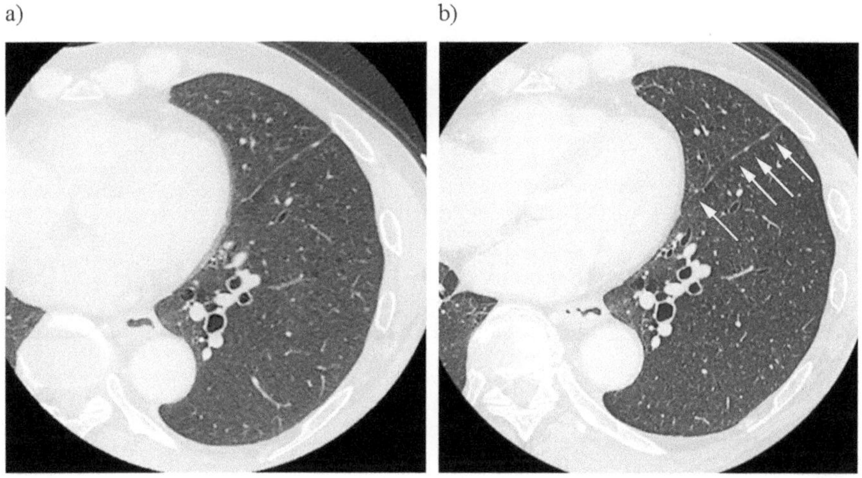

Fig. 5.16 CT images of a 73-year-old man diagnosed with pleural dissemination of left lung cancer. (**a**) EID-CT (512 × 512 matrix, 1.0 mm thickness), and (**b**) PCD-CT (1024 × 1024 matrix, 0.2 mm thickness, scanned 4 weeks later) images. (**a**) EID-CT shows tiny nodules along the left major fissure. (**b**) PCD-CT reveals more tiny nodules clearly delineated, suggesting pleural dissemination

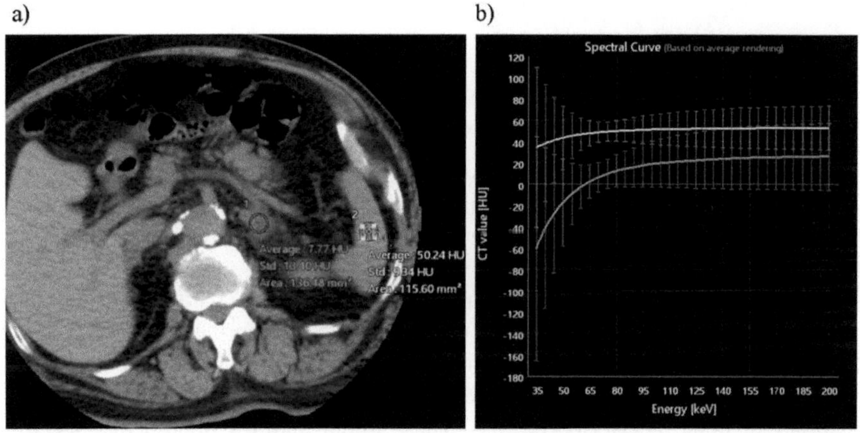

Fig. 5.17 (**a**) Abdominal PCD-CT image at 70 keV shows a small low-density nodule (mean CT value 7.77 HU) in the left adrenal gland of a 78-year-old man (same patient with Fig. 5.5). (**b**) The blue line reflecting the adrenal nodule appears lower at low kilo-electron volt, suggestive of lipid component, compared to the yellow line, which is a spectral curve of the spleen as a control. As a result, PCD-CT may have the potential to assess adrenal gland metastasis with high diagnostic accuracy

content, from metastasis [57, 58] (Fig. 5.17). For skeletal metastasis, FDG-PET/CT may carry a risk of overdiagnosis due to false positives as fractures, arthropathies, degenerative changes, inflammation, and other tumors [59]. Early studies suggest that spectral imaging techniques, such as an iodine density map, may improve

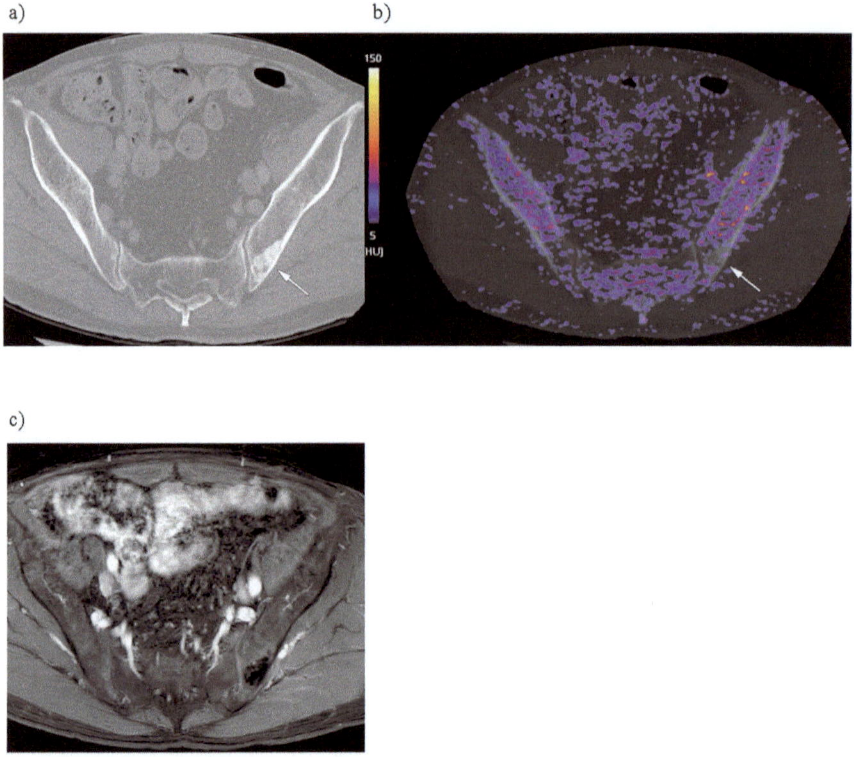

Fig. 5.18 (**a**) Pelvic PCD-CT image at 70 keV shows a high-density area (arrow) in the left iliac bone in a 77-year-old man diagnosed with lung cancer (bone window). (**b**) PCD-CT image at 70 keV with an overlay of iodine map shows less iodine uptake in the high-density area, which is suggestive of less vascularity. (**c**) Gadolinium contrast-enhanced fat-sat T1 weighted image shows less enhancement, which is suggestive of benign origin

specificity [60–62], although there may be limitations in separating iodine from bone tissue using current techniques [63, 64] (Fig. 5.18). Furthermore, other spectral imaging techniques, such as virtual non-calcium or hydroxyapatite-water decomposition imaging, have also been reported to potentially improve diagnostic accuracy of bone metastasis [62, 63]. These initial findings suggest that spectral imaging may be a promising tool for improving diagnostic accuracy by applying specific image analysis to target tissues, potentially complementing FDG-PET/CT in the assessment of distant metastasis.

5.6 Conclusion

In conclusion, PCD-CT shows great potential across various aspects of lung cancer imaging, including screening, diagnosis, and staging. The potential benefits of improved spatial resolution, spectral imaging capabilities, and reduced radiation dose could enhance our ability to detect, characterize, and manage lung cancer. However, it is important to note that many of these advantages remain theoretical or are based on preliminary studies and require further validation.

References

1. Sartoretti T, Wildberger JE, Flohr T, Alkadhi H. Photon-counting detector CT: early clinical experience review. Br J Radiol. 2023;96:20220544. https://doi.org/10.1259/bjr.20220544.
2. Nehra AK, Rajendran K, Baffour FI, Mileto A, Rajiah PS, Horst KK, et al. Seeing more with less: clinical benefits of photon-counting detector CT. Radiographics. 2023;43(5):e220158. https://doi.org/10.1148/rg.220158.
3. Nakamura Y, Higaki T, Kondo S, Kawashita I, Takahashi I, Awai K. An introduction to photon-counting detector CT (PCD CT) for radiologists. Jpn J Radiol. 2023;41(3):266–82. https://doi.org/10.1007/s11604-022-01350-6.
4. McCollough CH, Rajendran K, Leng S, Yu L, Fletcher JG, Stierstorfer K, et al. The technical development of photon-counting detector CT. Eur Radiol. 2023;33:5321–30. https://doi.org/10.1007/s00330-023-09545-9.
5. Molwitz I, Campbell GM, Yamamura J, Knopp T, Toedter K, Fischer R, et al. Fat quantification in dual-layer detector spectral computed tomography: experimental development and first in-patient validation. Invest Radiol. 2022;57(7):463–9. https://doi.org/10.1097/RLI.0000000000000858.
6. Liu LP, Shapira N, Chen AA, Shinohara RT, Sahbaee P, Schnall M, et al. First-generation clinical dual-source photon-counting CT: ultra-low-dose quantitative spectral imaging. Eur Radiol. 2022;32:8579. https://doi.org/10.1007/s00330-022-08933-x.
7. Esquivel A, Ferrero A, Mileto A, Baffour F, Horst K, Rajiah PS, et al. Photon-counting detector CT: key points radiologists should know. Korean J Radiol. 2022;23(9):854–65. https://doi.org/10.3348/kjr.2022.0377.
8. Willemink MJ, Persson M, Pourmorteza A, Pelc NJ, Fleischmann D. Photon-counting CT: technical principles and clinical prospects. Radiology. 2018;289(2):293–312. https://doi.org/10.1148/radiol.2018172656.
9. Zhan X, Zhang R, Niu X, Hein I, Budden B, Wu S, et al. Comprehensive evaluations of a prototype full field-of-view photon counting CT system through phantom studies. Phys Med Biol. 2023;68(17):175007. https://doi.org/10.1088/1361-6560/acebb3.
10. Van Ballaer V, Dubbeldam A, Muscogiuri E, Cockmartin L, Bosmans H, Coudyzer W, et al. Impact of ultra-high-resolution imaging of the lungs on perceived diagnostic image quality using photon-counting CT. Eur Radiol. 2024;34:1895–904. https://doi.org/10.1007/s00330-023-10174-5.
11. Milos RI, Röhrich S, Prayer F, Strassl A, Beer L, Heidinger BH, et al. Ultrahigh-resolution photon-counting detector CT of the lungs: association of reconstruction kernel and slice thickness with image quality. AJR Am J Roentgenol. 2023;220(5):672–80. https://doi.org/10.2214/ajr.22.28515.
12. Leng S, Bruesewitz M, Tao S, Rajendran K, Halaweish AF, Campeau NG, et al. Photon-counting detector CT: system design and clinical applications of an emerging technology. RadioGraphics. 2019;39(3):729–43. https://doi.org/10.1148/rg.2019180115.
13. McCollough CH, Rajendran K, Baffour FI, Diehn FE, Ferrero A, Glazebrook KN, et al. Clinical applications of photon counting detector CT. Eur Radiol. 2023;33(8):5309–20. https://doi.org/10.1007/s00330-023-09596-y.
14. Flohr T, Schmidt B. Technical basics and clinical benefits of photon-counting CT. Invest Radiol. 2023;58(7):441–50. https://doi.org/10.1097/RLI.0000000000000980.

15. Siegel MJ, Bugenhagen SM, Sanchez A, Kim S, Abadia A, Ramirez-Giraldo JC. Comparison of radiation dose and image quality of pediatric high-resolution chest CT between photon-counting detector CT and energy-integrated detector CT: a matched study. AJR Am J Roentgenol. 2023;221(3):363–71. https://doi.org/10.2214/AJR.23.29077.
16. Scharm SC, Schaefer-Prokop C, Winther HB, Huisinga C, Werncke T, Vogel-Claussen J, et al. Regional pulmonary morphology and function: photon-counting CT assessment. Radiology. 2023;308(1):e230318. https://doi.org/10.1148/radiol.230318.
17. Gaillandre Y, Duhamel A, Flohr T, Faivre JB, Khung S, Hutt A, et al. Ultra-high resolution CT imaging of interstitial lung disease: impact of photon-counting CT in 112 patients. Eur Radiol. 2023;33(8):5528–39. https://doi.org/10.1007/s00330-023-09616-x.
18. Inoue A, Johnson TF, White D, Cox CW, Hartman TE, Thorne JE, et al. Estimating the clinical impact of photon-counting-detector CT in diagnosing usual interstitial pneumonia. Invest Radiol. 2022;57(11):734–41. https://doi.org/10.1097/rli.0000000000000888.
19. Remy-Jardin M, Oufriche I, Guiffault L, Duhamel A, Flohr T, Schmidt B, et al. Diagnosis of acute pulmonary embolism: when photon-counting-detector CT replaces energy-integrating-detector CT in daily routine. Eur Radiol. 2024;34(10):6544–55. https://doi.org/10.1007/s00330-024-10724-5.
20. Pannenbecker P, Huflage H, Grunz JP, Gruschwitz P, Patzer TS, Weng AM, et al. Photon-counting CT for diagnosis of acute pulmonary embolism: potential for contrast medium and radiation dose reduction. Eur Radiol. 2023;33(11):7830–9. https://doi.org/10.1007/s00330-023-09777-9.
21. Inoue A, Johnson TF, Walkoff LA, Levin DL, Hartman TE, Burke KA, et al. Lung cancer screening using clinical photon-counting detector computed tomography and energy-integrating-detector computed tomography: a prospective patient study. J Comput Assist Tomogr. 2023;47(2):229–35. https://doi.org/10.1097/rct.0000000000001419.
22. Network NCC. Non-small cell lung cancer (Version 2.2023). https://www.nccn.org/professionals/Physician_gls/pdf/nscl.pdf. Accessed 2023 Mar 22.
23. de Koning HJ, van der Aalst CM, de Jong PA, Scholten ET, Nackaerts K, Heuvelmans MA, Lammers JJ, et al. Reduced lung-cancer mortality with volume CT screening in a randomized trial. N Engl J Med. 2020;382(6):503–13. https://doi.org/10.1056/NEJMoa1911793.
24. Aberle DR, Adams AM, Berg CD, Black WC, Clapp JD, Fagerstrom RM, et al. Reduced lung-cancer mortality with low-dose computed tomographic screening. N Engl J Med. 2011;365(5):395–409. https://doi.org/10.1056/NEJMoa1102873.
25. Lung-RADS® v2022. https://www.acr.org/-/media/ACR/Files/RADS/Lung-RADS/Lung-RADS-2022.pdf. Accessed 2024 Sept 9.
26. Christensen J, Prosper AE, Wu CC, Chung J, Lee E, Elicker B, et al. ACR lung-RADS v2022: assessment categories and management recommendations. J Am Coll Radiol. 2024;21(3):473–88. https://doi.org/10.1016/j.jacr.2023.09.009.
27. de Groot PM, Chung JH, Ackman JB, Berry MF, Carter BW, Colletti PM, et al. ACR appropriateness criteria(®) noninvasive clinical staging of primary lung cancer. J Am Coll Radiol. 2019;16(5s):S184–95. https://doi.org/10.1016/j.jacr.2019.02.008.
28. Mazzone PJ, Lam L. Evaluating the patient with a pulmonary nodule: a review. JAMA. 2022;327(3):264–73. https://doi.org/10.1001/jama.2021.24287.
29. Martin MD, Kanne JP, Broderick LS, Kazerooni EA, Meyer CA. RadioGraphics update: lung-RADS 2022. RadioGraphics. 2023;43(11):e230037. https://doi.org/10.1148/rg.230037.
30. de Hoop B, van Ginneken B, Gietema H, Prokop M. Pulmonary perifissural nodules on CT scans: rapid growth is not a predictor of malignancy. Radiology. 2012;265(2):611–6. https://doi.org/10.1148/radiol.12112351.
31. Snoeckx A, Reyntiens P, Desbuquoit D, Spinhoven MJ, Van Schil PE, van Meerbeeck JP, Parizel PM. Evaluation of the solitary pulmonary nodule: size matters, but do not ignore the power of morphology. Insights Imaging. 2018;9(1):73–86. https://doi.org/10.1007/s13244-017-0581-2.

32. Miyoshi T, Aokage K, Katsumata S, Tane K, Ishii G, Tsuboi M. Ground-glass opacity is a strong prognosticator for pathologic stage IA lung adenocarcinoma. Ann Thorac Surg. 2019;108(1):249–55. https://doi.org/10.1016/j.athoracsur.2019.01.079.
33. Aoki T, Nakata H, Watanabe H, Nakamura K, Kasai T, Hashimoto H, et al. Evolution of peripheral lung adenocarcinomas: CT findings correlated with histology and tumor doubling time. AJR Am J Roentgenol. 2000;174(3):763–8. https://doi.org/10.2214/ajr.174.3.1740763.
34. Lababede O, Meziane MA. The eighth edition of TNM staging of lung cancer: reference chart and diagrams. Oncologist. 2018;23(7):844–8. https://doi.org/10.1634/theoncologist.2017-0659.
35. Imai K, Minamiya Y, Ishiyama K, Hashimoto M, Saito H, Motoyama S, et al. Use of CT to evaluate pleural invasion in non-small cell lung cancer: measurement of the ratio of the interface between tumor and neighboring structures to maximum tumor diameter. Radiology. 2013;267(2):619–26. https://doi.org/10.1148/radiol.12120864.
36. Choong CK, Pasricha SS, Li X, Briggs P, Ramdave S, Crossett M, et al. Dynamic four-dimensional computed tomography for preoperative assessment of lung cancer invasion into adjacent structures†. Eur J Cardiothorac Surg. 2015;47(2):239–43.; discussion 243. https://doi.org/10.1093/ejcts/ezu265.
37. Volpi S, Ali JM, Tasker A, Peryt A, Aresu G, Coonar AS. The role of positron emission tomography in the diagnosis, staging and response assessment of non-small cell lung cancer. Ann Transl Med. 2018;6(5):95. https://doi.org/10.21037/atm.2018.01.25.
38. Qu X, Huang X, Yan W, Wu L, Dai K. A meta-analysis of ^{18}FDG-PET-CT, ^{18}FDG-PET, MRI and bone scintigraphy for diagnosis of bone metastases in patients with lung cancer. Eur J Radiol. 2012;81(5):1007–15. https://doi.org/10.1016/j.ejrad.2011.01.126.
39. Schneider BJ, Ismaila N, Aerts J, Chiles C, Daly ME, Detterbeck FC, et al. Lung cancer surveillance after definitive curative-intent therapy: ASCO guideline. J Clin Oncol. 2020;38(7):753–66. https://doi.org/10.1200/JCO.19.02748.
40. Huda W, Abrahams RB. X-ray-based medical imaging and resolution. AJR Am J Roentgenol. 2015;204(4):W393–7. https://doi.org/10.2214/AJR.14.13126.
41. Eisen Y, Shor A, Mardor I. CdTe and CdZnTe gamma ray detectors for medical and industrial imaging systems. Nucl Instrum Methods Phys Res A Accel Spectr Detect Assoc Equip. 1999;428(1):158–70.
42. McCollough CH, Boedeker K, Cody D, Duan X, Flohr T, Halliburton SS, et al. Principles and applications of multienergy CT: report of AAPM Task Group 291. Med Phys. 2020;47(7):e881–912.
43. Shefer E, Altman A, Behling R, Goshen R, Gregorian L, Roterman Y, et al. State of the art of CT detectors and sources: a literature review. Curr Radiol Rep. 2013;1(1):76–91.
44. Johnson T, Fink C, Schönberg SO, Reiser MF, editors. Dual energy CT in clinical practice. Berlin: Springer; 2011.
45. Haase V, Noo F, Stierstorfer K, Maier A, McNitt-Gray M. Reproducibility in dual energy CT: the impact of a projection domain material decomposition method. In: 7th International conference on image formation in X-ray computed tomography 2022 Oct 17, vol. 12304. SPIE. p. 101–6.
46. Yanagawa M, Tsubamoto M, Satoh Y, Hata A, Miyata T, Yoshida Y, et al. Lung adenocarcinoma at CT with 0.25-mm section thickness and a 2048 matrix: high-spatial-resolution imaging for predicting invasiveness. Radiology. 2020;297(2):462–71. https://doi.org/10.1148/radiol.2020201911.
47. Tomori Y, Yamashiro T, Tomita H, Tsubakimoto M, Ishigami K, Atsumi E, et al. CT radiomics analysis of lung cancers: differentiation of squamous cell carcinoma from adenocarcinoma, a correlative study with FDG uptake. Eur J Radiol. 2020;128:109032. https://doi.org/10.1016/j.ejrad.2020.109032.
48. Han R, Arjal R, Dong J, Jiang H, Liu H, Zhang D, et al. Three dimensional texture analysis of noncontrast chest CT in differentiating solitary solid lung squamous cell carcinoma from adenocarcinoma and correlation to immunohistochemical markers. Thorac Cancer. 2020;11(11):3099–106. https://doi.org/10.1111/1759-7714.13592.

49. Shimamoto H, Iwano S, Umakoshi H, Kawaguchi K, Naganawa S. Evaluation of locoregional invasiveness of small-sized non-small cell lung cancers by enhanced dual-energy computed tomography. Cancer Imaging. 2016;16(1):18. https://doi.org/10.1186/s40644-016-0077-1.
50. Iwano S, Ito R, Umakoshi H, Ito S, Naganawa S. Evaluation of lung cancer by enhanced dual-energy CT: association between three-dimensional iodine concentration and tumour differentiation. Br J Radiol. 2015;88(1055):20150224. https://doi.org/10.1259/bjr.20150224.
51. Ito R, Iwano S, Shimamoto H, Umakoshi H, Kawaguchi K, Ito S, et al. A comparative analysis of dual-phase dual-energy CT and FDG-PET/CT for the prediction of histopathological invasiveness of non-small cell lung cancer. Eur J Radiol. 2017;95:186–91. https://doi.org/10.1016/j.ejrad.2017.08.010.
52. Sato Y, Ishiyama M, Nakano S, Nakao M, Mun M, Ninomiya H, et al. Ringlike peripheral increased iodine concentration for the differentiation of primary lung cancer and pulmonary metastases on contrast-enhanced dual-energy CT. AJR Am J Roentgenol. 2023;220(6):828–37. https://doi.org/10.2214/ajr.22.28654.
53. Xu JJ, Boesen MR, Hansen SL, Ulriksen PS, Holm S, Lönn L, et al. Assessment of liver fat: dual-energy CT versus conventional CT with and without contrast. Diagnostics (Basel). 2022;12(3) https://doi.org/10.3390/diagnostics12030708.
54. Molwitz I, Leiderer M, McDonough R, Fischer R, Ozga AK, Ozden C, et al. Skeletal muscle fat quantification by dual-energy computed tomography in comparison with 3T MR imaging. Eur Radiol. 2021;31(10):7529–39. https://doi.org/10.1007/s00330-021-07820-1.
55. Kuno H, Onaya H, Iwata R, Kobayashi T, Fujii S, Hayashi R, et al. Evaluation of cartilage invasion by laryngeal and hypopharyngeal squamous cell carcinoma with dual-energy CT. Radiology. 2012;265(2):488–96. https://doi.org/10.1148/radiol.12111719.
56. Wu Q, Luo W, Zhao Y, Xu F, Zhou Q. The utility of 18F-FDG PET/CT for the diagnosis of adrenal metastasis in lung cancer: a PRISMA-compliant meta-analysis. Nucl Med Commun. 2017;38(12):1117–24. https://doi.org/10.1097/mnm.0000000000000757.
57. Shi JW, Dai HZ, Shen L, Xu DF. Dual-energy CT: clinical application in differentiating an adrenal adenoma from a metastasis. Acta Radiol. 2014;55(4):505–12. https://doi.org/10.1177/0284185113501660.
58. Kim YK, Park BK, Kim CK, Park SY. Adenoma characterization: adrenal protocol with dual-energy CT. Radiology. 2013;267(1):155–63. https://doi.org/10.1148/radiol.12112735.
59. Costelloe CM, William A, Murphy J, Chasen BA. Musculoskeletal pitfalls in 18F-FDG PET/CT: pictorial review. AJR Am J Roentgenol. 2009;193(3_supplement):WS1–WS13. https://doi.org/10.2214/ajr.07.7138.
60. Wu H, Dong S, Li X, Shi L, Shao D, Zhang Q, et al. Clinical utility of dual-energy CT used as an add-on to 18F FDG PET/CT in the preoperative staging of resectable NSCLC with suspected single osteolytic metastases. Lung Cancer. 2020;140:80–6. https://doi.org/10.1016/j.lungcan.2019.10.032.
61. Lim CH, Ahn TR, Moon SH, Cho YS, Choi JY, Kim BT, et al. PET/CT features discriminate risk of metastasis among single-bone FDG lesions detected in newly diagnosed non-small-cell lung cancer patients. Eur Radiol. 2019;29(4):1903–11. https://doi.org/10.1007/s00330-018-5764-9.
62. Tan MT, Lloyd TB. Utility of dual energy computed tomography in the evaluation of infiltrative skeletal lesions and metastasis: a literature review. Skeletal Radiol. 2022;51(9):1731–41. https://doi.org/10.1007/s00256-022-04032-6.
63. Borggrefe J, Neuhaus VF, Le Blanc M, Grosse Hokamp N, Maus V, Mpotsaris A, et al. Accuracy of iodine density thresholds for the separation of vertebral bone metastases from healthy-appearing trabecular bone in spectral detector computed tomography. Eur Radiol. 2019;29(6):3253–61. https://doi.org/10.1007/s00330-018-5843-y.
64. Hiyama T, Kuno H. Letter to editor "Dual-energy CT for the detection of skull base invasion in nasopharyngeal carcinoma: comparison of simulated single-energy CT and MRI". Insights Into Imaging. 2023;14(1):218. https://doi.org/10.1186/s13244-023-01510-w.

Chapter 6
Photon-Counting CT in Emergency Radiology

Fides R. Schwartz

Contents

6.1 Introduction to Photon-Counting CT for Emergency Applications

Photon-counting computed tomography (PCCT) is a novel imaging technology that offers numerous advantages over conventional energy-integrating detector (EID) CT systems. Its core mechanism is fundamentally different at the X-ray detector level, providing improved spatial resolution, spectral separation at the detector level, faster scan speeds without sacrificing spectral information, and other benefits that hold great potential for emergency radiology [1, 2].

F. R. Schwartz (✉)
Department of Radiology, Brigham and Women's Hospital, Harvard Medical School, Boston, MA, USA
e-mail: frschwartz@bwh.harvard.edu

S. A. Si-Mohamed, K. K. Iniewski (eds.), *Clinical Applications of Spectral Photon Counting Computed Tomography Technology*,
https://doi.org/10.1007/978-3-031-85078-3_6

Spatial resolution is improved, in large part because photon-counting detectors do not require the intermediate step of converting X-rays into visible light (as done with scintillator-based detectors) [3]. This means that the pixels in photon-counting detectors can be smaller than those in conventional CT because no physical separation (septa between the detector elements) is needed to isolate light impulses to the correct detector element [4]. This is particularly advantageous when imaging small structures such as trabecular bone, small blood vessels, or minute pathological changes in soft tissues [5, 6]. In the setting of emergency radiology, this may be useful for making confident diagnoses on initial imaging, instead of having to recommend follow-up imaging or additional clinical testing.

A photon-counting detector (PCD) operates by directly counting individual X-ray photons and measuring their energy, in contrast to the traditional EID systems that integrate energy from several photons that arrive at the same detector element to produce a signal [7]. This process allows PCCT to discriminate between different X-ray photon energies. Each photon is assigned a specific energy bin depending on its energy (between 4 and 8 bins, depending on the vendor), enabling continuous acquisition of multi-energy data, often referred to as spectral imaging. This capability allows for the differentiation of materials based on their atomic number, leading to more detailed tissue characterization, which can be helpful in emergency radiology when trying to identify bleeding sources (iodine) or discerning the composition of renal stones [8].

This method of acquiring spectral information at the detector level is a step forward from conventional dual-energy CT as offered by several vendors for existing EID systems. In most cases, specific protocols must be implemented to acquire spectral data when using dual-source scanners or rapid kVp switching techniques. The continuous access to spectral data that PCCT offers can improve lesion detection, enhance contrast resolution, and offer opportunities for advanced material decomposition, enabling better differentiation of tissue types such as fat, calcium, and iodine, even for unexpected findings.

Another advancement with photon-counting CT technology is the potential for extremely rapid image acquisition while acquiring spectral data. Utilizing a mode often referred to as "flash mode," PCCT can perform whole-body scans at up to 3.2 m/s table speed. The combination of high-speed acquisition and spectral capabilities offers significant advantages in acute care settings, particularly in patients who may not be able to cooperate with their imaging examination, as scanning can be performed without the need for breathing instructions [9]. This increased speed does not compromise image quality or spectral information [10].

The ability to obtain spectral data with every scan can also be used to improve tissue contrast and aid in differentiating materials such as hemorrhage, calcification, and iodine-based contrast agents. In emergency situations, particularly in trauma and acute stroke cases, PCCT can enable quicker and more accurate diagnosis of critical conditions like suspected intracranial hemorrhage, gastrointestinal bleeding, or other acute organ injuries. In addition, in acute chest pain assessment, PCCT can more easily differentiate between calcified plaques and contrast-enhanced blood vessels, improving the accuracy of coronary artery disease evaluation [11]. In cases of suspected pulmonary embolism, spectral data can enhance the ability to detect and classify thrombi or reduce the need for high volumes of contrast agents, especially in patients with impaired renal function [12].

Photon-counting detectors offer superior radiation dose efficiency compared to conventional CT systems [13]. They can operate at lower radiation doses while maintaining or even improving image quality [14]. This feature is especially beneficial in emergency settings, where patients often undergo multiple scans within a short time frame, such as during short-term follow-up evaluations in trauma care or in patients who are in need of emergency imaging repeatedly.

6.2 Image Quality

In the high-stakes environment of emergency radiology, image quality is critical for making rapid, accurate diagnoses. PCCT brings notable improvements to image quality that address the challenges inherent to emergency imaging, particularly in trauma, acute vascular events, and chest pain assessment [15].

A key challenge in emergency radiology is imaging patients under suboptimal conditions. Many emergency patients are in motion due to pain, discomfort, or agitation, and this movement can degrade image quality in conventional CT systems, leading to non-diagnostic scans. PCCT, with its ability to perform fast scans without sacrificing image quality, reduces the motion artifacts commonly seen in such cases. Faster scan acquisition times are particularly important in trauma imaging, where whole-body CT is often performed to assess for multiple injuries in unstable patients and in lung imaging for patients that cannot comply with breathing instructions [10].

Another critical advantage of PCCT in emergency radiology is its improved contrast resolution. In cases such as acute chest pain, where rapid assessment for aortic dissection, pulmonary embolism, or coronary artery disease is necessary, the enhanced contrast resolution of PCCT can improve diagnostic qualities, especially in situations where the contrast bolus timing may have been suboptimal [16]. Furthermore, in stroke imaging, where time-sensitive decisions about reperfusion therapy depend on the precise identification of ischemic regions, PCCT may provide improved visualization of subtle parenchymal changes and perfusion abnormalities when combined with virtual monoenergetic images or iodine maps [17, 18].

6.3 Contrast Agent Utilization

In emergency radiology, the use of intravenous contrast agents is essential for many diagnostic procedures, particularly in assessing vascular, abdominal, and traumatic injuries. However, the necessity of contrast agents must be balanced against the risks they pose, especially in patients with pre-existing renal impairment or those at risk of contrast-induced nephropathy (CIN) [19]. PCCT represents a significant advance in this context, as its ability to generate high-quality images with lower contrast agent volumes offers substantial benefits for patient safety and clinical outcomes [20].

This is primarily due to the photon-counting detector‘s ability to efficiently differentiate between photon energies, resulting in improved contrast-to-noise ratios even at reduced contrast agent volumes. In conventional CT systems, increasing contrast agent dose is often necessary to enhance the attenuation differences between tissues, especially in low radiation-dose settings. Another approach to increase contrast conspicuity in EID CT is to lower the tube voltage to achieve a higher number of low-energy photons in the X-ray beam, which improves soft tissue contrast [21]. This approach is, however, limited by individual patients' body habitus as the image noise increases with larger patient diameters. PCCT acquisitions can be reconstructed at lower virtual monoenergetic energy levels, essentially simulating a CT acquired with a lower tube voltage on an EID CT. This is not limited by patient size and thus all patients can benefit from better image quality [22, 23].

In vascular emergencies, such as suspected aortic dissection, pulmonary embolism, or acute coronary syndromes, the ease of assessment of contrast-enhanced vascular structures is crucial for diagnosis and treatment planning. PCCT's efficient photon-counting detectors allow for enhanced visualization of contrast-enhanced vessels even at lower contrast doses, minimizing the renal burden on patients who may have underlying comorbidities, such as diabetes or chronic kidney disease [24]. In these cases, where timely and precise imaging is essential, the reduced contrast requirements of PCCT do not compromise the ability to accurately assess the vascular system for critical pathologies like thrombi, dissection, or aneurysms [25].

Beyond the immediate reduction in contrast agent volume, PCCT also offers the potential for more optimized protocols that can further minimize contrast use. The spectral imaging capability of PCCT allows for the differentiation of materials based on their attenuation properties at different energy levels, meaning that certain tissues, such as iodine-enhanced blood vessels or contrast-enhanced organs, can be more effectively imaged, especially when taking advantage of lower keV virtual monoenergetic image creation. These images can be automatically created on the scanner console or as part of a vendor-neutral server-based reconstruction logic, which can save time for radiologists, who do not need to access a separate thin-client solution for spectral image post-processing.

6.4 Radiation Dose Management

PCCT excels in low-dose imaging scenarios, which are often necessary in emergency settings, particularly in vulnerable populations such as pediatric or polytrauma patients requiring multiple follow-up scans [26]. Conventional CT scans performed at lower radiation doses frequently suffer from increased noise, reducing image quality and diagnostic confidence. PCCT, however, maintains high image quality even at reduced radiation doses, with minimal noise interference [27]. By offering high image quality at lower radiation doses, PCCT reduces the cumulative radiation burden on patients without sacrificing diagnostic accuracy.

PCCT can reduce radiation exposure by 30% to 60% compared to conventional EID CT, depending on the clinical scenario, without compromising the diagnostic value of the scans [28]. This reduction in dose is especially important for vulnerable populations, such as pediatric patients, polytrauma patients, and those requiring serial imaging for ongoing conditions, where minimizing cumulative radiation is key for long-term health outcomes [29, 30].

The primary mechanism by which PCCT achieves such significant radiation dose reductions lies, once again, in its unique detectors, which allow for a more efficient use of the X-ray beam [7]. PCD can filter out low-energy photons that contribute to patient dose but do not improve image quality. By setting a threshold, typically between 20 and 30 keV, PCCT effectively eliminates low-energy photons that are prone to be absorbed in superficial tissues without contributing to diagnostic information, contributing to a better image noise profile at lower radiation doses [31, 32].

In emergency radiology, where rapid decision-making is often necessary, the ability to perform high-quality imaging at lower radiation doses is invaluable. In trauma patients, for example, whole-body CT is frequently used to assess for fractures, internal injuries, and active bleeding in one exam. These patients, particularly if polytrauma is present, may require multiple follow-up scans to monitor for changes in their condition, which increases their overall radiation exposure. PCCT reduces the cumulative radiation burden in such patients, making serial imaging safer, particularly for young adults or children who are more sensitive to long-term effects of radiation. By offering lower doses without sacrificing diagnostic accuracy, PCCT helps emergency radiologists provide the necessary imaging to guide patient care while minimizing long-term risks associated with radiation exposure [33].

For pediatric patients, the benefits of dose reduction with PCCT are even more pronounced. Children are more radiosensitive than adults due to their rapidly dividing cells and longer life expectancy, increasing their risk of developing radiation-induced cancers over time [29]. In pediatric emergency radiology, where the need for rapid imaging often outweighs concerns about radiation exposure, PCCT provides a safer alternative to conventional CT.

The ability of PCCT to provide high-quality images at lower radiation doses is not just relevant in pediatric or polytrauma cases. It also offers significant advantages in more routine emergency scenarios, such as imaging patients with chest pain, suspected stroke, or abdominal pain. For example, in the evaluation of pulmonary embolism or aortic dissection, high-quality imaging is critical for detecting life-threatening conditions. PCCT allows for these diagnoses to be made at a lower radiation dose, especially if whole acquisitions can be removed from protocols (e.g., when using virtual monoenergetic reconstructions from contrast-enhanced series instead of acquiring a true noncontrast scan [34]), reducing patient risk without compromising the detail or accuracy of the images. In patients with chronic conditions who may present to the emergency department repeatedly, such as those with recurrent kidney stones or bowel obstructions, the cumulative radiation exposure from serial CT scans can be significant. PCCT's dose-sparing capabilities help mitigate this risk, allowing these patients to receive the necessary imaging at every visit while reducing the potential for radiation-related complications down the line.

The ability to maintain image quality at lower radiation doses also supports the growing trend toward low-dose protocols in emergency imaging. While conventional CT systems can reduce radiation dose, this often comes at the cost of increased image noise, which reduces diagnostic confidence, particularly in low-contrast situations [35]. PCCT, however, maintains high contrast-to-noise ratios even at reduced radiation doses, which is especially beneficial in scenarios where low-dose imaging is critical, such as in assessing lung parenchyma for pulmonary contusions or early lung infections [36].

6.5 Spectral Capabilities

PCCT provides a significant advancement in spectral imaging over many EID systems by inherently offering spectral data in every scan. This sets PCCT apart from most traditional dual-energy CT (DECT) systems, where dual-energy scanning must be specifically activated, limiting the availability of spectral information, often especially in emergency imaging. In PCCT, all photons are measured across a range of energy levels, and the spectral information is automatically embedded in every scan (archived in the spectral post-processing (SPP) file). This feature is particularly valuable in the fast-paced environment of emergency radiology, where there may be no way to anticipate the need for dual-energy imaging ahead of scanning. The availability of spectral data in every PCCT scan enhances diagnostic capabilities and opens up new possibilities for tissue characterization, lesion differentiation, and improved diagnostic confidence, all of which are especially important in emergency settings where incidental findings often occur [37].

One of the key strengths of PCCT's spectral imaging capability is that it allows for material decomposition, meaning different materials within the body can be differentiated based on their attenuation properties at varying energy levels. This ability mirrors the capabilities of tube-based dual-energy CT, where scans from two X-ray tubes operating at two different energy levels or one tube rapidly switching its kVp output are used to differentiate between tissue types, contrast agents, and other materials [38]. However, with PCCT performing the energy separation at the detector level, these spectral images are available by default in every scan, eliminating the need to set specific dual-energy protocols. This can be a game-changer in emergency radiology, where patients often present with complex and unexpected findings that require further characterization. In these cases, the immediate availability of spectral data provides clinicians with additional diagnostic information that can help clarify incidental or ambiguous findings, reducing the need for follow-up imaging and additional clinical diagnostics, potentially saving resources [37].

For example, incidental lesions are commonly encountered in emergency radiology. While the primary reason for imaging may be to assess trauma, acute abdominal pain, or chest pain, radiologists frequently discover incidental findings such as renal cysts, adrenal masses, or pulmonary nodules. In conventional CT, these incidental lesions often require further characterization through follow-up imaging or other diagnostic tests to determine whether they are benign or malignant. However, with PCCT's spectral imaging capability, many of these lesions may be more confidently classified during the initial scan. For instance, spectral imaging can help differentiate between benign cystic renal lesions and more suspicious solid masses based on their iodine content or attenuation properties at different energy levels [39]. By doing so, PCCT enables radiologists to make more confident diagnoses without the need for additional imaging, reducing patient anxiety, decreasing the number of unnecessary follow-up exams, and streamlining patient management [40].

In cases where incidental lesions are discovered in the liver, pancreas, or adrenal glands, spectral imaging can further aid in differentiating between benign and malignant pathologies. For example, incidental liver lesions may be classified as hemangiomas, cysts, or potentially malignant tumors using spectral data to analyze their contrast uptake characteristics [41]. PCCT's material decomposition capability can distinguish between different types of tissue, enabling radiologists to determine whether a lesion contains fat, calcium, or hemorrhage and helping to distinguish between benign findings and those that may warrant further investigation. This ability is particularly useful in the emergency department, where incidental findings may not be immediately addressed, leading to delays in care or unnecessary worry for the patient.

Additionally, PCCT's spectral imaging is valuable for differentiating between contrast-enhanced structures and pathological tissue, particularly in vascular emergencies. For example, in suspected pulmonary embolism, spectral data can be used to distinguish between an acute thrombus and enhancing soft tissue, increasing the diagnostic confidence in cases where the presence of a clot might otherwise be ambiguous (Fig. 6.1). Similarly, in patients with suspected aortic dissection, PCCT can aid in the detection of subtle dissections or intramural hematomas, especially when making use of virtual non-contrast images [42]. This immediate access to spectral data can speed up decision-making and prevent unnecessary delays in initiating life-saving treatments.

6.6 Diagnostic Workflows

Photon-counting CT (PCCT) brings notable improvements to diagnostic workflows in the emergency department (ED). One of the most immediate impacts of PCCT on workflow is its ability to streamline the imaging process by eliminating the need for separate protocols, particularly dual-energy CT (DECT) scans. In conventional systems, specific imaging protocols must be selected based on clinical indications—such as dual-energy modes for enhanced tissue characterization or vascular

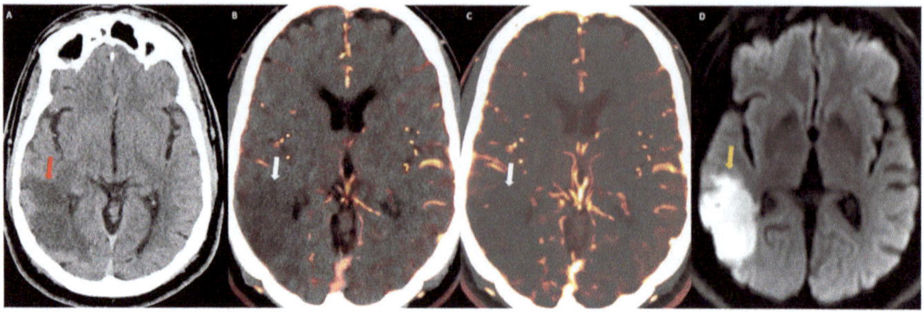

Fig. 6.1 Shows a subacute right temporal ischemic infarct in a 65-year-old patient. (**a**) Shows the standard reconstruction at 67 keV with a clearly delineated hypodense area in the right temporal lobe (red arrow). The (**b**) CTA and (**c**) CTV show no iodine enhancement in the area surrounding the hypodense zone (red arrows). An MRI (**d**), acquired 2 h later, shows diffusion restriction in the same distribution (yellow arrow)

imaging—often requiring technologists or radiologists to anticipate diagnostic needs before scanning. PCCT, however, automatically captures spectral data in every scan, removing the need for these decisions upfront. This simplification reduces the time spent setting up protocols, minimizes errors associated with incorrect protocol selection, and ensures that all relevant diagnostic data is available after a single scan.

Another key advantage of PCCT in the ED is its potential to reduce repeat imaging, which is frequently necessary for conventional CT when findings are inconclusive or incidental lesions require further characterization. With PCCT, the spectral data captured during the initial scan can often provide enough information to fully characterize these findings without the need for additional scans or other imaging modalities. For example, a patient undergoing evaluation for trauma may have an incidental adrenal lesion or renal mass detected [43, 44]. In traditional CT, this might prompt a follow-up scan with a different contrast protocol or even a different imaging modality such as MRI. With PCCT, material differentiation—such as the ability to assess iodine uptake or differentiate between calcifications and iodine—can often provide the necessary diagnostic clarity, enabling radiologists to classify the lesion as benign or malignant with greater confidence during the initial assessment. This not only expedites patient care if the emergency department can refer patients directly to relevant medical sub-specialties for further care but also reduces the strain on imaging resources.

In acute clinical scenarios, the availability of spectral data in all PCCT scans also enhances diagnostic accuracy and decision-making speed. For instance, in suspected pulmonary embolism or aortic dissection cases, where rapid and accurate imaging is vital, PCCT can provide detailed information about blood flow and distribution without requiring multiphase acquisitions [22, 34]. This can result in faster interpretations and more confident diagnoses, reducing time-to-treatment for critical conditions. By offering higher-quality images and spectral data in a single acquisition, PCCT reduces the need for radiologists to request additional imaging, thus accelerating the diagnostic process.

In terms of workflow, the high-quality images produced by PCCT—coupled with its spectral capabilities—allow radiologists to make more confident, definitive diagnoses during the first scan, reducing the back-and-forth between departments, cutting down on consultation times, and minimizing delays. This improvement in efficiency ultimately translates to faster patient disposition, whether that means discharge, hospital admission, or surgical intervention. PCCT, by integrating advanced imaging capabilities into routine CT scans, can enhance both the speed and precision of diagnostic workflows in the emergency department, improving patient outcomes while optimizing resource use.

6.7 Stroke Imaging

PCCT offers several advantages for stroke imaging, especially because most practices have never used dual-energy acquisitions for these protocols, but this spectral information is available through PCCT [45]. In ischemic stroke, the ability to differentiate between salvageable brain tissue (penumbra) and irreversibly damaged tissue (core) is crucial for timely intervention [46]. PCCT's enhanced contrast-to-noise ratio allows for better detection of subtle tissue changes, improving the visualization of ischemic regions [18]. The spectral data captured during every PCCT scan can also assist in distinguishing between blood products and calcifications in hemorrhagic stroke cases [47]. Furthermore, PCCT may optimize contrast-enhanced brain perfusion imaging, which plays a critical role in assessing the extent of brain ischemia, all while minimizing contrast volume and radiation dose—an important consideration in stroke patients who often present with comorbid conditions like renal impairment (Fig. 6.2) [48]. These features help streamline the stroke diagnosis and treatment pathway, leading to faster decision-making for clinical teams.

6.8 Pulmonary Embolism Detection

For detecting pulmonary embolism (PE), PCCT enhances both the sensitivity and specificity of imaging by improving the visualization of pulmonary arteries and thrombi. In conventional CT angiography (CTA), diagnosing PE can be challenging, especially in subsegmental vessels or in cases with extensive clot burden [49]. PCCT's superior resolution and spectral imaging capabilities allow for better delineation of vascular structures, even at lower contrast doses or when the optimal contrast bolus was missed during the scan (Fig. 6.3). Additionally, spectral images help assess the iodine content within the vessels, and lung parenchyma, giving the option to routinely assess lung perfusion [50]. This reduces the need for repeat imaging or follow-up scans, enabling quicker diagnoses and faster treatment initiation, whether

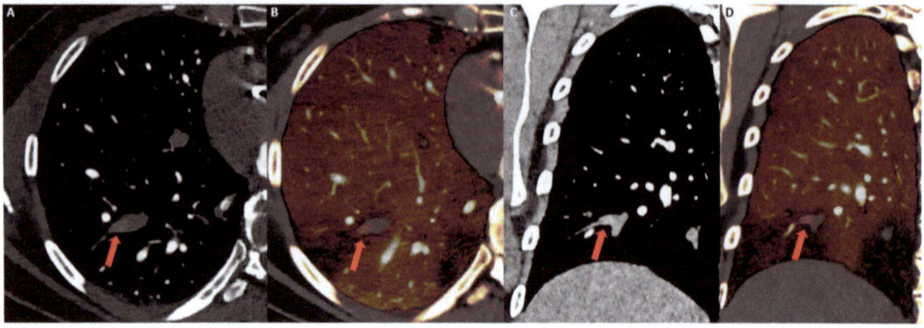

Fig. 6.2 Shows contrast enhancing thrombus inside a pulmonary vein (red arrows) in a 48-year-old patient with known pancreas adenocarcinoma. This was confirmed to be metastatic tissue via pathology. (**a** and **c**) show the standard contrasted images of a pulmonary-embolism detection study at 40 keV virtual monoenergetic reconstruction, with the metastatic thrombus showing a higher density than the adjacent pulmonary vein. (**b** and **d**) show the corresponding iodine maps that demonstrate a clear iodine uptake in the metastatic tissue

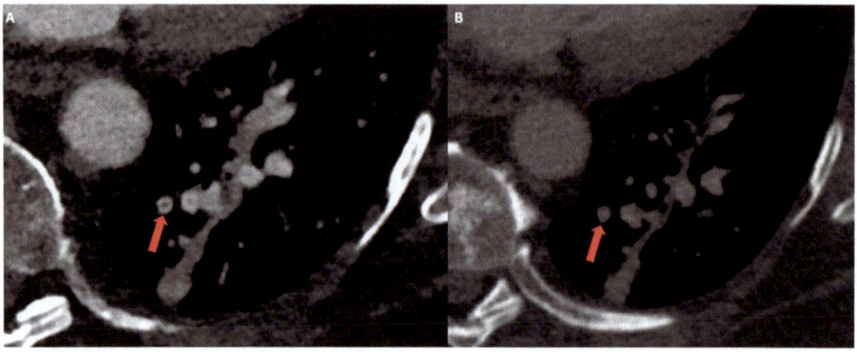

Fig. 6.3 Shows a subsegmental thrombus in a pulmonary artery in a 59-year-old patient, who had received 50 mL of intravenous contrast. The 40 keV monoenergetic reconstruction in (**a**) shows the filling defect clearly (red arrow), while the (**b**) reconstruction at 67 keV (similar to an acquisition at 120 kVp or dual-energy blended images) does not show the thrombus

through anticoagulation or thrombolysis, thus improving patient outcomes in a life-threatening condition like PE.

6.9 Acute Abdominal CT

PCCT offers improvements for acute abdominal imaging, which is commonly used in the emergency department to evaluate conditions such as appendicitis, bowel obstruction, or mesenteric ischemia. One of the major benefits of PCCT in this context is its ability to enhance the visualization of soft tissues and vasculature at lower radiation doses and with reduced contrast agent volumes (Fig. 6.4) [51]. The

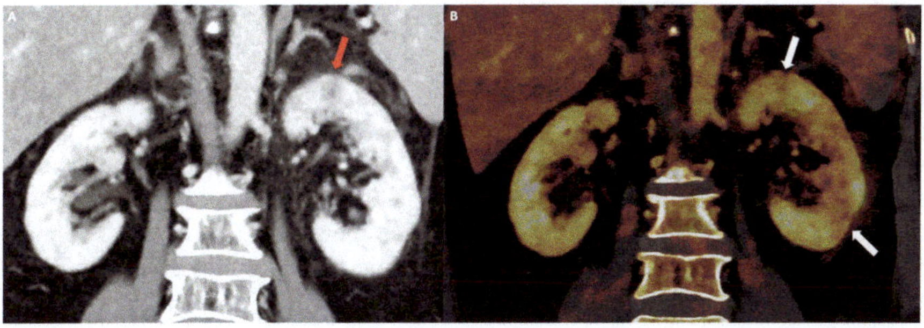

Fig. 6.4 Shows pyelonephritis in the left kidney of a 61-year-old patient. The (**a**) standard 67 keV reconstruction shows a zone of potential hypoenhancement (red arrow), which is confirmed in the (**b**) iodine map as a wedge-shaped perfusion deficit, in conjunction with additional inhomogeneities of iodine uptake in the mid-section and lower pole of the left kidney (white arrows), confirming the diagnosis of a pyelonephritis

spectral data provided by PCCT allows for better characterization of vascular abnormalities and inflammation, distinguishing between different tissue types and enhancing the detection of small, subtle lesions that may be missed on conventional CT [52–54]. This way, PCCT streamlines the diagnostic process, reducing the need for repeat imaging and ensuring timely intervention for acute abdominal emergencies.

6.10 Trauma Imaging

In trauma imaging, where rapid whole-body CT is often required to assess multiple injuries, PCCT provides high-resolution images that are invaluable for detecting fractures, organ injuries, and active hemorrhage. One of the primary advantages in this setting is PCCT's ability to acquire high-quality images at lower radiation doses and with a lower overall number of acquisitions, which is particularly important for trauma patients who may require multiple follow-up scans [55]. Additionally, PCCT's spectral imaging capabilities improve the visualization of hemorrhages, differentiate between active bleeding and other high-density materials, and enhance the detection of small, often subtle injuries in complex trauma cases [56]. By improving the accuracy of initial imaging, PCCT helps reduce the need for repeat scans, allowing for faster decision-making regarding surgical or interventional treatments.

6.11 Incidental Findings Management

In emergency radiology, incidental findings—such as lung nodules, adrenal masses, or renal cysts—are frequently encountered, often unrelated to the acute clinical presentation. PCCT offers a significant advantage in the management of these incidental lesions through its inherent spectral imaging capabilities, which enable tissue characterization during the initial scan. For example, PCCT can differentiate between benign cystic structures and more suspicious solid masses by analyzing their contrast uptake and tissue components, such as fat [57]. This ability to confidently characterize lesions reduces the need for follow-up imaging and additional diagnostics, lowering patient anxiety and preventing unnecessary healthcare utilization. With PCCT, many incidental findings can be definitively assessed in the emergency department, streamlining patient care and improving workflow efficiency.

6.12 Opportunistic Screening

Opportunistic screening is another area where PCCT can provide added value in the emergency setting. During routine or trauma-related imaging, PCCT's spectral data can be leveraged to assess a variety of conditions that might not be the primary focus of the scan. For instance, bone mineral density can be evaluated for osteoporosis screening, vascular calcifications can be assessed for cardiovascular risk, and liver fat content can be quantified, all from a single scan performed for a different diagnostic purpose [11, 58–62]. This capability allows radiologists to gather important health information opportunistically, potentially identifying conditions like early atherosclerosis or osteoporosis without the need for additional imaging or clinical appointments. PCCT's ability to provide comprehensive data during routine emergency scans opens up new possibilities for preventive care, even in acute care settings.

6.13 Potential for Functional Imaging

With its spectral capabilities, photon-counting CT (PCCT) opens the door to functional imaging in emergency settings, providing valuable insights beyond traditional anatomical visualization. This is particularly useful in acute conditions such as stroke or pulmonary embolisms, where rapid assessment of tissue viability and perfusion is crucial for determining appropriate treatment strategies. In stroke imaging, one of PCCT's key advantages is its ability to differentiate between various materials, such as iodine-based contrast agents and hemorrhage [47]. By using the spectral data inherent in every scan, PCCT can help to identify areas of older hemorrhage in contrast with acute contrast extravasation, which can be relevant for the triaging of these patients [63].

In ischemic stroke, iodine-based contrast agents are commonly used in perfusion studies to assess blood flow to the brain [64]. PCCT's spectral imaging capabilities may allow for the assessment of iodine uptake and distribution, which are critical in evaluating whether tissue is salvageable (penumbra) or irreversibly damaged (infarct core), without the need for the multiple imaging volumes that traditionally need to be acquired for perfusion studies. This may provide the opportunity to reduce the number of perfusion studies and instead rely on the iodine maps generated from standard CT angiography and venography performed during standard stroke workups.

Beyond iodine, the potential to use multiple contrast agents with distinct energy signatures in PCCT could, in the future, offer new possibilities for functional imaging in the emergency setting. PCCT's energy discrimination ability allows it to detect and differentiate contrast agents based on their unique energy profiles, making it theoretically possible to use agents such as gadolinium, hafnium, tantalum, or gold in conjunction with traditional iodine-based agents [65, 66]. Each of these elements has different attenuation characteristics at various energy levels, which may be exploited by PCCT to provide a more comprehensive assessment of vascular, soft tissue, and functional processes in the body.

For example, gadolinium, commonly used in magnetic resonance imaging (MRI), could be employed in PCCT to complement iodine-based contrast agents. There have been animal studies showing the possibility for this [67, 68], but it is not clear that it will be achievable using typical contrast doses for human imaging [69]. Similarly, materials such as hafnium or tantalum could be used for specific applications where enhanced contrast or differentiation of certain structures is needed, for example, as oral contrast agents in conjunction with intravenous contrast agents, when there is concern for bowel obstruction and non-perfused bowel [70–72].

In trauma settings, where the rapid assessment of multiple organ systems is required, the use of multiple contrast agents could offer advantages. PCCT could simultaneously assess perfusion, organ integrity, and blood loss by utilizing different contrast agents tailored for specific tissues or functions. The ability to gather functional and anatomical data in a single scan without the need for multiple imaging modalities could significantly reduce the time to diagnosis and treatment in critically ill patients.

Another promising application of multiple contrast agents in PCCT is the evaluation of tumor perfusion and angiogenesis in emergency oncology cases, such as patients presenting with acute complications of cancer. Using a combination of iodine and a secondary contrast agent could offer a more nuanced assessment of tumor vascularity, enabling clinicians to differentiate between necrotic and viable tumor tissue, assess the extent of tumor invasion, and determine the appropriate course of action. Even though oncological imaging is not a standard for emergency imaging, these patients are frequently seen in the emergency department and would benefit from not needing additional imaging after an initial investigation in the emergency setting, if enough information can be provided using PCCT capabilities.

In summary, PCCT's imaging capabilities allow for a combination of high spatial and contrast resolution, at low radiation doses, with optimized contrast agent dosing and the potential for functional imaging that goes beyond mere anatomical assessment, offering the ability to differentiate iodine from blood,, and evaluate tissue viability in emergency situations such as stroke, pulmonary embolism or trauma. Moreover, the possibility of using spectral data retrospectively presents new opportunities for more detailed assessments of incidental findings that can be encountered in a variety of emergency settings. This innovation holds the potential to significantly enhance diagnostic workflows and accuracy, reduce the need for multiple imaging modalities where incidental findings are discovered, and optimize patient outcomes in critical care scenarios.

References

1. Almqvist H, Crotty D, Nyren S, Yu J, Arnberg-Sandor F, Brismar T, et al. Initial clinical images from a second-generation prototype silicon-based photon-counting computed tomography system. Academic Radiology. 2023;31(23):572–81. https://doi.org/10.1016/j.acra.2023.06.031.
2. Douek PC, Boccalini S, Oei EHG, Cormode DP, Pourmorteza A, Boussel L, et al. Clinical applications of photon-counting CT: a review of pioneer studies and a glimpse into the future. Radiology. 2023;309(1):e222432. https://doi.org/10.1148/radiol.222432.
3. Wang J, Fleischmann D. Improving spatial resolution at CT: development, benefits, and pitfalls. Radiology. 2018;289(1):261–2. https://doi.org/10.1148/radiol.2018181156.
4. Danielsson M, Persson M, Sjölin M. Photon-counting x-ray detectors for CT. Phys Med Biol. 2021;66(3):03TR01. https://doi.org/10.1088/1361-6560/abc5a5.
5. Halfmann MC, Bockius S, Emrich T, Hell M, Schoepf UJ, Laux GS, et al. Ultrahigh-spatial-resolution photon-counting detector CT angiography of coronary artery disease for stenosis assessment. Radiology. 2024;310(2):e231956. https://doi.org/10.1148/radiol.231956.
6. Rajagopal JR, Schwartz FR, Solomon JB, Enterline DS, Samei E. High spatial-resolution skull base imaging with photon-counting computed tomography and energy-integrating computed tomography: a comparative phantom study. Reconstruction. 2023;50:32.
7. Flohr T, Schmidt B. Technical basics and clinical benefits of photon-counting CT. Invest Radiol. 2023;58(7):441. https://doi.org/10.1097/RLI.0000000000000980.
8. Marcus RP, Fletcher JG, Ferrero A, Leng S, Halaweish AF, Gutjahr R, et al. Detection and characterization of renal stones by using photon-counting-based CT. Radiology. 2018;289(2):436–42. https://doi.org/10.1148/radiol.2018180126.
9. Euler A, Higashigaito K, Mergen V, Sartoretti T, Zanini B, Schmidt B, et al. High-pitch photon-counting detector computed tomography angiography of the aorta: intraindividual comparison to energy-integrating detector computed tomography at equal radiation dose. Invest Radiol. 2022;57(2):115–21. https://doi.org/10.1097/RLI.0000000000000816.
10. Bunch PM, Fulwadhva UP, Wortman JR, Primak AN, Madan R, Steigner ML, et al. Motion artifact reduction from high-pitch dual-source computed tomography pulmonary angiography. J Comput Assist Tomogr. 2018;42(4):623–9. https://doi.org/10.1097/RCT.0000000000000736.
11. Boussel L, Coulon P, Thran A. Photon counting spectral CT component analysis of coronary artery atherosclerotic plaque samples. Br J Radiol. 2014;87(1040):20130798.
12. Kopp FK, Daerr H, Si-Mohamed S. Evaluation of a preclinical photoncounting CT prototype for pulmonary imaging. Sci Rep. 2018;8(1):17386.

13. Dettmer S, Werncke T, Mitkovska VN, Brod T, Joean O, Vogel-Claussen J, et al. Photon counting computed tomography with the radiation dose of a chest X-ray: feasibility and diagnostic yield. Respiration. 2024;103(2):88–94. https://doi.org/10.1159/000536065.
14. Schwartz FR, Vinson EN, Spritzer CE, Colglazier R, Samei E, French RJ, et al. Prospective multireader evaluation of photon-counting CT for multiple myeloma screening. Radiol Imaging Cancer. 2022;4(6):e220073. https://doi.org/10.1148/rycan.220073.
15. Flohr T, Petersilka M, Henning A, Ulzheimer S, Ferda J, Schmidt B. Photon-counting CT review. Phys Med. 2020;79:126–36. https://doi.org/10.1016/j.ejmp.2020.10.030.
16. Bae K, Jeon KN, Cho SB. Improved opacification of a suboptimally enhanced pulmonary artery in chest CT: experience using a dual-layer detector spectral CT. AJR Am J Roentgenol. 2018;210(4):734–41.
17. Fransson V, Mellander H, Wasselius J, Ydström K. Detection of perfusion deficits in multiphase computed tomography angiography—a stroke imaging technique based on iodine mapping on spectral computed tomography: initial findings. J Comput Assist Tomogr. 2021;45(4):618. https://doi.org/10.1097/RCT.0000000000001173.
18. Taguchi K, Itoh T, Fuld MK, Fournie E, Lee O, Noguchi K. "X-Map 2.0" for edema signal enhancement for acute ischemic stroke using non–contrast-enhanced dual-energy computed tomography. Invest Radiol. 2018;53(7):432. https://doi.org/10.1097/RLI.0000000000000461.
19. Isaka Y, Hayashi H, Aonuma K, Horio M, Terada Y, Doi K, et al. Guideline on the use of iodinated contrast media in patients with kidney disease 2018. Jpn J Radiol. 2020;38(1):3–46. https://doi.org/10.1007/s11604-019-00850-2.
20. Alobeidi H, Alshamari M, Widell J, Eriksson T, Lidén M. Minimizing contrast media dose in CT pulmonary angiography with high-pitch technique. Br J Radiol. 2020;93(1111):20190995. https://doi.org/10.1259/bjr.20190995.
21. Apfaltrer P, Sudarski S, Schneider D, Nance JW Jr, Haubenreisser H, Fink C, et al. Value of monoenergetic low-kV dual energy CT datasets for improved image quality of CT pulmonary angiography. Eur J Radiol. 2014;83(2):322–8. https://doi.org/10.1016/j.ejrad.2013.11.005.
22. Dane B, Freedman D, Qian K, Ginocchio L, Smereka P, Megibow A. Photon-counting CT urogram: optimal acquisition potential (kV) determination for virtual noncontrast creation. Abdom Radiol. 2024;49(3):868–74. https://doi.org/10.1007/s00261-023-04113-7.
23. Marth AA, Goller SS, Kajdi GW, Marcus RP, Sutter R. Photon-counting detector CT: clinical utility of virtual monoenergetic imaging combined with tin prefiltration to reduce metal artifacts in the postoperative ankle. Invest Radiol. 2024;59(8):545–53. https://doi.org/10.1097/RLI.0000000000001058.
24. Noro T, Ojio Y, Urano M, Ohta K, Suzuki K, Sato T, et al. Photon-counting detector CT with an ultra-low-dose contrast media to diagnose a renal pseudoaneurysm: a case report. Radiol Case Rep. 2024;19(9):3618–21. https://doi.org/10.1016/j.radcr.2024.05.077.
25. Emrich T, O'Doherty J, Schoepf UJ, Suranyi P, Aquino G, Kloeckner R, et al. Reduced iodinated contrast media administration in coronary CT angiography on a clinical photon-counting detector CT system: a phantom study using a dynamic circulation model. Invest Radiol. 2023;58(2):148–55. https://doi.org/10.1097/RLI.0000000000000911.
26. Cao J, Bache S, Schwartz FR, Frush D. Pediatric applications of photon-counting detector CT. Am J Roentgenol. 2023;220(4):580–9. https://doi.org/10.2214/AJR.22.28391.
27. Rajagopal JR, Farhadi F, Saboury B, Sahbaee P, Negussie AH, Pritchard WF, et al. Multivariate signal-to-noise ratio as a metric for characterizing spectral computed tomography. Phys Med Biol. 2024;69(14):145005. https://doi.org/10.1088/1361-6560/ad5d4a.
28. Jungblut L, Euler A, von Spiczak J, Sartoretti T, Mergen V, Englmaier V, et al. Potential of photon-counting detector CT for radiation dose reduction for the assessment of interstitial lung disease in patients with systemic sclerosis. Invest Radiol. 2022;57(12):773–9. https://doi.org/10.1097/RLI.0000000000000895.
29. Frush D. The cumulative radiation dose paradigm in pediatric imaging. Br J Radiol. 2021;94(1126):20210478. https://doi.org/10.1259/bjr.20210478.

30. Goodman TR, Mustafa A, Rowe E. Pediatric CT radiation exposure: where we were, and where we are now. Pediatr Radiol. 2019;49(4):469–78. https://doi.org/10.1007/s00247-018-4281-y.
31. Duan X, Wang J, Leng S, Schmidt B, Allmendinger T, Grant K, et al. Electronic noise in CT detectors: impact on image noise and artifacts. AJR Am J Roentgenol. 2013;201(4):W626–32. https://doi.org/10.2214/AJR.12.10234.
32. Kappler S, Henning A, Kreisler B, Schoeck F, Stierstorfer K, Flohr T. Photon counting CT at elevated X-ray tube currents: contrast stability, image noise and multi-energy performance. In: Medical imaging 2014: physics of medical imaging. SPIE; 2014. p. 368–75. https://doi.org/10.1117/12.2043511.
33. Newman B, Callahan MJ. ALARA (as low as reasonably achievable) CT 2011—executive summary. Pediatr Radiol. 2011;41(Suppl 2):453–5. https://doi.org/10.1007/s00247-011-2154-8.
34. Decker JA, Bette S, Scheurig-Muenkler C. Virtual non-contrast reconstructions of photon-counting detector CT angiography datasets as substitutes for true non-contrast acquisitions in patients after EVAR—performance of a novel calcium-preserving reconstruction algorithm. Diagnostics https://doi.org/10.3390/diagnostics12030558
35. Chang KJ, Collins S, Li B, Mayo-Smith WW. Optimizing CT technique to reduce radiation dose: effect of changes in kVp, iterative reconstruction, and noise index on dose and noise in a human cadaver. Radiol Phys Technol. 2017;10(2):180–8. https://doi.org/10.1007/s12194-016-0382-1.
36. Chen M, Mattar G, Abdulkarim JA. Computed tomography pulmonary angiography using a 20% reduction in contrast medium dose delivered in a multiphasic injection. World J Radiol. 2017;9(3):143–7. https://doi.org/10.4329/wjr.v9.i3.143.
37. Wortman JR, Shyu JY, Fulwadhva UP, Sodickson AD. Impact analysis of the routine use of dual-energy computed tomography for characterization of incidental renal lesions. J Comput Assist Tomogr. 2019;43(2):176. https://doi.org/10.1097/RCT.0000000000000828.
38. Johnson TRCC. Dual-energy CT: general principles. AJR Am J Roentgenol. 2012;199(5 Suppl):S3.
39. Homayounieh F, Gopal N, Firouzabadi FD, et al. A prospective study of the diagnostic performance of photon-counting CT compared with MRI in the characterization of renal masses. Invest Radiol. 2023;59:774. https://doi.org/10.1097/RLI.0000000000001087.
40. Schwartz FR, Roth CJ, Boardwine B, Hardister L, Thomas-Campbell S, Lander K, et al. Electronic health record closed-loop communication program for unexpected nonemergent findings. Radiology. 2021;301(1):123–30. https://doi.org/10.1148/radiol.2021210057.
41. Bette S, Decker JA, Braun FM, Becker J, Haerting M, Haeckel T, et al. Optimal conspicuity of liver metastases in virtual monochromatic imaging reconstructions on a novel photon-counting detector CT—effect of keV settings and BMI. Diagnostics. 2022;12(5):1231. https://doi.org/10.3390/diagnostics12051231.
42. Sandfort V, Persson M, Pourmorteza A, Noël PB, Fleischmann D, Willemink MJ. Spectral photon-counting CT in cardiovascular imaging. J Cardiovasc Comput Tomogr. 2021;15(3):218–25. https://doi.org/10.1016/j.jcct.2020.12.005.
43. Bette S, Risch F, Canalini L, Becker J, Leithner EV, Huber A, et al. Diagnostic performance of photon-counting detector CT for differentiation between adrenal adenomas and metastases. Eur Radiol. 2024;34:5944. https://doi.org/10.1007/s00330-024-10675-x.
44. Mileto A, Nelson RC, Marin D, Roy Choudhury K, Ho LM. Dual-energy multidetector CT for the characterization of incidental adrenal nodules: diagnostic performance of contrast-enhanced material density analysis. Radiology. 2015;274(2):445–54. https://doi.org/10.1148/radiol.14140876.
45. Abel F, Schubert T, Winklhofer S. Advanced neuroimaging with photon-counting detector CT. Invest Radiol. 2023;58(7):472–81.
46. Sarraj A, Hassan AE, Abraham MG, Ortega-Gutierrez S, Kasner SE, Hussain MS, et al. Endovascular thrombectomy for large ischemic stroke across ischemic injury and penumbra profiles. JAMA. 2024;331(9):750–63. https://doi.org/10.1001/jama.2024.0572.

47. Wiggins WF, Potter CA, Sodickson AD. Dual-energy CT to differentiate small foci of intracranial hemorrhage from calcium. Radiology. 2020;294(1):129–38. https://doi.org/10.1148/radiol.2019190792.
48. Review of perfusion imaging in acute ischemic stroke | Stroke. . https://www.ahajournals.org/doi/full/10.1161/STROKEAHA.119.028337. Accessed 30 Sept 2024.
49. Costa AF, Basseri H, Sheikh A, Stiell I, Dennie C. The yield of CT pulmonary angiograms to exclude acute pulmonary embolism. Emerg Radiol. 2014;21(2):133–41. https://doi.org/10.1007/s10140-013-1169-x.
50. Li K, Li Y, Qi Z, Garrett JW, Grist TM, Chen GH. Quantitative lung perfusion blood volume using dual energy CT-based effective atomic number. Med Phys. 2021;48(11):6658–72.
51. Schwartz FR, Samei E, Marin D. Exploiting the potential of photon-counting CT in abdominal imaging. Invest Radiol. 2023;58(7):488–98.
52. Turrion Gomollon AM, Mergen V, Sartoretti T, Polacin M, Nakhostin D, Puippe G, et al. Photon-counting detector CT angiography for endoleak detection after endovascular aortic repair: triphasic CT with true noncontrast versus biphasic CT with virtual noniodine imaging. Invest Radiol. 2023;58(11):816–21. https://doi.org/10.1097/RLI.0000000000000993.
53. Dane B, Qian K, Soni R, Megibow A. Crohn's disease inflammation severity assessment with iodine density from photon counting CT enterography: comparison with endoscopic histopathology. Abdom Radiol. 2024;49(1):271–8. https://doi.org/10.1007/s00261-023-04060-3.
54. Breiding PS, Gomollon AMT, Martini K, Nakhostin D, Alkadhi H, Euler A. Photon-counting detector CT for kidney stone detection in excretory phase CT—Comparison between virtual non-contrast and virtual non-iodine reconstructions in a 3D printed kidney phantom. Acad Radiol. 2024;31(9):3650–6. https://doi.org/10.1016/j.acra.2024.04.002.
55. Sonnow L, Salimova N, Behrendt L, Wacker FK, Örgel M, Plagge J, et al. Photon-counting CT of elbow joint fractures: image quality in a simulated post-trauma setting with off-center positioning. Eur Radiol Exp. 2023;7(1):15. https://doi.org/10.1186/s41747-023-00329-w.
56. Dillinger D, Overhoff D, Froelich MF, Kaatsch HL, Booz C, Hagen A, et al. Photon-counting detector CT virtual monoenergetic images in cervical trauma imaging—optimization of dental metal artifacts and image quality. Diagnostics. 2024;14(6):626. https://doi.org/10.3390/diagnostics14060626.
57. Schade KA, Mergen V, Sartoretti T, Alkadhi H, Euler A. Pseudoenhancement in cystic renal lesions–impact of virtual monoenergetic images of photon-counting detector CT on lesion classification. Acad Radiol. 2023;30:S305. https://www.sciencedirect.com/science/article/pii/S1076633223001873. Accessed 14 Nov 2023.
58. Schwartz FR, Ashton J, Wildman-Tobriner B, Molvin L, Ramirez-Giraldo JC, Samei E, et al. Liver fat quantification in photon counting CT in head to head comparison with clinical MRI–first experience. Eur J Radiol. 2023;161:110734.
59. Niehoff JH, Woeltjen MM, Saeed S, Michael AE, Boriesosdick J, Borggrefe J, et al. Assessment of hepatic steatosis based on virtual non-contrast computed tomography: Initial experiences with a photon counting scanner approved for clinical use. Eur J Radiol. 2022;149:110185. https://doi.org/10.1016/j.ejrad.2022.110185.
60. Lin H, Xu X, Deng R, Xu Z, Cai X, Dong H, et al. Photon-counting detector CT for liver fat quantification: validation across protocols in metabolic dysfunction–associated Steatotic liver disease. Radiology. 2024;312(3):e240038. https://doi.org/10.1148/radiol.240038.
61. Healy J, Searle E, Panta RK, Chernoglazov A, Roake J, Butler P, et al. Ex-vivo atherosclerotic plaque characterization using spectral photon-counting CT: comparing material quantification to histology. Atherosclerosis. 2023;378:117160. https://doi.org/10.1016/j.atherosclerosis.2023.06.007.
62. Pickhardt PJ, Graffy PM, Zea R, Lee SJ, Liu J, Sandfort V, et al. Automated abdominal CT imaging biomarkers for opportunistic prediction of future major osteoporotic fractures in asymptomatic adults. Radiology. 2020;297(1):64–72. https://doi.org/10.1148/radiol.2020200466.

63. Computed tomography angiography spot sign, hematoma expansion, and functional outcome in spontaneous cerebellar intracerebral hemorrhage | Stroke. https://www.ahajournals.org/doi/full/10.1161/STROKEAHA.120.033297. Accessed 30 Sept 2024.
64. Automated brain perfusion imaging in acute ischemic stroke: interpretation pearls and pitfalls | Stroke. https://www.ahajournals.org/doi/full/10.1161/STROKEAHA.121.035049. Accessed 30 Sept 2024.
65. Lambert JW, Sun Y, Stillson C, Li Z, Kumar R, Wang S, et al. An intravascular tantalum oxide-based CT contrast agent: preclinical evaluation emulating overweight and obese patient size. Radiology. 2018;289(1):103–10. https://doi.org/10.1148/radiol.2018172381.
66. Moghiseh M, Searle E, Dixit D, Kim J, Dong YC, Cormode DP, et al. Spectral photon-counting CT imaging of gold nanoparticle labelled monocytes for detection of atherosclerosis: a preclinical study. Diagnostics (Basel). 2023;13(3):499. https://doi.org/10.3390/diagnostics13030499.
67. Boccalini S, Dessouky R, Rodesch P-A, Lacombe H, Yagil Y, Lahoud E, et al. Gadolinium K-edge angiography with a spectral photon counting CT in atherosclerotic rabbits. Diagn Interv Imaging. 2023;104(10):490–9. https://doi.org/10.1016/j.diii.2023.05.002.
68. Ren L, Huber N, Rajendran K, Fletcher JG, McCollough CH, Yu L. Dual-contrast biphasic liver imaging with iodine and gadolinium using photon-counting detector computed tomography: an exploratory animal study. Invest Radiol. 2022;57(2):122. https://doi.org/10.1097/RLI.0000000000000815.
69. Baubeta E, Laurin Gadsböll E, Will L, Holmquist F, Aurumskjöld M-L. No gadolinium K-edge detected on the first clinical photon-counting computed tomography scanner. J Appl Clin Med Phys. 2024;25(4):e14324. https://doi.org/10.1002/acm2.14324.
70. Lipengolts AA, Skribitsky VA, Finogenova YA, Shulyak AT, Abakumov MA, Bykov AY, et al. Hafnium complexes as contrast media and dose enhancing agents for radiology and contrast-enhanced radiotherapy. Biophysics. 2024;69(1):148–55. https://doi.org/10.1134/S0006350924700167.
71. Safari A, Mahdavi M, Fardid R, Oveisi A, Jalli R, Haghani M. Evaluation of hafnium oxide nanoparticles imaging characteristics as a contrast agent in X-ray computed tomography. Radiol Phys Technol. 2024;17(2):441–50. https://doi.org/10.1007/s12194-024-00797-8.
72. Chakravarty S, Hix JML, Wiewiora KA, Volk MC, Kenyon E, Shuboni-Mulligan DD, et al. Tantalum oxide nanoparticles as versatile contrast agents for X-ray computed tomography. Nanoscale. 2020;12(14):7720–34. https://doi.org/10.1039/D0NR01234C.

Chapter 7
Clinical Applications of Photon-Counting CT in Adult and Pediatric Oncology

Corinne Balleyguier, Rémy Barbé, Salma Moalla, Gabriel Garcia, Samy Ammari, François Bidault, Yves Menu, and Nathalie Lassau

Contents

C. Balleyguier (✉) · S. Ammari · F. Bidault · N. Lassau
Service d'Imagerie Diagnostique, Gustave Roussy, Villejuif, France

BIOMAPS, UMR 1281 INSERM, CEA, CNRS, Université Paris-Saclay, Villejuif, France
e-mail: corinne.balleyguier@gustaveroussy.fr

R. Barbé · S. Moalla · G. Garcia · Y. Menu
Service d'Imagerie Diagnostique, Gustave Roussy, Villejuif, France

S. A. Si-Mohamed, K. K. Iniewski (eds.), *Clinical Applications of Spectral Photon Counting Computed Tomography Technology*,
https://doi.org/10.1007/978-3-031-85078-3_7

7.1 Introduction

Photon-counting detector computed tomography (PCD-CT) represents a major technological advance in CT imaging. Unlike conventional energy-integrating detectors, PCD-CT systems can count individual photons and measure their energy, enabling improved spatial resolution, reduced electronic noise, and spectral imaging capabilities [1, 2]. These technical advantages translate to several potential clinical benefits in oncologic imaging, including enhanced lesion detection and characterization, reduced radiation dose, and new quantitative imaging biomarkers [3, 4].

This chapter focuses on the emerging clinical applications of PCD-CT in adult and pediatric oncology. We review current evidence and discuss future directions for using this technology to improve cancer detection, staging, and treatment response assessment.

7.2 General Oncologic Applications

7.2.1 Improved Lesion Detection

The higher spatial resolution, especially when additional slices are in ultra-high-resolution mode, and reduced image noise of PCD-CT can improve the detection of small lesions across multiple organ systems. In a prospective study of 27 patients with multiple myeloma, Baffour et al. found that PCD-CT detected significantly more lytic bone lesions compared to conventional CT, especially for lesions <5 mm [5]. Similar improvements in lesion detection have been demonstrated for pulmonary nodules, hepatic metastases, and pancreatic tumors [6, 7].

7.2.2 Enhanced Lesion Characterization

The spectral capabilities of PCD-CT allow for improved material decomposition and characterization of lesions. Virtual monoenergetic imaging at low keV levels can increase lesion contrast and conspicuity [8]. Virtual non-contrast images have been proven to be similar to standard non-contrast CT images; thus, this reduces the need to acquire additional non-contrast slices. This is interesting to decrease the radiation dose, especially in emergencies in oncology, such as acute bleeding. Iodine quantification enables a better assessment of tumor vascularity and perfusion [3]. These techniques may help differentiate benign from malignant lesions and characterize tumor heterogeneity.

7.2.3 Reduced Radiation Dose

PCD-CT can achieve similar or improved image quality at lower radiation doses compared to conventional CT. This is particularly beneficial for cancer patients who often undergo multiple CT exams for initial staging and treatment response assessment. Studies have demonstrated dose reductions of 30–60% with PCD-CT while maintaining high diagnostic image quality [4].

7.2.4 Quantitative Imaging Biomarkers

The quantitative capabilities of PCD-CT may provide new imaging biomarkers for assessing tumor biology and treatment response. Iodine concentration measurements can serve as a surrogate for tumor vascularity and angiogenesis [9]. Changes in these quantitative metrics over time may allow for an earlier and more accurate assessment of treatment response compared to conventional size-based criteria. In addition, the evaluation of new drugs or contrast agents based on nanoparticles may be a future research area [10].

7.2.5 Reduced Contrast Media Requirements

PCD-CT demonstrates significant potential for reducing contrast media volume while maintaining or even improving image quality compared to conventional CT. This advantage stems from the technology's enhanced detection of iodine signals and improved contrast-to-noise ratio, particularly when using low keV virtual monoenergetic reconstructions. Higashigaito et al. demonstrated that PCD-CT allows for a 25% reduction in contrast media volume for CT aortography while maintaining diagnostic image quality [11]. Similar benefits have been shown in abdominal imaging, where Hoeijmakers et al. achieved a 20% reduction in total iodine load while maintaining superior signal-to-noise and contrast-to-noise ratios compared to conventional CT [12]. This capability is particularly valuable for patients with renal impairment or those requiring multiple contrast-enhanced studies, as it helps reduce the cumulative contrast burden while maintaining diagnostic confidence.

7.3 Applications in Specific Organ Systems

7.3.1 Thoracic Oncology

PCD-CT shows promise for improving lung cancer screening and nodule characterization.

PCD-CT shows considerable promise for improving lung cancer screening and nodule characterization. The higher spatial resolution and reduced image noise of this technology offer several advantages for thoracic imaging (Fig. 7.1).

In a prospective study comparing PCD-CT with conventional CT for lung cancer screening, Inoue et al. demonstrated that PCD-CT provided superior visualization of small pulmonary nodules while maintaining lower radiation dose levels. The authors found improved reader confidence in nodule detection and characterization, particularly for ground-glass and part-solid nodules [13].

The technical advantages of PCD-CT are particularly evident in the evaluation of lung parenchyma. Milos et al. investigated ultra-high-resolution PCD-CT imaging of the lungs and found that using an appropriate reconstruction kernel (Bl64), peripheral bronchial branches, lobular fissures, and bronchial walls could be better visualized without compromising the evaluation of pulmonary nodules [14].

The spectral capabilities of PCD-CT also offer potential advantages for characterizing pulmonary nodules. By analyzing iodine uptake patterns and generating virtual monoenergetic images at different energy levels, PCD-CT may help differentiate benign from malignant lesions, though further research is needed to validate these applications.

For mediastinal staging, the improved contrast resolution and reduced image noise of PCD-CT may enhance the detection and characterization of lymph nodes, particularly in areas near vascular structures where beam hardening artifacts can limit conventional CT evaluation.

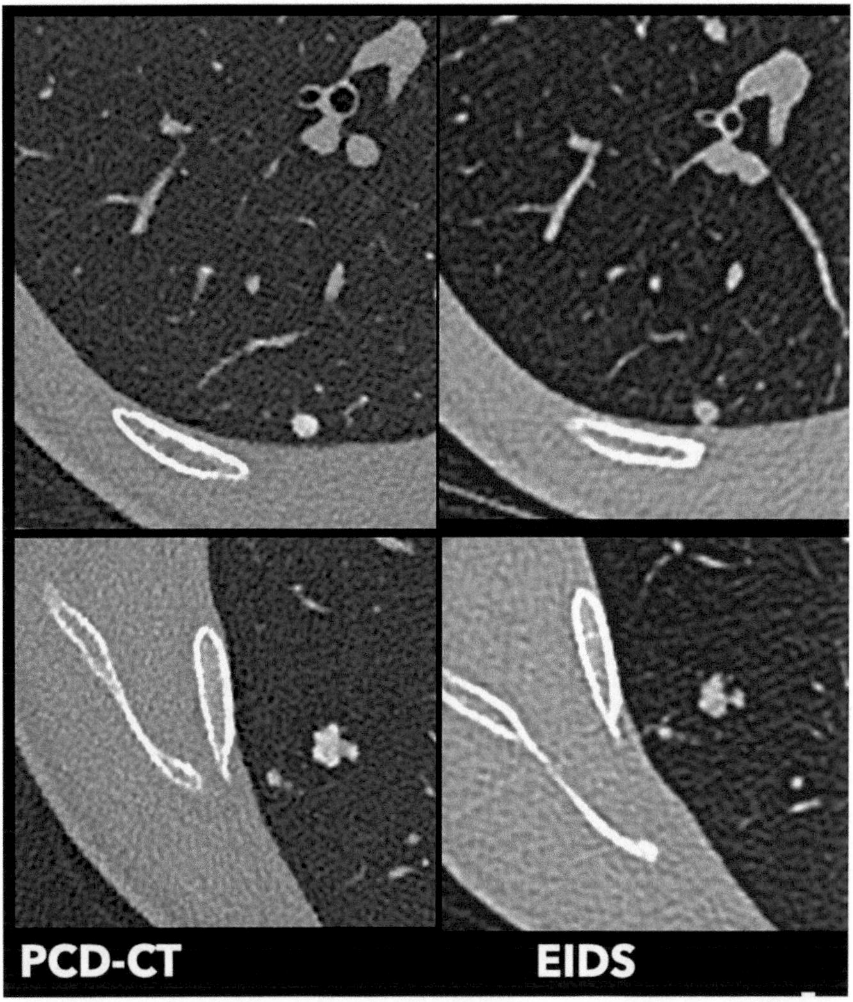

Fig. 7.1 Pulmonary metastases from tibial osteosarcoma in an 18-year-old male. Comparison of conventional CT (EIDS) versus PCD-CT demonstrates superior delineation of two lung metastases on PCD-CT, with enhanced visualization of irregular margins, improved internal contrast, and reduced noise

7.3.2 Abdominal Oncology

7.3.2.1 Pancreatic Cancer and Neuroendocrine Tumors

PCD-CT offers significant advantages in pancreatic imaging, particularly for pancreatic ductal adenocarcinoma (PDAC) and neuroendocrine tumors (NETs). The improved spatial resolution allows for better delineation of tumor margins and vascular involvement, crucial for accurate staging and surgical planning (Figs. 7.2 and 7.3).

For PDAC, PCD-CT shows promise in detecting subtle changes in pancreatic parenchymal attenuation, potentially enabling earlier diagnosis. The study by Decker et al. found that tumor conspicuity was significantly improved at lower keV levels ($\leq$ 70 keV) in both arterial and portal venous phases. Contrast-to-noise ratio (CNR) was highest at 40 keV, with an 18.6% increase compared to conventional CT [15].

Regarding NETs, the spectral capabilities of PCD-CT offer enhanced lesion detection and characterization. The ability to generate virtual monoenergetic images (VMI) at low keV levels increases the contrast between NETs and surrounding pancreatic tissue. This is particularly beneficial for small, hypervascular NETs or hypervascular pancreatic metastases that may be difficult to detect on conventional CT [6] (Fig. 7.4). Additionally, the iodine quantification possible with PCD-CT can help differentiate NETs from other pancreatic lesions based on their enhancement patterns.

7.3.2.2 Liver Imaging

In hepatic imaging, PCD-CT can improve the detection and characterization of focal liver lesions. Bette et al. demonstrated higher contrast-to-noise ratios for hypovascular liver metastases on PCD-CT compared to conventional CT, especially at low keV levels [16]. Same considerations are also noticed for liver NET metastases, which can be depicted more accurately at low keV images (Fig. 7.5a, b). The quantitative iodine maps generated by PCD-CT may help differentiate hepatocellular carcinoma from other focal liver lesions based on arterial phase hyperenhancement patterns [15].

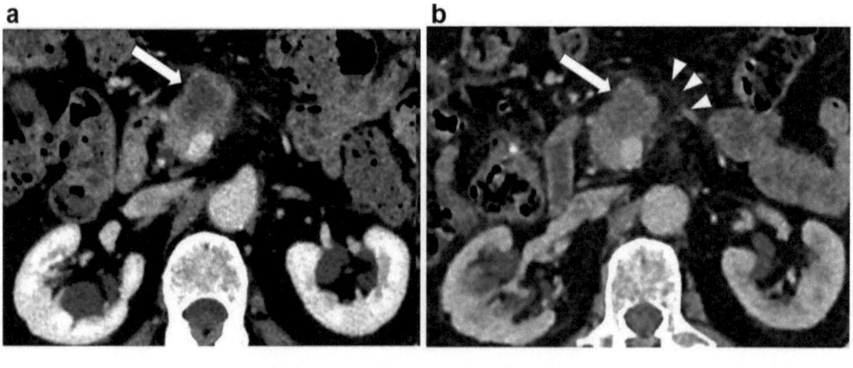

Fig. 7.2 Pancreatic cancer staging in a 53-year-old male. (**a**) PCD-CT reveals a well-defined hypodense tumor with irregular margins in the pancreatic head. (**b**) EIDS shows reduced tumor-to-parenchyma contrast differentiation and increased peritumoral fat noise, limiting vascular staging assessment

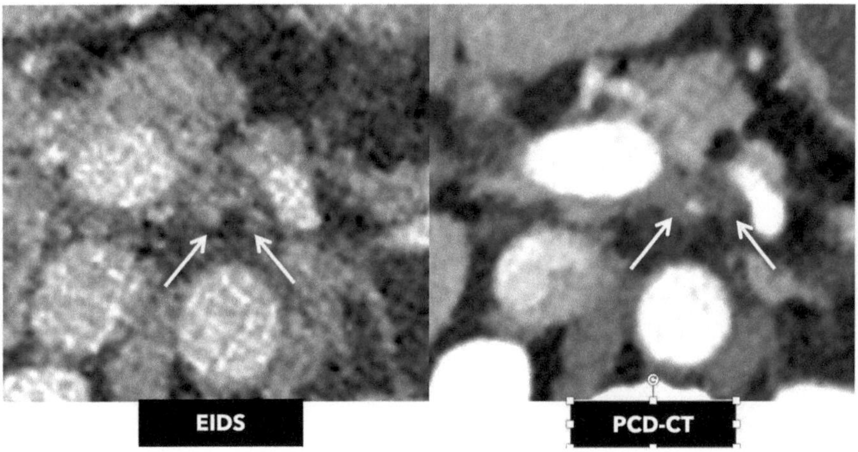

Fig. 7.3 Pancreatic cancer with mesenteric artery involvement in a 67-year-old male. PCD-CT demonstrates superior visualization of near-circumferential tumor spread around the mesenteric artery compared to EIDS, with improved assessment of vascular invasion due to reduced noise

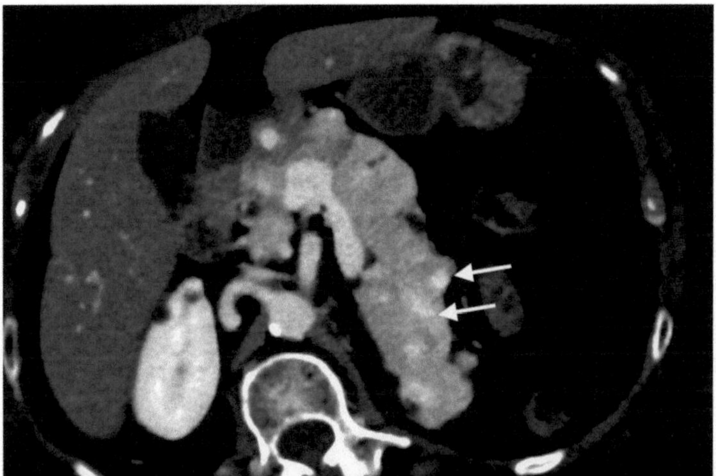

Fig. 7.4 Hypervascular pancreatic metastases from renal cell carcinoma. Arterial phase PCD-CT at 50 keV showing well-defined round pancreatic metastases with enhanced contrast relative to normal pancreatic tissue

The improved lesion detection and characterization capabilities of PCD-CT in liver imaging have important clinical implications. For patients with colorectal cancer, more accurate detection of small liver metastases could alter staging and

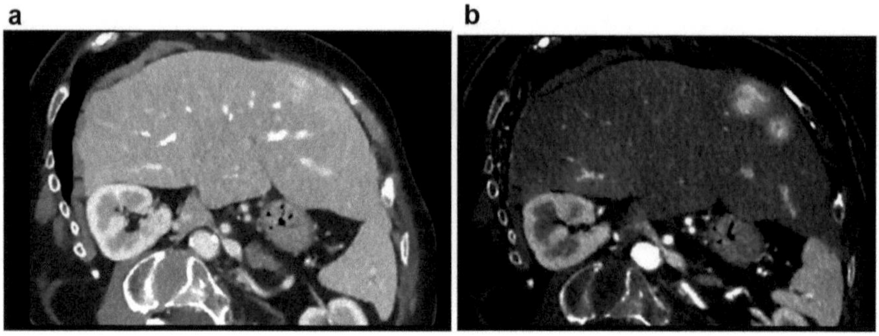

Fig. 7.5 Neuroendocrine liver metastases: multi-energy comparison. Late arterial phase PCD-CT at (**a**) 100 keV and (**b**) 40 keV demonstrates improved lesion detection at lower energy, revealing two left hepatic metastases compared to single lesion visibility at standard energy

treatment decisions. In hepatocellular carcinoma surveillance, the ability to better characterize small lesions might reduce the need for follow-up MRI or biopsy.

7.3.2.3 Renal Imaging

In renal imaging, PCD-CT shows potential for improved characterization of small renal masses. The ability to perform spectral analysis may help differentiate benign renal cysts from solid renal cell carcinomas without the need for multiphase imaging [17]. This could reduce radiation exposure and improve workflow efficiency in the evaluation of incidental renal lesions.

7.3.2.4 Peritoneal Carcinomatosis

PCD-CT has shown promising results in the evaluation of peritoneal carcinomatosis, a condition that can be challenging to assess on conventional CT. The improved spatial resolution and spectral capabilities of PCD-CT offer several advantages in this context:

1. Enhanced detection of small peritoneal implants: The higher spatial resolution of PCD-CT allows for better visualization of small peritoneal nodules, which can be crucial for accurate staging and treatment planning.
2. Improved characterization of peritoneal lesions: The spectral imaging capabilities of PCD-CT enable better differentiation between peritoneal implants and benign peritoneal thickening or fibrosis. Low keV virtual monoenergetic images can enhance the conspicuity of small implants, while iodine maps can provide information about lesion vascularity [18].
3. Quantitative assessment of tumor burden: PCD-CT allows for more accurate quantification of peritoneal tumor volume, which can be valuable for assessing treatment response and predicting outcomes in patients undergoing cytoreductive

surgery and hyperthermic intraperitoneal chemotherapy (HIPEC) [18] (Fig. 7.6a, b).
4. Reduced artifacts: The improved image quality and reduced beam hardening artifacts with PCD-CT can enhance the evaluation of peritoneal surfaces adjacent to high-density structures like bone or contrast-filled bowel [8].
5. Lower radiation dose: The dose reduction capabilities of PCD-CT are particularly beneficial for patients with peritoneal carcinomatosis, who often require multiple follow-up scans to monitor disease progression or treatment response [4].

These advancements in peritoneal carcinomatosis imaging with PCD-CT have the potential to improve patient selection for surgical intervention, enhance the accuracy of preoperative planning, and provide a more precise assessment of treatment response.

7.3.3 *Musculoskeletal Oncology*

The high spatial resolution of PCD-CT is particularly advantageous for evaluating bone lesions and assessing cortical involvement. In multiple myeloma imaging, PCD-CT has demonstrated superior detection of lytic bone lesions compared to conventional CT [5]. The improved trabecular bone visualization may also allow for earlier detection of marrow infiltration.

For primary bone tumors, PCD-CT provides exquisite detail of tumor matrix mineralization and cortical destruction. This can aid in differentiating benign from malignant bone lesions and assessing local tumor extent [19]. The reduced beam

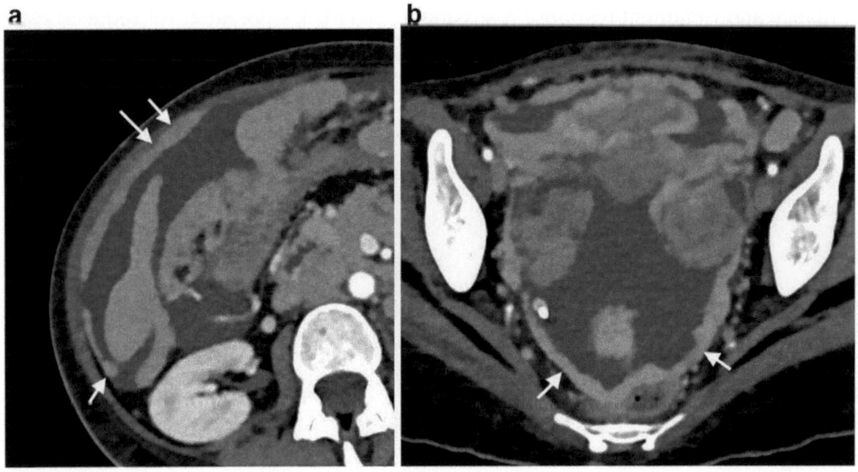

Fig. 7.6 Peritoneal carcinomatosis in post-treatment Ewing sarcoma. Portal phase PCD-CT showing (**a**) distinct peritoneal infiltration along the right paracolic gutter with associated ascites, and (**b**) clear visualization of pelvic carcinomatosis

hardening artifacts with PCD-CT also improve the evaluation of tumors adjacent to metallic implants.

7.3.4 Head and Neck Oncology

Photon-counting detector CT (PCD-CT) represents a promising advance for head and neck cancer imaging, where precise delineation of tumor extent and relationships to adjacent critical structures is essential for treatment planning. The technology's improved spatial resolution, reduced image noise, and spectral imaging capabilities offer potential advantages over conventional CT in this anatomically complex region.

In the evaluation of head and neck vasculature, which is crucial for tumor staging and surgical planning, PCD-CT has demonstrated significant benefits. Symons et al. performed the first in-vivo human study of PCD-CT for cervical vascular imaging and demonstrated superior vessel visualization compared to conventional CT. Their research showed improved contrast-to-noise ratios and better delineation of small vessels, particularly beneficial when assessing tumor relationships to critical vascular structures [20].

The spectral imaging capabilities of PCD-CT enable additional post-processing opportunities that may benefit head and neck cancer imaging. Virtual monoenergetic images can be generated at varying energy levels to optimize contrast for different tissue types. The ability to perform spectral imaging with a single exposure also allows for material decomposition analysis without requiring additional radiation exposure [21]. Thus, PCD-CT may be more accurate than conventional CT in assessing thyroïd cartilage invasion in thyroid tumors (Fig. 7.7a, b). For tumor surveillance and treatment response assessment, PCD-CT's potential for reduced radiation dose while maintaining image quality becomes particularly relevant. As head

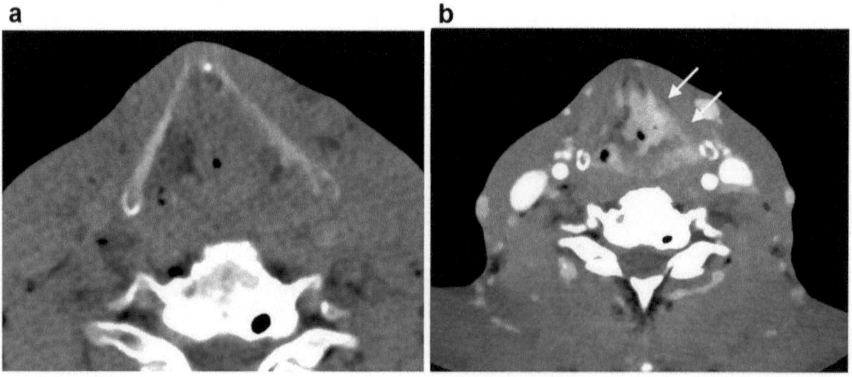

Fig. 7.7 Laryngeal cancer: dual-energy assessment. Virtual Monoenergetic Imaging (VMI) at (**a**) 190 keV showing intact hyperdense thyroid cartilage and (**b**) 40 keV highlighting tumor iodine enhancement with preserved cartilage interface

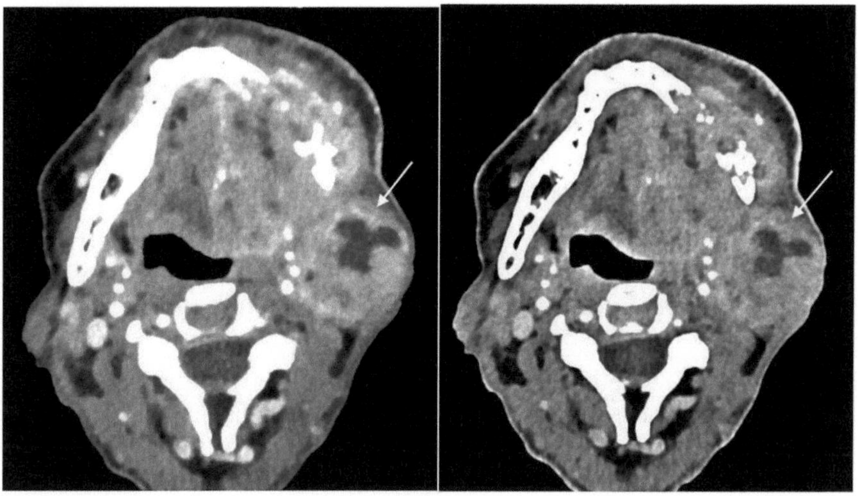

Fig. 7.8 Advanced pharyngeal cancer with nodal metastasis. PCD-CT comparison at (**a**) 40 keV and (**b**) 90 keV demonstrates enhanced tumor and lymph node contrast at lower energy, with improved margin and necrosis definition

and neck cancer patients often require multiple follow-up scans, the ability to achieve diagnostic image quality at lower radiation doses could represent a significant advantage for long-term patient care [4].

While early clinical experience suggests that PCD-CT may offer advantages in reducing dental artifacts and improving soft tissue contrast in the head and neck region (Fig. 7.8a, b), further research is needed to quantify these benefits specifically in oncologic imaging. Studies are currently underway to evaluate the technology's impact on tumor staging accuracy, treatment planning, and patient outcomes.

7.4 Pediatric Oncology Applications

PCD-CT has significant potential in pediatric oncology due to its dose reduction capabilities and improved image quality. Children with cancer often undergo multiple CT exams throughout their treatment course, making radiation dose reduction particularly important in this population.

7.4.1 Dose Reduction

Several studies have demonstrated substantial dose reductions with PCD-CT in pediatric imaging while maintaining diagnostic image quality. Dose reduction may reach 60% with PCD-CT in comparison with conventional CT without compromising image quality and diagnostic certainty [22, 23].

7.4.2 Improved Lesion Detection and Characterization

The higher spatial resolution and reduced image noise of PCD-CT may improve the detection of small metastases in pediatric patients. This is particularly relevant for diseases like neuroblastoma and Wilms tumor, where accurate staging is crucial for treatment planning (Fig. 7.9).

For pediatric bone and soft tissue sarcomas, the high spatial resolution of PCD-CT allows for better delineation of tumor extent and improved surgical planning [23]. The reduced beam hardening artifacts with PCD-CT also improve the

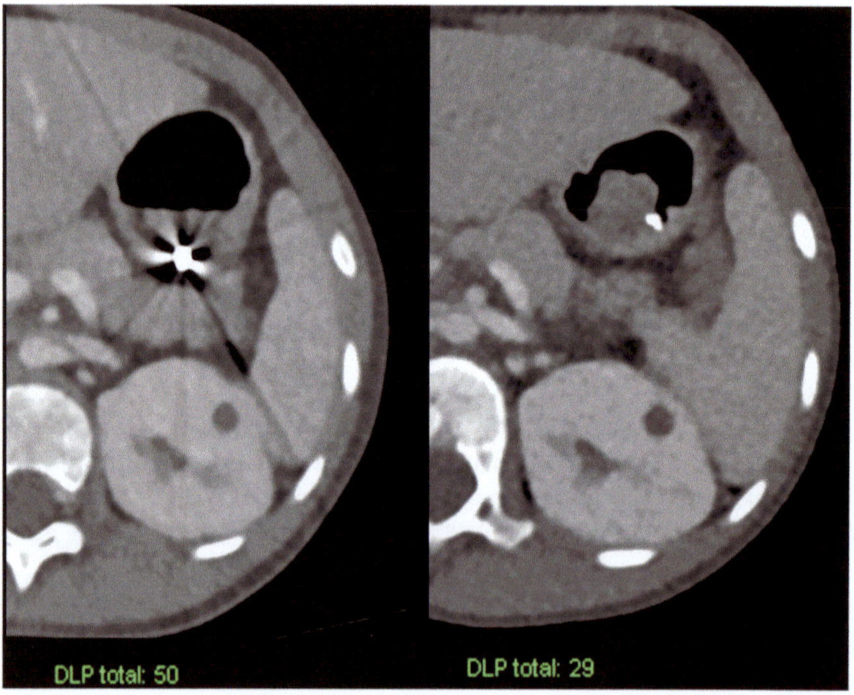

Fig. 7.9 Post-treatment neuroblastoma follow-up in a 5-year-old child. Comparison of standard EIDS (DLP 50 mGycm) versus PCD-CT (DLP 29 mGycm) showing superior image quality with reduced noise and metallic artifact at lower radiation dose

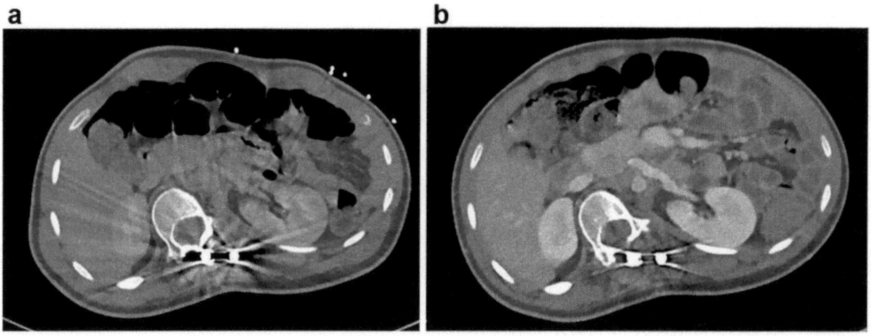

Fig. 7.10 Post-surgical imaging in metastatic poorly differentiated adenocarcinoma in a 20-year-old patient. (**a**) Conventional CT versus (**b**) PCD-CT with spectral imaging and IMAR algorithm demonstrating improved metal artifact reduction and image quality

evaluation of tumors adjacent to metallic implants, which is particularly relevant for patients with limb-salvage prostheses.

As in adult patients, the reduction of metallic artifacts in the case of implants or equivalent may improve tumor visibility and tumor changes (Fig. 7.10a, b).

7.4.3 Reduced Need for Sedation

The faster scan times possible with PCD-CT may reduce the need for sedation in young children. This is particularly beneficial for pediatric oncology patients who may require frequent imaging studies.

7.5 Clinical Implementation and Workflow Considerations

7.5.1 Protocol Optimization

Implementing PCD-CT in clinical oncology practice requires careful optimization of acquisition and reconstruction protocols. While PCD-CT offers numerous advantages, it also introduces new variables that must be considered:

- Selection of optimal energy thresholds for spectral imaging
- Choice of virtual monoenergetic image (VMI) energies for different applications
- Optimization of iterative reconstruction parameters

Radiologists and medical physicists should work together to develop standardized protocols that maximize the benefits of PCD-CT while maintaining workflow efficiency [24].

7.5.2 Data Management and Storage

PCD-CT generates large datasets, particularly when multiple energy bins or VMI series are reconstructed. Institutions implementing PCD-CT should ensure their picture archiving and communication systems (PACS) can handle the increased data volume. Cloud-based storage solutions may be necessary for long-term archiving of spectral datasets.

7.5.3 Radiologist Training

Interpretation of PCD-CT images, particularly spectral datasets, requires additional training for radiologists. Familiarity with iodine maps, VMI, and other advanced post-processing techniques is essential. Institutions should develop educational programs to ensure radiologists are comfortable with these new imaging tools.

7.6 Emerging Applications and Future Directions

7.6.1 Multi-Material Decomposition

PCD-CT enables simultaneous imaging of multiple contrast agents, opening new possibilities for functional and molecular imaging. For example, simultaneous imaging of iodine and gadolinium-based contrast agents could allow for combined assessment of vascular permeability and extracellular volume in tumors [25].

7.6.2 Nanoparticle Imaging

The energy-resolving capabilities of PCD-CT make it well-suited for imaging nanoparticle-based contrast agents. This could enable more specific targeting of tumors and improved characterization of tumor biology [10].

7.6.3 Radiomics and Artificial Intelligence

The rich spectral information provided by PCD-CT offers new opportunities for radiomics and artificial intelligence applications in oncology. Quantitative features extracted from PCD-CT datasets may provide novel biomarkers for tumor characterization and treatment response prediction.

7.6.4 Theranostics

PCD-CT may play a role in theranostic applications, where imaging is used to guide targeted therapies. For example, the ability to quantify iodine concentrations could be used to optimize dosing in radioiodine therapy for thyroid cancer.

7.7 Economic Considerations

While PCD-CT offers numerous clinical benefits, the higher cost of these systems compared to conventional CT scanners may pose a barrier to widespread adoption. However, several factors should be considered when evaluating the economic impact of PCD-CT.

7.7.1 Reduced Need for Multiple Acquisitions

The spectral capabilities of PCD-CT may reduce the need for multiple acquisitions or follow-up imaging studies, potentially offsetting the higher initial cost.

7.7.2 Improved Diagnostic Accuracy

Enhanced lesion detection and characterization with PCD-CT could lead to earlier and more accurate diagnoses, potentially reducing downstream healthcare costs.

7.7.3 Dose Reduction

The ability to perform low-dose examinations may reduce the long-term health costs associated with radiation exposure, particularly in pediatric patients.

7.8 Conclusion

PCD-CT represents a significant advancement in CT technology with numerous applications in adult and pediatric oncology. The improved spatial resolution, reduced image noise, and spectral imaging capabilities offer opportunities for enhanced lesion detection and characterization, reduced radiation dose, and new quantitative imaging biomarkers. While challenges remain in terms of clinical implementation and economic considerations, PCD-CT has the potential to significantly impact cancer imaging and improve patient care.

As the technology continues to evolve and more clinical experience is gained, PCD-CT is likely to become an essential tool in oncologic imaging. Radiologists should familiarize themselves with the principles and applications of this innovative technology to fully leverage its capabilities in clinical practice.

References

1. McCollough CH, Leng S, Yu L, Fletcher JG. Dual- and multi-energy CT: principles, technical approaches, and clinical applications. Radiology. 2015;276(3):637–53.
2. Willemink MJ, Persson M, Pourmorteza A, Pelc NJ, Fleischmann D. Photon-counting CT: technical principles and clinical prospects. Radiology. 2018;289(2):293–312.
3. Pourmorteza A, Symons R, Sandfort V, Mallek M, Fuld MK, Henderson G, et al. Abdominal imaging with contrast-enhanced photon-counting CT: first human experience. Radiology. 2016;279(1):239–45.
4. Rajendran K, Petersilka M, Henning A, Shanblatt ER, Schmidt B, Flohr TG, et al. First clinical photon-counting detector CT system: technical evaluation. Radiology. 2022;303(1):130–8.
5. Baffour FI, Huber NR, Ferrero A, Rajendran K, Glazebrook KN, Larson NB, et al. Photon-counting detector CT with deep learning noise reduction to detect multiple myeloma. Radiology. 2023;306(1):229–36.
6. Woeltjen MM, Niehoff JH, Roggel R, Michael AE, Gerdes B, Surov A, et al. Pancreatic cancer in photon-counting CT: low keV virtual monoenergetic images improve tumor conspicuity. Eur J Radiol. 2024;173:111374.
7. Racine D, Mergen V, Viry A, Frauenfelder T, Alkadhi H, Vitzthum V, et al. Photon-counting detector CT for liver lesion detection-optimal virtual monoenergetic energy for different simulated patient sizes and radiation doses. Invest Radiol. 2024;59(8):554–60.
8. Pelgrim GJ, van Hamersvelt RW, Willemink MJ, Schmidt BT, Flohr T, Schilham A, et al. Accuracy of iodine quantification using dual energy CT in latest generation dual source and dual layer CT. Eur Radiol. 2017;27(9):3904–12.
9. Grosse Hokamp N, Gilkeson R, Jordan MK, Laukamp KR, Neuhaus VF, Haneder S, et al. Virtual monoenergetic images from spectral detector CT as a surrogate for conventional CT images: unaltered attenuation characteristics with reduced image noise. Eur J Radiol. 2019;117:49–55.
10. Amato C, Susenburger M, Lehr S, Kuntz J, Gehrke N, Franke D, et al. Dual-contrast photon-counting micro-CT using iodine and a novel bismuth-based contrast agent. Phys Med Biol. 2023;68(13):135001.
11. Higashigaito K, Mergen V, Eberhard M, Jungblut L, Hebeisen M, Ratzer S, et al. CT angiography of the aorta using photon-counting detector CT with reduced contrast media volume. Radiol Cardiothorac Imaging. 2023;5(1):e220140.
12. Hoeijmakers EJI, Stammen L, Wildberger JE, Eijsvoogel NG, Hersbach JM, Pernot J, et al. PCD-CT enables contrast media reduction in abdominal imaging compared to an individualized kV-adapted contrast media injection protocol on EID-CT. Eur J Radiol. 2024;179:111680.
13. Inoue A, Johnson TF, Walkoff LA, Levin DL, Hartman TE, Burke KA, et al. Lung cancer screening using clinical photon-counting detector computed tomography and energy-integrating-detector computed tomography: a prospective patient study. J Comput Assist Tomogr. 2023;47(2):229–35.

14. Milos RI, Rohrich S, Prayer F, Strassl A, Beer L, Heidinger BH, et al. Ultrahigh-resolution photon-counting detector CT of the lungs: association of reconstruction kernel and slice thickness with image quality. AJR Am J Roentgenol. 2023;220(5):672–80.
15. Decker JA, Becker J, Harting M, Jehs B, Risch F, Canalini L, et al. Optimal conspicuity of pancreatic ductal adenocarcinoma in virtual monochromatic imaging reconstructions on a photon-counting detector CT: comparison to conventional MDCT. Abdom Radiol (NY). 2024;49(1):103–16.
16. Bette S, Decker JA, Braun FM, Becker J, Haerting M, Haeckel T, et al. Optimal conspicuity of liver metastases in virtual monochromatic imaging reconstructions on a novel photon-counting detector CT-effect of keV settings and BMI. Diagnostics (Basel). 2022;12(5):1231.
17. Grosse Hokamp N, Maintz D, Shapira N, Chang H, Noel PB. Technical background of a novel detector-based approach to dual-energy computed tomography. Diagn Interv Radiol. 2020;26(1):68–71.
18. Nehra AK, Dane B, Yeh BM, Fletcher JG, Leng S, Mileto A. Dual-energy, spectral and photon counting computed tomography for evaluation of the gastrointestinal tract. Radiol Clin North Am. 2023;61(6):1031–49.
19. Gillet R, Boubaker F, Hossu G, Thay A, Gillet P, Blum A, et al. Computed tomography bone imaging: pushing the boundaries in clinical practice. Semin Musculoskelet Radiol. 2023;27(4):397–410.
20. Symons R, Reich DS, Bagheri M, Cork TE, Krauss B, Ulzheimer S, et al. Photon-counting computed tomography for vascular imaging of the head and neck: first in vivo human results. Invest Radiol. 2018;53(3):135–42.
21. Si-Mohamed SA, Greffier J, Miailhes J, Boccalini S, Rodesch PA, Vuillod A, et al. Comparison of image quality between spectral photon-counting CT and dual-layer CT for the evaluation of lung nodules: a phantom study. Eur Radiol. 2022;32(1):524–32.
22. Hagen F, Soschynski M, Weis M, Hagar MT, Krumm P, Ayx I, et al. Photon-counting computed tomography—clinical application in oncological, cardiovascular, and pediatric radiology. Rofo. 2024;196(1):25–35.
23. Mese I, Altintas Mese C, Demirsoy U, Anik Y. Innovative advances in pediatric radiology: computed tomography reconstruction techniques, photon-counting detector computed tomography, and beyond. Pediatr Radiol. 2024;54(1):1–11.
24. Rajendran K, Voss BA, Zhou W, Tao S, DeLone DR, Lane JI, et al. Dose reduction for sinus and temporal bone imaging using photon-counting detector CT with an additional tin filter. Invest Radiol. 2020;55(2):91–100.
25. Muenzel D, Bar-Ness D, Roessl E, Blevis I, Bartels M, Fingerle AA, et al. Spectral photon-counting CT: initial experience with dual-contrast agent K-edge colonography. Radiology. 2017;283(3):723–8.

Chapter 8
Photon-Counting Breast CT: System Requirements, Development, Current Clinical Applications, and Possible Future Applications

Veikko Ruth, Christian Steiding, David Heinemann, Felix Althoff, and Daniel Kolditz

Contents

V. Ruth (✉) · C. Steiding · D. Heinemann · F. Althoff · D. Kolditz
AB-CT—Advanced Breast-CT GmbH, Erlangen, Germany
e-mail: veikko.ruth@ab-ct.com

S. A. Si-Mohamed, K. K. Iniewski (eds.), *Clinical Applications of Spectral Photon Counting Computed Tomography Technology*,
https://doi.org/10.1007/978-3-031-85078-3_8

8.1 State of the Art

8.1.1 The First Photon Counting Detectors in Medical Imaging

Although photon counting detectors (PCD) are considered to be one of the latest approaches in CT imaging, they have been around for quite a while. In the mid-1990s, the potential of photon-counting detectors for X-ray imaging was already assessed to be very promising. The materials used by today's manufactures, such as Si, GaAs, CdTe, and CdZnTe, were already discussed as "photon detectors for the 21st century" [1]. Nevertheless, it took nearly a decade for the first PCD to be integrated into a clinical system. A silicon detector was the first to be used in the innovative full-field digital mammography system *Sectra MicroDose* in 2003. Compared to other semiconductors, silicon has relatively low attenuation properties. To overcome this fact, the detector modules were mounted with their edges facing incident X-rays. Several lines of detector modules are used in a multi-slit arrangement, which is moved along the object to be imaged. The final image is assembled by a dedicated data processing system [2]. The diagnostic performance of the system was comparable to those of existing modalities, and at the same time, the dose could be reduced to less than 50% by utilizing a PCD [3]. Despite good initial results, the photon-counting mammography system gradually disappeared from the market. One possible reason for this could be the non-uniform resolution in the two spatial directions of the detector.

Perhaps because of the low attenuation properties of silicon, but possibly also simply for the reason of doing it differently, other manufacturers developing PCDs focused on materials with higher atomic number such as GaAs, CdTe, and CdZnTe. Producing these materials in high purity and processing them into X-ray detectors is not yet as well understood as the production of silicon detectors. Therefore, it took the manufacturers another decade before the first system using a CdTe detector was introduced: in 2018 the first CT system, the dedicated Breast CT (BCT) scanner *nu:view* (AB-CT—Advanced Breast-CT GmbH, Erlangen, Germany), using PCDs was launched on the European market [4].

But what makes the development of PCDs with high atomic number so difficult and which challenges need to be overcome?

The first task is the production of crystals with high homogeneity and thus good charge transport properties. Only a small number of vendors are currently able to produce materials with sufficient characteristics, but only for crystal sizes of a few centimeters. When a large-area detector is to be produced, it must be assembled from many individual elements, which poses further problems for manufacturers. Another obstacle, especially for CT applications, is high X-ray fluxes, especially if there are crystal defects or impurities in the sensor material [5]. If that weren't enough of a challenge, the design of the application-specific integrated circuit (ASIC) and the connection of this ASIC to the sensor material is another highly complex problem in its very own right.

So why bother with all these problems when there are already excellent X-ray detectors available? From the perspective of a physicist and engineer, there is always the urge to achieve the best possible technical feasibility. Focusing on PCDs brings advantages for clinicians and radiologists. The diagnostic images can be created with higher contrast and higher spatial resolution at a patient dose similar to that applied in already existing modalities. Furthermore, the detectors provide information about the energy of the detected X-rays, which in turn could answer specific clinical questions that arise during the diagnostic workup. Above all, however, these images should enable radiologists to make better diagnoses with less X-ray radiation, which in turn also benefits the patient.

8.1.2 Photon Counting Detectors and Breast Imaging

It is no coincidence that the first photon-counting detectors were used for breast imaging. In the early days, when the technology was just becoming available, the options for using it at high photon fluxes were much more limited than they are today. However, the requirements necessary for sufficient breast imaging made them the ideal choice for this application. Since most breast imaging is performed on healthy women for screening purposes, low doses and low photon fluxes are a precondition. Furthermore, breast tissue has only slight differences in density, which results in minor differences in the photon flux behind the object. The photon energies used are low (<50 keV), so the detectors do not have to be very thick to ensure good dose efficiency, and materials with a lower atomic number, such as silicon, are sufficient. Therefore, most limitations of PCDs do not apply to breast imaging. On the other hand, the strengths of PCDs can improve breast imaging.

PCDs are very dose-efficient as they can operate without septa between the pixels and they offer nearly 100% geometric and absorption efficiency if the thickness is suitable for the X-ray spectrum used. This makes the PCD an ideal candidate to be used for screening purposes.

In breast imaging pathological changes like the occurrence of microcalcifications (<0.5 mm in diameter) can be an indication of malignant lesions. To detect those calcifications, high spatial resolution is required. The charge carriers, which are generated by the X-rays in the detector material, can be directed by an electric field. Thus, their detection with PCDs can be done with much greater spatial precision than with conventional scintillation detectors. The increased spatial resolution might allow the detection of microcalcifications at an earlier stage, which has the potential to improve long-term survival rates.

In conventional scintillating detectors, X-rays of higher energy produce more visible light and contribute more information to the final image. But it is particularly the low-energy photons that carry information about soft tissue attenuation. In PCDs photons of all X-ray energies make an equal contribution to the image, which improves the soft tissue contrast in PCD images. The effect is comparable to switching to a lower kV value on your imaging system, without the disadvantage of increased noise or delivering a higher dose. Subtle soft tissue differences or small deposits of contrast agents can thus be better characterized by using PCDs.

In summary, none of the major limitations of PCTs applies to breast imaging, so the full potential of PCT in terms of improved resolution and better contrast at higher dose efficiency can be utilized.

8.1.3 Taking the Leap from Mammography to Breast CT

Screening mammography is the backbone of the early detection of breast cancer. It is a cheap, easy-to-operate, and highly available modality. It is the only imaging method that has been proven to reduce mortality. While being very good at detecting in-situ carcinomas associated with microcalcification, its overall sensitivity depends highly on breast density and drops from 85.7% for non-dense breasts to 61.0% for dense breasts, due to superimposing structures [6]. For this reason, sonography is often used adjunct to mammography in screening and the diagnosis of women with dense breasts with significant success in terms of sensitivity improvements [7]. Even though mammography in combination with sonography shows high sensitivity, this only applies to patients who actually take part in the screening program. However, many women refuse to undergo mammography screening due to painful breast compression. It has been shown that 25–46% of women attending a screening program refused repeated examination, citing pain as the reason [8]. Unfortunately, the reduction in mortality rates through screening will always be limited if the modality used is not fully accepted by patients. Currently screening alternatives to mammography are limited.

Despite being marketed as an innovative technique, digital breast tomosynthesis is effectively a mammography product update with increased performance but with similar drawbacks as breast compression and tissue superimpositions [9]. Of course, the superimpositions are reduced by the underlying imaging concept of limited-angle tomography, but adjacent structures will always impair tissue contrast.

While sonography is the first choice for diagnosis in young women under 40, as breast cancer incidence is low, breasts are predominantly dense and the use of X-rays does not outweigh the risk of not using them, it is not as appropriate as a stand-alone screening method [10].

Used for screening women with a high risk of breast cancer, MRI is the most sensitive but also the most expensive and time-consuming modality used in breast imaging. Its superior sensitivity can be attributed to the use of contrast agents. It basically detects all cancers with relevant contrast-media uptake resp. ascularization independent of breast density [11]. But it cannot detect microcalcification and thus might miss early ductal carcinoma in situ (DCIS). However, the high cost, limited availability, long examination times, and the need to apply contrast agents are substantial limitations of using MRI as a primary screening modality.

Table 8.1 Requirements for breast imaging devices [12]

1. Full 3D capabilities
2. Good soft tissue differentiation
3. Dynamic imaging capabilities
4. High isotropic spatial resolution of about 100 μm
5. Low patient dose
6. High patient comfort (fast acquisition and no breast compression)
7. Low costs

In summary, there are sufficient screening options available, but they are either not fully accepted by patients, have limited sensitivity in dense breasts, or are costly and not widely available.

With the advantages and disadvantages of the existing modalities elaborated, from a development perspective, it is now very tempting to design a system that meets all those requirements (see Table 8.1).

As the technical possibilities seemed promising, an ambitious project was launched under the leadership of Prof. Willi Kalender to implement all requirements in one novel imaging system. There were good reasons to assume that all these demands could be met by a *photon-counting breast CT* system.

8.1.4 Cone Beam Breast CT Systems

The first efforts to develop a dedicated BCT started in the 1970s. At the time, it was more or less a demonstration that the manufacturer, who had no experience in the field of CT at the time, was able to develop such a device in a very short time. However, the available analog-to-digital converters only had a bit depth of 8 bits. The object to be examined was therefore only allowed to have very small differences in attenuation. Again, it was no coincidence that the female breast was chosen. The scanner was highly innovative and was one of the first third-generation CTs using a "rotation only" approach [13]. But due to its low sensitivity, poor spatial resolution and unreasonably high patient dose, its use was discontinued after the initial trials.

It was only when high-resolution digital flat panel detectors were introduced that the development of new BCT systems became feasible. Like in conventional CT, dedicated BCT images are reconstructed from a plurality of projection images acquired by a rotating imaging system. And yet the patient lies prone on the patient table, with their breast hanging freely through an opening into the imaging system rotating around a vertical axis underneath it (see Fig. 8.1). One breast is examined after the other. BCT utilizes X-rays but there is no need for breast compression. This setup limits the ionizing radiation to the examined organ. Eponymous for cone beam BCTs is the considerably large cone angle of the applied X-ray beam. The "height" of the beam and the flat-panel detector of these systems are large enough to image the entire breast, from chest wall to nipple. The imaging system does one rotation around the breast during image acquisition.

In the early 2000s, two US-based groups started the development of two independent systems. One system, the UC Davis research device, was launched by Prof. John Boone and developed over four generations of prototypes [14]. The other system driven by Prof. Ruola Ning became a commercialized product known as the "Koning Breast CT" and is currently sold under the name "Vera" [15].

Both groups used off-the-shelf components where possible which is why it was inevitable to make compromises in some aspects. The probably biggest one had to be made when selecting the X-ray detector. The flat panel detectors used offer high spatial resolution but only at slow readout rates (<10 frames per second (fps)). When they were operated with the detection speed required for CT, the use of pixel binning and thus reduced spatial resolution was inevitable. The Koning BCT and the third generation of the UC Davis, Cambria, are very similar in their main components and design. Therefore, they offer a similar spatial resolution. The effective detector element size of the two devices is 0.388 mm. This allows a spatial resolution in the CT images—10% MTF values—of only 1.7 mm^{-1} for the third generation of the UC Davis BCT, Cambria, and 1.3 mm^{-1} for the Koning BCT [16, 17]. This limits the detection of calcifications to a size of 250–300 μm. In the fourth generation of the UC Davis BCT, Doheny, the detector was changed, allowing higher readout rates with an effective detector element size of 0.15 mm, offering an improved spatial resolution of 3.4 mm^{-1} (10% MTF) [17]. With this setup calcifications of 150 μm are expected to be visible. Acquisition times vary from 7 to 16.6 s, depending on the spatial sampling. The Koning BCT acquires images with angle increments of 1.7° and the latest UC Davis device with 0.45°. A smaller angle increment allows for better spatial sampling and improved spatial resolution in the periphery of the CT image.

Several studies showed that the Koning BCT such as the third generation of the UC Davis BCT, Cambria, may be superior to mammography for mass visualization and characterization, but the ability to detect and characterize microcalcifications was still lagging behind the capabilities of mammography [18]. Unfortunately, studies demonstrating the clinical performance of the latest and most evolved UC Davis device, Doheny, are still pending.

8.2 Photon-Counting Breast CT

8.2.1 Design and Technical Specifications

The demands for a breast imaging system that can close the gap between the existing modalities have already been listed in Table 8.1. The design and components of the system were carefully developed in line with these requirements. Figure 8.1 shows how a patient is positioned on the system while in Fig. 8.2 the position of the main components is indicated.

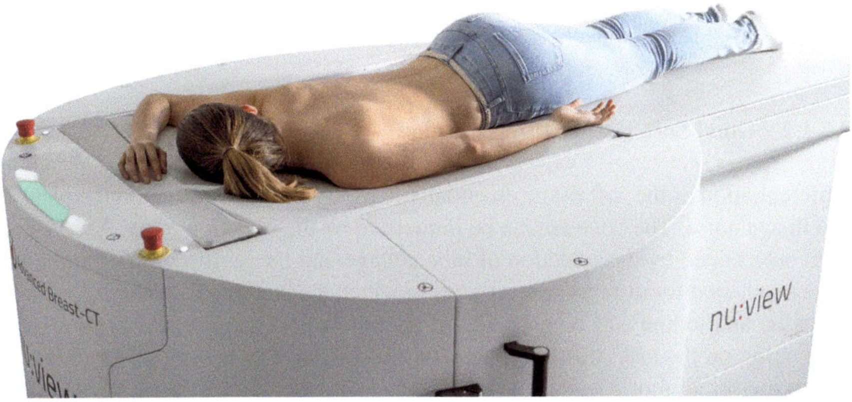

Fig. 8.1 Patient positioning in breast CT. The woman lies prone on the table and the breast is inserted in an opening in the center

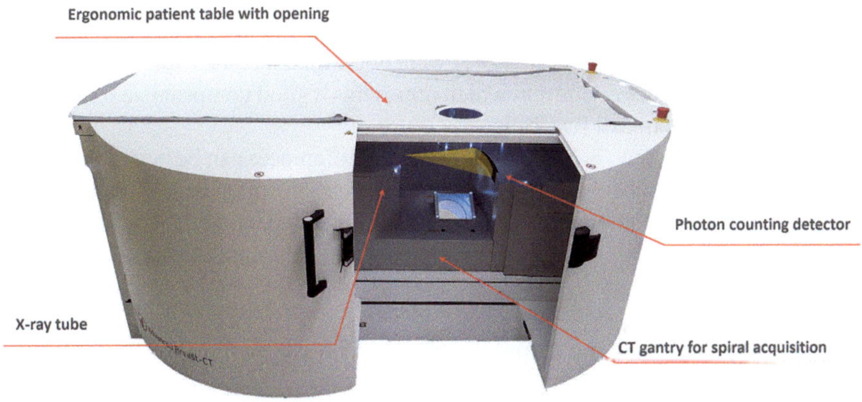

Fig. 8.2 System design with the position of main components indicated. The correct positioning of the patient can be checked by opening the side doors. In future versions of the system the breast can be accessed to perform a biopsy

8.2.1.1 Full 3D Capabilities

The required complete 3D representation will be realized by a spiral CT approach. This has the advantage that the cone beam of the X-ray beam can be kept small, thus minimizing cone beam artifacts and the occurrence of scattered radiation. This means that the height of the detector can also be chosen to be equally small, which will keep costs down. To execute the spiral trajectory, the imaging system requires a translational degree of freedom in addition to the rotational degree of freedom. Ideally, the rotation should be continuous, i.e., with any number of revolutions. This requires the use of a rotational transducer to transfer energy and signals between the stator and rotor. A special requirement for breast CT is that it is necessary to first perform a circular or partial circular scan of at least 180° + fan angle, before starting the spiral, in the position closest to the patient in order to be able to perform CT reconstruction in the area close to the chest wall.

8.2.1.2 Good Soft Tissue Differentiation

Differentiation of the soft tissue essentially depends on the resulting image contrast and image noise. The contrast can be influenced by the selection of the X-ray spectrum determined by the selection of tube voltage and the prefiltration. Simulations were conducted for different breast sizes in which the contrast between fat, glandular tissue and iodine was investigated in relation to image noise and applied dose. Depending on the task, these simulations showed that a tube voltage of 30–54 kV with a prefiltration of 3 mm Al should be chosen. For example, for a breast with a mean diameter of 12 cm, the ideal tube voltage for visualizing iodine is 53 kV, for visualizing calcifications 34 kV and for differentiating soft tissue 43 kV [19]. However, it must be noted that a higher tube power is required when increasing the prefiltration or reducing the tube voltage. X-ray tubes and the matching generator were only available in a very limited selection during the design phase of the system. Since models with sufficient power were not available, 60 kV tube voltage with 3 mm Al prefiltration was chosen, which represents a good compromise for clinical applications and technical feasibility.

As mentioned in Sect. 8.1.2, subtle soft tissue differences can be better characterized using PCDs since low-energy photons carry more information about differences in attenuation and make a higher contribution to the image. Therefore, for use in BCT a photon-counting detector is better suited than a scintillating detector.

8.2.1.3 Dynamic Imaging Capabilities

Dynamic or functional imaging uses contrast agents and visualizes the behavior of these agents in the patient's body. The additional information obtained can increase sensitivity substantially, which is reflected in the increased number of findings in breast MRI, where contrast agents are generally in use.

In order to ensure the best possible visibility of the contrast agent, often subtraction images are calculated after the examination. Ideally, the subtraction images should only show the contrast agent. They allow small differences in contrast to be better visualized by removing irrelevant and distracting background structures. To obtain the subtraction images, the general approach is to subtract the pre-contrast images from the post-contrast images. Since both images are acquired at different times, artifacts are usually visible in the subtraction images, which can be attributed to slight patient movement between the two images. For small deviations, the artifacts can be addressed by image registration algorithms. If PCDs with at least two energy thresholds are used, spectral subtraction can be utilized. Thereby, a high-energy and a low-energy image are acquired simultaneously from a single X-ray exposure. Using spectral subtraction techniques, artifacts due to patient movement can be eliminated. The algorithm generates not only a spectral subtraction image but also a virtual image without a contrast agent (VNC image), which corresponds to a pre-contrast image. In addition, as the pre-contrast image is no longer essential, both patient dose and examination time can be reduced. If the best benefit is to be utilized from contrast-enhanced X-ray imaging, PCDs must be used for this application.

8.2.1.4 High Isotropic Spatial Resolution

The maximum spatial resolution of a CT system is determined not only by the focal spot size and the detector pixel pitch but also by the imaging geometry and the number of projections. If, for example, it is not possible to obtain detectors with a small pixel pitch, this can be compensated for by a geometry with sufficient magnification. On the other hand, a large focal spot can be counteracted by adjusting the ratio of the "isocenter-detector distance" to the "isocenter-focus distance." For best results the "isocenter-focus distance" must be larger than the "isocenter-detector distance."

In order to detect cancers associated with microcalcifications as early as possible, it is important to detect the particles as small as possible. For this reason, a limiting resolution of 5 mm^{-1} (10% MTF) was specified. If this requirement is met, the detection of 100 μm calcifications should be possible and detection properties similar to mammography are achieved.

Simulations were performed to define the necessary system geometry, focal spot size and detector element size [20]. These simulations indicated that a resolution of 7 mm^{-1} (10% MTF) can be achieved with the following system parameters. Focal spot size of 300 μm, "isocenter-detector distance" of 150 mm, "isocenter-focus distance" of 300 mm, and a detector pixel pitch of 100 μm. Since the actual size of the focal spot can vary and depends on the operating parameters of the X-ray tube, additional safety measures should be considered. Therefore, the "isocenter-focus distance" was increased to 500 mm; thus, variations of the focal spot size of 40% are tolerable without compromising the limiting system resolution.

It must also be taken into account that the relative movement between the imaging system and the object during acquisition leads to motion blur and thus losses in spatial resolution. This is a particular problem for systems with high-resolution requirements. If a sufficient number of projections cannot be recorded, there will be a drop in spatial resolution in the periphery of the field of measurement. This effect can be reduced by using pulsed X-rays, as this results in an acquisition resembling a "step-and-shoot" procedure [17]. However, pulsed X-ray beams generate very high photon fluxes, which can have a negative effect on the detection properties of PCDs. For this reason, a continuous acquisition with a high sampling rate of 2000 projections per rotation was specified and validated by simulations. This approach represents a good compromise between increased data rate and sufficient spatial resolution of 5 mm^{-1} (10% MTF) even in the peripheral field of measurement [21].

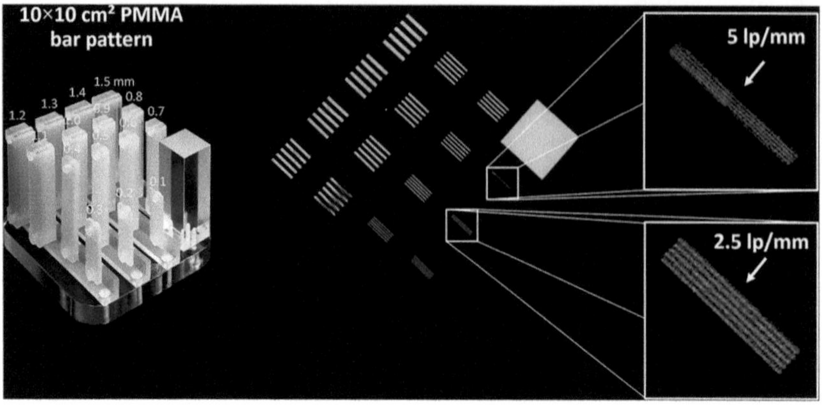

Fig. 8.3 Left: bar pattern phantom. Center: CT image of phantom. Right: magnification of the smallest patterns. The detail "5 lp/mm" visualizes 100 μm thick PMMA slabs, which can be easily separated

Based on these results, the first prototypes were built, and it was demonstrated that the desired limiting resolution of 5 mm^{-1} (10% MTF) and higher could be obtained in all three spatial directions [22]. An image of a resolution phantom obtained with the first prototype is shown in Fig. 8.3.

8.2.1.5 Low Patient Dose

Dose reduction is a constantly relevant topic in X-ray imaging. However, the focus of this topic must be shifted from the reduction itself to the optimization of utilization. From a developer's point of view, this can be achieved by adapting the spectrum and the intensity of the X-ray beam to the object and by using geometrically- and absorption-efficient detectors.

As discussed earlier in the Sect. 8.2.1.2 the X-ray spectrum was already optimized using simulation studies [19]. A comparative evaluation of conventional and photon-counting detector technologies indicates that a high potential dose saving appears possible by using PCDs while still providing the highest spatial resolution and the highest low-contrast detectability [23]. Due to the design of conventional scintillation detectors, it is very difficult to achieve 100% geometric efficiency. Based on the technical implementation, there are non-sensitive areas between the individual detector elements; the smaller the size of the detector elements, the more edge area exists and consequently the more non-sensitive area. In the most recent implementations, the non-sensitive area is already minimal but cannot be completely removed. Therefore, the detectors of the two breast CT devices presented in Sect. 8.1.4 have a limited geometrical efficiency of 70% (Koning BCT) and 84% (UC Davis BCT). In PCDs, the detector pixels are defined by the arrangement of the readout electrodes and the electric field lines that lead to the corresponding electrodes. Since there is no non-sensitive area between the neighboring field lines, these detectors achieve almost 100% geometric efficiency. 100% absorption efficiency can be achieved simply by using the right sensor material and the appropriate thickness. Due to its good availability and good attenuation properties, CdTe was selected. When this material is used a detector thickness of 750 μm is sufficient to ensure the desired absorption efficiency.

It has been shown that in breast CT an object-dependent modulation of the spectrum and intensity of the X-rays can reduce the dose by more than a third without compromising image quality [24]. This modulation is realized by applying so-called shaped or bowtie filters.

The use of tube current modulation to reduce the dose is also conceivable, but since breasts are quite homogeneous, the potential for dose reduction is limited.

Thus, from a design point of view, most of the options for dose reduction have been exploited.

8.2.1.6 High Patient Comfort

Increased patient comfort does not increase the radiological value of a modality, and yet it is of fundamental importance when it comes to BCT. As already mentioned in Sect. 8.1.3, a large number of women refrain from mammography screening due to pain. If only half of these women decide to take part in screening thanks to the possibility of a compression-free alternative, several million women in the US alone would benefit from it.

The concept of compression-free modality is already fulfilled by the choice of CT. To make the examination as pleasant as possible, the acquisition time should be as short as possible and should not exceed 10 s, even for larger breasts. A maximal breast length of 16 cm and a maximal diameter of 20 cm was assumed according to initial investigations [25]. As mentioned in Sect. 8.2.1.1, a circle-spiral trajectory was chosen.

The requirement resulting from these specifications and the desired system geometry is the detector size. With a pitch factor of 1 and a rotation time of 2 s, a detector size of 5 cm in height and 28 cm in width will be necessary.

A detector of the desired size cannot be produced from just one crystal but must be composed of several tiles. Since the price of tiles increases disproportionately with the size, a compromise must be made between the number of tiles and the overall cost. A tile size of 25.6 mm width and 12.8 mm height was chosen, which resulted in a total arrangement of 44 tiles (11 × 4 array). In addition, the separation of the total detector area into individual tiles makes it possible to deviate from a suboptimal flat detector design. In order to keep the detection of the X rays and their interaction length in the semiconductor crystal as uniform as possible, the tiles are arranged along an arc around the tube focus. Assuming that a gap of 100 μm exists at each tile joint, a geometric efficiency of approximately 99% can be achieved.

With the resulting detector geometry and a circle-spiral trajectory, the targeted acquisition time of 10 s can be achieved even for large breasts.

8.2.1.7 Low Costs

It goes without saying that the requirement for low costs cannot always be met when a new modality is being introduced. Many custom-made parts increase the purchase costs in the initial phase of a product's launch, which cannot entirely be covered by the manufacturer. In the longer term, especially when photon-counting detectors have reached a mature stage of development and production becomes more economical, it is anticipated that the purchase costs will not exceed those of a mammography or tomosynthesis device. An important cost factor that already applies today is that the operating costs of breast CT do not exceed those of a mammography or tomosynthesis device and are far below those of an MRI.

8.2.1.8 Additional Requirements

In addition to the requirements mentioned above, there are other interesting features that are worth considering for breast CT.

The advantages of computed tomography in 3D localization and navigation of lesions and calcification, in combination with the acquisition in pendant geometry, imply the implementation of a biopsy option. For realization, the use of proven equipment and familiar workflows of vacuum assisted breast biopsy, such as the ones used in MRI, is a good basis.

As described in Sects. 8.1.1 and 8.2.1.3, the photon-counting detector offers the possibility of measuring the spectral properties of the X-rays. In addition to spectral subtraction, the quantification and precise identification of various materials such as contrast agents, calcifications or breast implant material can be carried out. The information can also be used to determine tissue densities or to generate virtual monochromatic images. With the implementation of two energy thresholds in the detector design, it will be possible to fulfill these requirements.

In addition to the technical challenges, there is also a need for quality assurance. As BCT is a new modality, there are no established procedures for monitoring image quality. For photon-counting-BCT (PC-BCT), image quality guidelines based on whole-body CT are applied and they are evaluated regularly [26].

8.2.2 Clinical Implementation and Performance

8.2.2.1 Clinical Performance

The first clinical installation followed by the first patient scans were done at the University Hospital of Zurich (USZ), Switzerland, in 2018. Two independent readers evaluated the image quality of the first 12 examined patients. Both readers identified 11 out of 12 (91.7%) of the calcifications and 3 out of 3 (100%) masses [27]. Example images comparing the visibility of microcalcifications in mammography and PC-BCT are shown in Fig. 8.4.

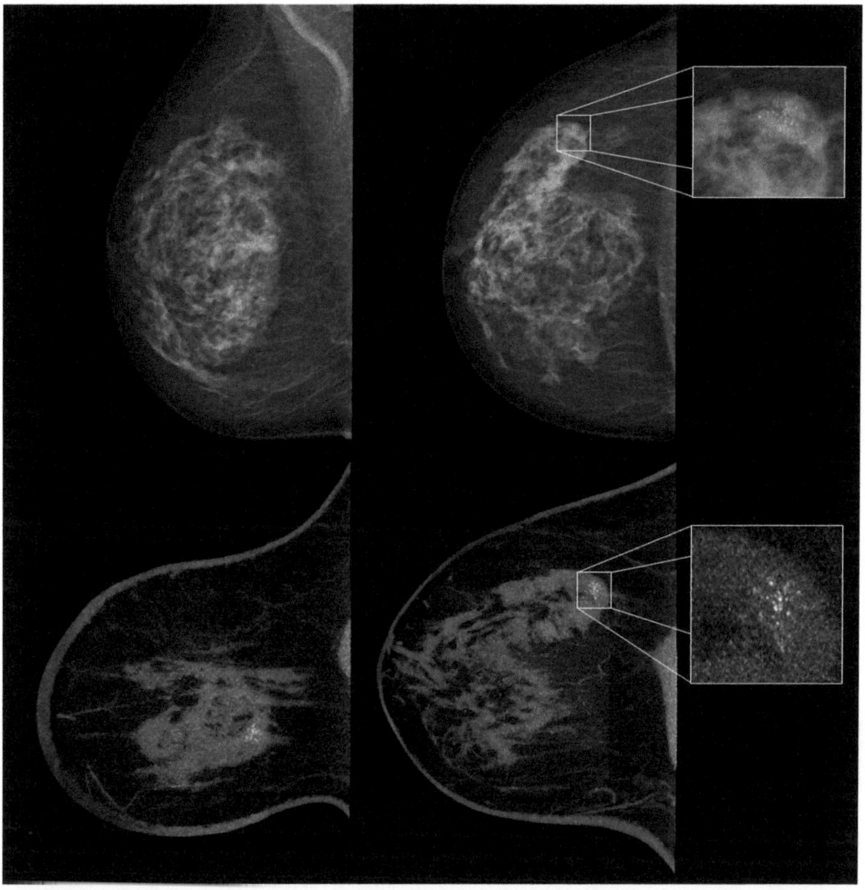

Fig. 8.4 Microcalcifications in mammography, top row and PC-BCT, bottom row. Left: MLO/sagittal view, center: CC/axial view, right: magnification of calcifications. (Images courtesy of Prof. Evelyn Wenkel)

Updated experiences were published in 2021. Retrospectively 160 women: 80 women with known breast imaging findings and 80 without, matched according to their age, underwent bilateral breast CT examinations. The diagnostic performance of maximum intensity projections (MIP) of 15 mm thickness was accessed by 2 readers regarding detectability of lesions and BI-RADS classification. Malignant lesions were diagnosed with a sensitivity of 88.9%; specificity was 99.3% resp. 100% for both readers. Inter-reader agreement was almost perfect ($\kappa = 0.94$) for the detection of malignant lesions and strong ($\kappa = 0.82$) for BI-RADS classification [28].

A comparison of PC-BCT to tomosynthesis (DBT) showed that PC-BCT allows the depiction of microcalcifications similar to DBT, in particular also including microcalcifications associated with malignant changes, as long as they are not located in the posterior part of the breast. PC-BCT allows the assessment of the distribution of microcalcifications equivalent to DBT, whereas the assessment of the individual morphology of microcalcifications is limited [29].

The first larger cohort of women was analyzed retrospectively in 2022. PC-BCT scans of 1222 women (mean age 57.7, mean body mass index (BMI) 24.3 kg/m^2) were performed at USZ and assessed image quality in terms of lesion contrast, detectability of calcification, and breast coverage using a Likert-Scale. In the study radiologists rated image quality with "no" or "negligible" complaints at 96.7% ($n = 1182$). Lesions were found in 525 patients, for which contrast was rated "optimal" or "good" in 92.6%. Ratings for lesion contrast were significantly better among heavier groups. For demonstration purposes, the visibility of 2 small cysts is shown in Fig. 8.5. BMI groups also demonstrated significantly different ratings of lesion contrast. Groups with higher BMI received ratings of optimal lesion contrast more frequently. Image quality, as well as delineation of soft tissue lesions, received high ratings overall from the radiologists [30].

Recent investigations showed that in PC-BCT malignant lesions have higher density values compared to cystic lesions and normal tissue, which may result in a better detection of lesions even in high breast density. The measurements of density values are highly reproducible between different readers [31].

The first investigations on patients with implants using PC-BCT were made by Ruby et al. [32]. They wrote that augmentation mammography comes with the

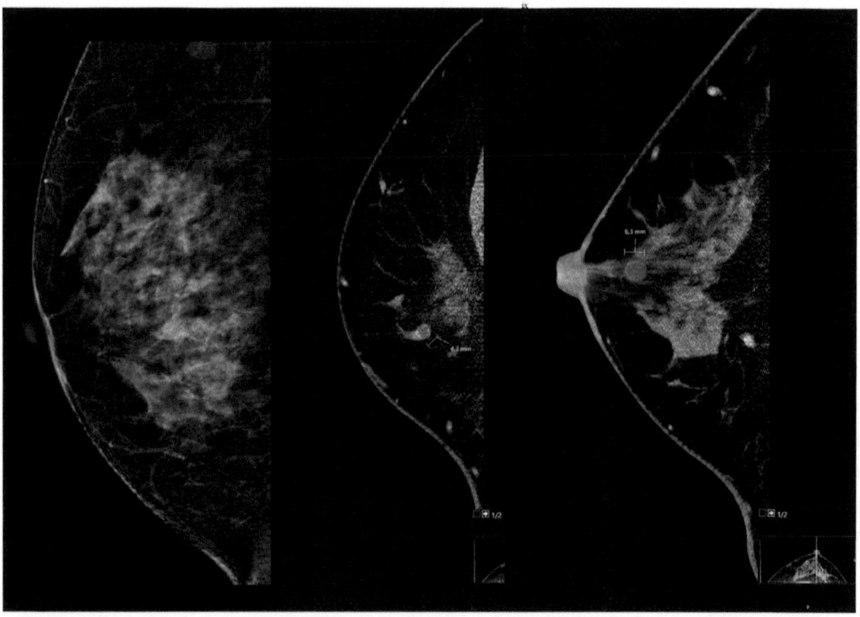

Fig. 8.5 Visibility of small cysts with a diameter of 4.3 mm (center) and 6.3 mm (right) in contrast-enhanced PC-BCT, both were not visible in the corresponding mammography (left). A contrast-enhanced rim is clearly visible in the outer perfused area of the cysts. (Images courtesy of Dr. Martin Wasser)

disadvantage of a higher mean glandular dose due to decreased compression as well as radiation absorption of the implant itself. The discomfort caused by compression during mammography can be particularly enhanced in women with implants. Mammography cannot, in most cases, detect intracapsular ruptures, which account for the majority of implant failures. In addition, although very rarely, mechanical pressure during mammography can lead to implant rupture. In women with implants, PC-BCT presents a promising modality for evaluating breast density, implant integrity, extensive capsular fibrosis, soft tissue lesions and micro–/macrocalcifications without the need for breast compression by utilizing a lower dose compared to doses reported for conventional four-view mammography. PC-BCT is the only modality providing both excellent depiction of implants and microcalcifications in a single examination without overlay effects [32]. Figure 8.6 shows PC-BCT images of a woman with a ruptured implant.

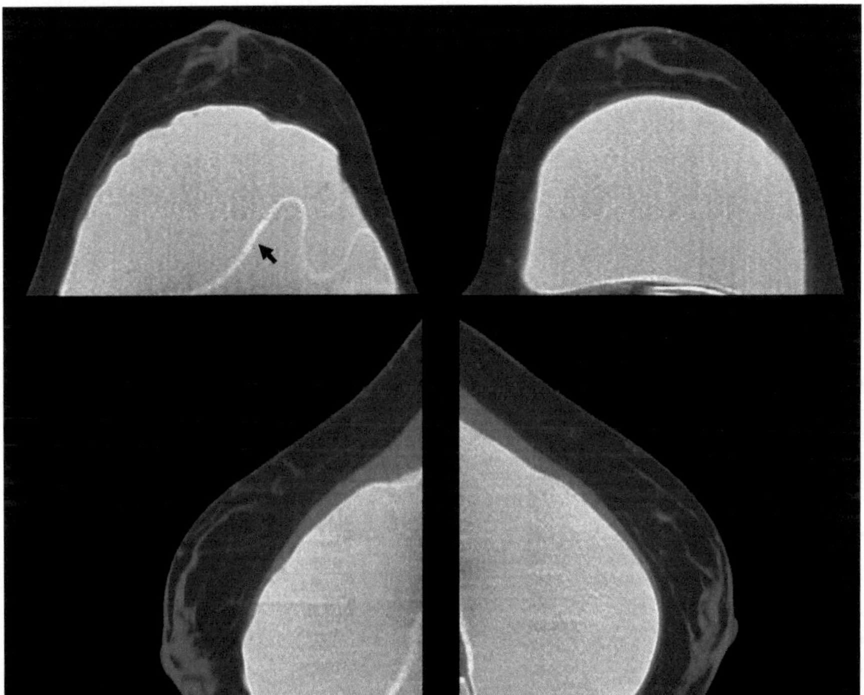

Fig. 8.6 PC-BCT images of an implant. Top row, axial view, bottom row, sagittal view. In the left images linguini signs (arrows) are visible, indicating an implant rupture. (Images courtesy of Prof. Andreas Boss)

PC-BCT can be performed with the administration of contrast media, but it is certainly not limited to it. As mentioned in Sect. 8.2.1.3, contrast-enhanced breast CT (CE-BCT) makes use of an administered contrast agent to visualize functional information derived from lesion enhancement additionally to the regular morphological features. With CE-BCT iodine-based contrast agents are used. Analog to contrast-enhanced spectral mammography, both breasts are examined sequentially after the contrast agent administration. Protocols with pre- and post-contrast examination and with post-contrast examination only can be used. Wetzl et al. [33] were the first to publish their experience with contrast-enhanced examinations acquired with a PC-BCT. They examined 93 patients, with one acquisition before and 2 min after contrast media injection for each of these examinations without problems with the clinical workflow. It was shown that no adverse events occurred during the examination that could have affected the quality of the examinations. Berger et al. [34] investigated CE-BCT by correlating the imaging findings to histology after surgical resection. They were able to demonstrate that the CE-BCT is capable of depicting both the contrast-enhancing mass component of a breast cancer lesion and associated microcalcifications of surrounding DCIS with high accuracy. All suspicious lesions substantially enhanced after contrast media application. Breast cancer lesions exhibit high contrast media uptake with an SNR above 100 and a CNR to surrounding glandular tissue above 10. Comparing the histologic size of the invasive tumors with the size of the mass component of the lesion with contrast media uptake at CE-BCT showed a strong and significant correlation with an R value of 0.77 ($P < 0.05$).

Another study with 90 women with 105 biopsy-proven malignant breast lesions showed that CE-BCT is far superior to conventional non-contrast mammography in terms of both detectability and diagnostic confidence [35]. Figure 8.7 shows a patient with a multicentric carcinoma. Contrast-enhanced MRI and contrast-enhanced PC-BCT images are displayed for comparison.

8.2.2.2 Patient Dose

The first comprehensive study on radiation dose in real patients was carried out at the USZ. The dose delivered in 1657 examinations to 829 women (mean age 57 years) between 2018 and 2021 was analyzed retrospectively. The average glandular dose (AGD) was determined using Monte Carlo simulations. Part of the dose simulation was a precise segmentation of the breast, which was carried out individually for each patient, thereby further increasing the accuracy of the simulation [36]. For small, medium and large breasts the mean AGD was 7.3 mGy, 6.5 mGy and 5.9 mGy, respectively. All women were examined with 32 mA tube current independent of the actual breast size and its density [37].

It is apparent that using a single tube current for all breast sizes and densities is not optimal in terms of radiation dose and image quality. Even simple adjustments like using 25 mA for small breasts and 40 mA for large ones instead of 32 mA, as

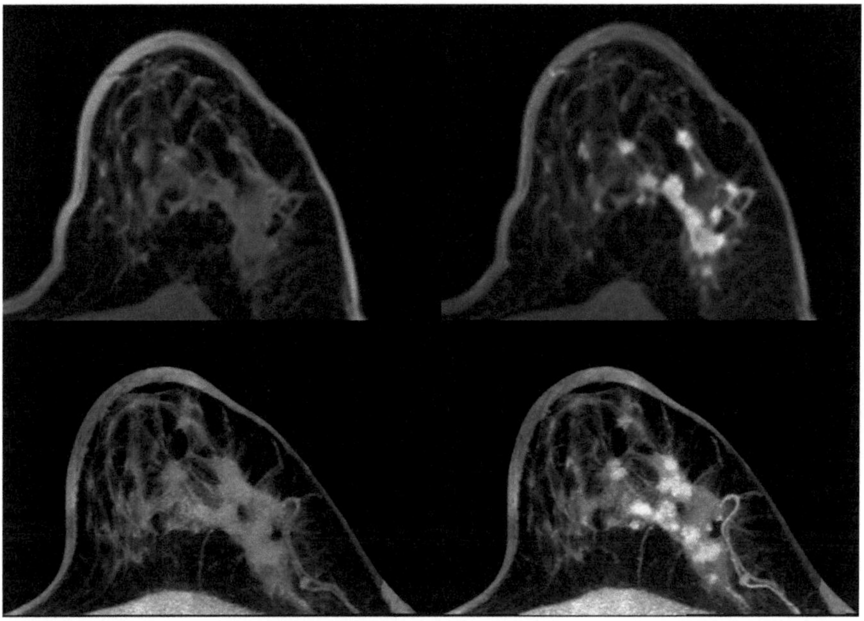

Fig. 8.7 Contrast-enhanced images of MRI top row, PC-BCT bottom row, left without and right with contrast agent applied. (Images courtesy of Prof. Evelyn Wenkel)

suggested in a phantom study, would reduce the overall dose to 4.3 mGy for small and 5.7 mGy for large breasts, respectively [38].

8.2.3 Future Developments in Photon Counting Breast CT

8.2.3.1 Integration of Biopsy Unit

A biopsy option is an urgent requirement of clinical routine. If a potentially malignant lesion is detected by any imaging procedure, a histological clarification of the finding is necessary. Therefore breast biopsy is available for several imaging modalities. There are already established biopsy solutions for ultrasound, MRI, mammography and tomosynthesis. Although these are suitable biopsy options, each modality has its own limitations. The spatial resolution is a limiting factor in breast MRI, which is why a biopsy of microcalcifications is typically not possible. In addition, there are the high costs and limited availability of an MRI examination. Although mammography and breast tomosynthesis have a high spatial resolution, they cannot achieve the soft tissue contrast of breast CT due to superpositions. As a result, the biopsy of soft tissue lesions is only possible to a limited extent. Furthermore, the precise positioning of the biopsy needle is impaired by the lack of three-dimensionality. Due to the limited resolution of sonography, a biopsy of microcalcifications is usually not feasible.

Biopsy techniques used in MRI should be transferrable to breast CT without major adaptations. Common procedures are needle biopsy and vacuum-assisted biopsy. So-called “grid and cube” or “post and pillar” devices can be used for navigation. The basic principle of these procedures and instruments can be retained with only minor adaptations for integration into BCT. This means that the biopsy procedures that clinical users are familiar with will remain unchanged. The added value of the application lies in the comprehensive information provided by the BCT data and its improved availability. Thus, a precise biopsy of calcifications, soft tissue lesions and enhancing areas is possible with only one device.

A biopsy solution is already available in the Koning Breast CT and is under development for the PC-BCT.

8.2.3.2 Spectral Applications

A major advantage of photon-counting detectors is that the spectral information of the detected X-ray quanta can be made available. In reality, however, this is linked to many challenges.

Depending on the area and thickness of the detector elements, the charge-sharing effect and fluorescence photons distort the correct detection of the X-ray spectrum. If the detector elements are larger, these effects can be counteracted, but at the price of lower spatial resolution and a reduced dynamic range; the ability to detect high photon fluxes decreases with a larger detector element size [39]. The charge-sharing effect for detector elements with an edge length of 100 μm, as specified in Sect. 8.2.1, is already severe. As a result, the detected spectral information is almost completely lost. Yet a large part of the spectral information can be restored by intelligent correction during the readout process, which already takes place on the ASIC. This correction is called “charge sharing correction” or “anti-coincidence mode” [40]. Nevertheless, the computational effort necessary for the correction further reduces the detector’s ability to detect high photon fluxes. For breast CT as a low-dose application, this constraint can be tolerated.

To distinguish between two energies, the separation of photons in at least two energy bins is necessary. Thereby the amount of created data is doubled, whereby the number of detected photons is divided between the two bins, which in turn results in more noise in the spectral information. This effect becomes worse when more spectral information and thus more energy bins are needed. The required fast readout rates have already been considered in the design of the PC-BCT. It is therefore possible to acquire projections with two energy bins at 1000 fps. The increased noise in the spectral information, caused by the low photon number, can only be reduced by increasing the dose or the application of sufficient post-processing algorithms. Since a dose increase is not an option, the development of effective algorithms for material decomposition and noise reduction is necessary.

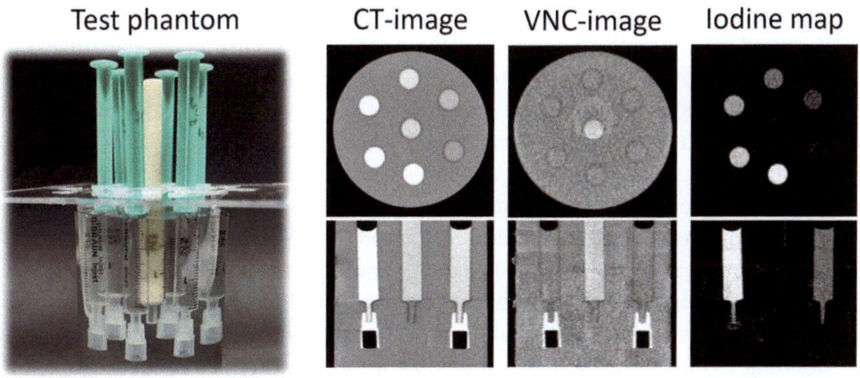

Fig. 8.8 From left to right: Syringes filled with different concentrations of iodine, non-iodine material in the center. CT image. VNC image and iodine map calculated from the spectral information

The first positive results that tested the feasibility of spectral applications have already been obtained (see Fig. 8.8). Utilizing the K-edge of iodine, accurate iodine quantification and generation of virtual non-contrast images can be achieved using contrast-enhanced PC-BCT from a single scan in the absence of temporal or spatial misalignment [41]. Nevertheless, challenges such as high noise levels and inhomogeneous detection of the spectrum still need to be addressed before spectral applications will reach market maturity.

8.3 Conclusion

Providing a comprehensive and unbiased summary of a modality will always be a challenge, as personal preferences, and the limitations of one's knowledge will invariably leave room for improvement. Nevertheless, it can be said with certainty that the availability of photon-counting detectors opens up new possibilities in the field of imaging. For example, the development of a CT device that aims to achieve a similar dose and spatial resolution as projection imaging offers all the advantages of a 3D modality. This idea alone sounds far-fetched. However, initial results show that even if not all the goals have been achieved, the idea can become a reality.

One problem, not only in BCT, that is becoming apparent is that the high spatial resolution of photon-counting detectors can often not be fully utilized in practice due to high noise values. In CT reconstructions, the maximum possible system resolution is often not pursued in order to keep the noise at an adequate level. Perhaps new reconstruction algorithms and the increased use of artificial intelligence will help to address this problem.

The possibility of utilizing the information of the energy spectrum seems appealing, but developing applications that really add value for radiologists is all the more difficult, especially since a change in established processes and the increased workload caused by more information slow down the workflow while the potential benefits are not yet fully known. In some areas of application, however, a benefit has already been recognized. Spectral applications for better visualization of contrast agents are essential in mammography and have also been tested successfully in PC-BCT. In contrast, other applications such as the spectral discrimination of malignant and benign calcifications are of great interest but have not been implemented yet due to the limited spectral resolution of the PCDs available.

An advantage and simultaneously a disadvantage of breast CT is the limited field of view. Although only the organ of interest is exposed to X-rays, the axillary margins of the breast tissue cannot always be imaged adequately. Improved patient positioning by adapting the patient table or a modified trajectory of the imaging system may overcome these limitations in the near future.

Despite the very good utilization of the applied X-ray radiation, mammography currently still has a dose advantage. However, this advantage will fade with the introduction of shaped filters and newer reconstruction techniques. And due to the numerous advantages that breast CT offers, it is possible that it can fulfill and even exceed the tasks currently performed by mammography, one day.

The quality of photon-counting detectors for imaging has just reached market maturity. There is still a lot of room for improvement in both the materials and the readout electronics. As demand increases, so will the number of manufacturers and the quality of photon-counting detectors, which will benefit BCT and all other X-ray-based imaging modalities.

The acquisition of further clinical data and constant efforts in the field of engineering are necessary to further improve the benefits of BCT and prove its effectiveness. There is still a long way to go to improve not only the detection of cancer but also the comfort of women and the workflow of radiologists.

Acknowledgments Potential conflicts of interest include the following: The authors V. Ruth, C. Steiding, D. Heinemann, F. Althoff and D. Kolditz are employees of AB-CT–Advanced Breast-CT GmbH.

References

1. Breskin A. Photon detectors for the 21st century. Nucl Instrum Methods Phys Res A Accel Spectr Detect Assoc Equip. 1997;387(1–2):1–18. https://doi.org/10.1016/s0168-9002(96)01203-x.
2. Lundqvist M, Danielsson M, Cederstroem B, Chmill V, Chuntonov A, Aslund M. Measurements on a full-field digital mammography system with a photon counting crystalline silicon detector. In: Yaffe MJ, Antonuk LE, editors. Medical imaging 2003: physics of medical imaging. Lesquin: SPIE; 2003. https://doi.org/10.1117/12.480251.
3. Weigel S, Berkemeyer S, Girnus R, Sommer A, Lenzen H, Heindel W. Digital mammography screening with photon-counting technique: can a high diagnostic performance be realized at low mean glandular dose? Radiology. 2014;271(2):345–55. https://doi.org/10.1148/radiol.13131181.

4. Berger N, Marcon M, Saltybaeva N, Kalender WA, Alkadhi H, Frauenfelder T, et al. Dedicated breast computed tomography with a photon-counting detector: initial results of clinical in vivo imaging. Invest Radiol. 2019;54(7):409–18. https://doi.org/10.1097/rli.0000000000000552.
5. Bale DS, Szeles C. Nature of polarization in wide-bandgap semiconductor detectors under high-flux irradiation: application to semi-insulating $cd_{1-x}zn_xTe$. Phys Rev B. 2008;77:035205. https://doi.org/10.1103/PhysRevB.77.035205.
6. Wanders JOP, Holland K, Veldhuis WB, Mann RM, Pijnappel RM, Peeters PHM, et al. Volumetric breast density affects performance of digital screening mammography. Breast Cancer Res Treat. 2016;162(1):95–103. https://doi.org/10.1007/s10549-016-4090-7.
7. Tagliafico AS, Calabrese M, Mariscotti G, Durando M, Tosto S, Monetti F, et al. Adjunct screening with tomosynthesis or ultrasound in women with mammography-negative dense breasts: interim report of a prospective comparative trial. J Clin Oncol. 2016;34(16):1882–8. https://doi.org/10.1200/jco.2015.63.4147.
8. Whelehan P, Evans A, Wells M, MacGillivray S. The effect of mammography pain on repeat participation in breast cancer screening: a systematic review. Breast. 2013;22(4):389–94. https://doi.org/10.1016/j.breast.2013.03.003.
9. Heindel W, Weigel S, Joachim GÃŸ, Hense H-W, Sommer A, Krischke M, et al. Digital breast tomosynthesis plus synthesised mammography versus digital screening mammography for the detection of invasive breast cancer (tosyma): a multicentre, open-label, randomised, controlled, superiority trial. Lancet Oncol. 2022;23(5):601–11. https://doi.org/10.1016/s1470-2045(22)00194-2.
10. Evans A, Trimboli RM, Athanasiou A, Balleyguier C, Baltzer PA, Bick U, et al. Breast ultrasound: recommendations for information to women and referring physicians by the european society of breast imaging. Insights Imaging. 2018;9(4):449–61. https://doi.org/10.1007/s13244-018-0636-z.
11. Kuhl CK, Strobel K, Bieling H, Leutner C, Schild HH, Schrading S. Supplemental breast mr imaging screening of women with average risk of breast cancer. Radiology. 2017;283(2):361–70. https://doi.org/10.1148/radiol.2016161444.
12. Kalender WA. Computed tomography: fundamentals, system technology, image quality, applications. 3rd ed. Erlangen: Publicis Publishing; 2011.
13. Chang CH, Sibala JL, Fritz SL, Gallagher JH, Dwyer SJ, Templeton AW. Computed tomographic evaluation of the breast. AJR Am J Roentgenol. 1978;131(3):459–64. https://www.ncbi.nlm.nih.gov/pubmed/98990
14. Lindfors KK, Boone JM, Nelson TR, Yang K, Kwan ALC, Miller DF. Dedicated breast CT: initial clinical experience. Radiology. 2008;246(3):725–33. https://www.ncbi.nlm.nih.gov/pubmed/18195383
15. O'Connell A, Conover DL, Zhang Y, Seifert P, Logan-Young W, Lin C-FL, et al. Cone-beam CT for breast imaging: radiation dose, breast coverage, and image quality. Am J Roentgenol. 2010;195(2):496–509. https://doi.org/10.2214/AJR.08.1017. https://www.ncbi.nlm.nih.gov/pubmed/20651210
16. Brombal L, Arfelli F, Delogu P, Donato S, Mettivier G, Michielsen K, et al. Image quality comparison between a phase-contrast synchrotron radiation breast ct and a clinical breast ct: a phantom based study. Sci Rep. 2019;9(1):17778. https://doi.org/10.1038/s41598-019-54131-z.
17. Gazi PM, Yang K, Burkett GW, Aminololama-Shakeri S, Seibert JA, Boone JM. Evolution of spatial resolution in breast CT at UC Davis. Med Phys. 2015;42(4):1973–81. https://www.ncbi.nlm.nih.gov/pubmed/25832088
18. Aminololama-Shakeri S, Hargreaves JB, Boone JM, Lindfors KK. Dedicated breast CT: screening technique of the future. Curr Breast Cancer Rep. 2016;8(4):242–7. https://doi.org/10.1007/s12609-016-0227-2.
19. Weigel M, Vollmar SV, Kalender WA. Spectral optimization for dedicated breast CT. Med Phys. 2011;38(1):114–24. https://doi.org/10.1118/1.3523599. https://www.ncbi.nlm.nih.gov/pubmed/21361181

20. Kalender WA, Beister M, Boone JM, Kolditz D, Vollmar SV, Weigel M. High-resolution spiral ct of the breast at very low dose: concept and feasibility considerations. Eur Radiol. 2012;22:1–8. https://doi.org/10.1007/s00330-011-2169-4.
21. Althoff F. Systemdesign und Patientenlagerung bei der dedizierten Brust-Computertomografie. PhD thesis, Institute of Medical Physics, Friedrich-Alexander-University Erlangen-Nuremberg, Erlangen, Germany; 2013.
22. Kalender WA, Kolditz D, Steiding C, Ruth V, Lück F, Rößler A-C, et al. Technical feasibility proof for high-resolution low-dose photon-counting ct of the breast. Eur Radiol. 2017;27(3):1081–6. https://doi.org/10.1007/s00330-016-4459-3.
23. Kuttig JD, Steiding C, Kolditz D, Hupfer M, Karolczak M, Kalender WA. Comparative investigation of the detective quantum efficiency of direct and indirect conversion detector technologies in dedicated breast CT. Phys Med. 2015;31(4):406–13. https://doi.org/10.1016/j.ejmp.2015.03.007.
24. Lück F, Kolditz D, Hupfer M, Steiding C, Kalender WA. Experimental validation of a single shaped filter approach for ct using variable source-to-filter distance for examination of arbitrary object diameters. Phys Med Biol. 2014;59(19):5691. https://doi.org/10.1088/0031-9155/59/19/5691.
25. Huang SY, Boone JM, Yang K, Packard NJ, McKenney SE, Prionas ND, Lindfors KK, et al. The characterization of breast anatomical metrics using dedicated breast CT. Med Phys. 2011;38(4):2180–91. https://www.ncbi.nlm.nih.gov/pubmed/21626952
26. Steiding C, Kolditz D, Kalender WA. A quality assurance framework for the fully automated and objective evaluation of image quality in cone-beam computed tomography. Med Phys. 2014;41(3):031901. https://doi.org/10.1118/1.4863507.
27. Berger N, Marcon M, Saltybaeva N, Kalender WA, Alkadhi H, Frauenfelder T, et al. Dedicated breast computed tomography with a photon-counting detector: initial results of clinical in vivo imaging. Invest Radiol. 2019;54(7):409–18. https://doi.org/10.1097/RLI.0000000000000552.
28. Zellweger C, Berger N, Wieler J, Cioni D, Neri E, Boss A, et al. Breast computed tomography: diagnostic performance of the maximum intensity projection reformations as a stand-alone method for the detection and characterization of breast findings. Invest Radiol. 2021;57(4):205–11. https://doi.org/10.1097/rli.0000000000000829.
29. Huck LC, Bode M, Zanderigo E, Wilpert C, Raaff V, Dethlefsen E, et al. Dedicated photon-counting ct for detection and classification of microcalcifications: an intraindividual comparison with digital breast tomosynthesis. Invest Radiol. 2024;59 https://doi.org/10.1097/rli.0000000000001097.
30. Schmidt CS, Zellweger C, Wieler J, Berger N, Marcon M, Frauenfelder T, et al. Clinical assessment of image quality, usability and patient comfort in dedicated spiral breast computed tomography. Clin Imaging. 2022;90:50–8. https://doi.org/10.1016/j.clinimag.2022.07.001.
31. Weber J, Zanetti G, Nikolova E, Frauenfelder T, Boss A, Wieler J, et al. Potential of non-contrast spiral breast ct to exploit lesion density and favor breast cancer detection: a pilot study. Eur J Radiol. 2024;178:111614. https://doi.org/10.1016/j.ejrad.2024.111614.
32. Ruby L, Shim S, Berger N, Marcon M, Frauenfelder T, Boss A. Diagnostic value of a spiral breast computed tomography system equipped with photon counting detector technology in patients with implants. Medicine. 2020;99(30):e20797. https://doi.org/10.1097/md.0000000000020797.
33. Wetzl M, Wenkel E, Dietzel M, Siegler L, Emons J, Dethlefsen E, et al. Potential of spiral breast computed tomography to increase patient comfort compared to dm. Eur J Radiol. 2021;145:110038. https://doi.org/10.1016/j.ejrad.2021.110038.
34. Berger N, Marcon M, Wieler J, Vorburger D, Dedes KJ, Frauenfelder T, et al. Contrast media enhanced breast computed tomography with a photon-counting detector: initial experiences on in vivo image quality and correlation to histology. Invest Radiol. 2022;57(10):704–9. https://doi.org/10.1097/rli.0000000000000863.

35. Wetzl M, Heilingbrunner T, Heindl F, Wenkel E, Uder M, Ohlmeyer S. Detectability of breast cancer in dedicated breast ct compared with mammography dependent on breast density. Invest Radiol. 2024;59 https://doi.org/10.1097/rli.0000000000001105.
36. Shim S, Cester D, Ruby L, Bluthgen C, Marcon M, Berger N, et al. Fully automated breast segmentation on novel breast computed tomography images. In Swiss Congress of Radiology (SCR) 2021, Online, 2021. https://congress.sgr-ssr.ch/file/2022/01/SCR-21_OAB_Final.pdf.
37. Shim S, Kolditz D, Steiding C, Ruth V, Hoetker AM, Unkelbach J, et al. Radiation dose estimates based on Monte Carlo simulation for spiral breast computed tomography imaging in a large cohort of patients. Med Phys. 2023;50(4):2417–28. https://doi.org/10.1002/mp.16211.
38. Shim S, Saltybaeva N, Berger N, Marcon M, Alkadhi H, Boss A. Lesion detectability and radiation dose in spiral breast CT with photon-counting detector technology: a phantom study. Invest Radiol. 2020;55(8):515–23. https://doi.org/10.1097/rli.0000000000000662.
39. Trueb P, Zambon P, Broennimann C. Assessment of the spectral performance of hybrid photon counting x-ray detectors. Med Phys. 2017;44(9) https://doi.org/10.1002/mp.12323.
40. Ullberg C, Eriksson C, Urech M, Stewart A, Weber N. Photon counting dual energy x-ray imaging at CT count rates: measurements and implications of in-pixel charge sharing correction. In: Chen G-H, Lo JY, Schmidt TG, editors. SPIE medical imaging 2018: physics of medical imaging. Lesquin: SPIE; 2018. https://doi.org/10.1117/12.2293591.
41. Ruth V, Kolditz D, Steiding C, Kalender WA. Investigation of spectral performance for single-scan contrast-enhanced breast CT using photon-counting technology: a phantom study. Med Phys. 2020;47(7):2826–37. https://doi.org/10.1002/mp.14133.

Chapter 9
Clinical Applications of MARS Spectral Photon-Counting CT

Jennifer A. Clark, Steven Gieseg, Maya R. Amma, Anthony P. H. Butler, Devyani Dixit, and The MARS Collaboration

Contents

J. A. Clark (✉)
MARS Bioimaging Limited, Christchurch, New Zealand

University of Otago Christchurch, Christchurch, New Zealand
e-mail: jenn.clark@marsbioimaging.com

S. Gieseg · D. Dixit
MARS Bioimaging Limited, Christchurch, New Zealand

University of Canterbury, Christchurch, New Zealand

M. R. Amma
MARS Bioimaging Limited, Christchurch, New Zealand

A. P. H. Butler
MARS Bioimaging Limited, Christchurch, New Zealand

University of Otago Christchurch, Christchurch, New Zealand

University of Canterbury, Christchurch, New Zealand

European Organisation for Nuclear Research (CERN), Geneva, Switzerland

The MARS Collaboration
European Organisation for Nuclear Research (CERN), Geneva, Switzerland

S. A. Si-Mohamed, K. K. Iniewski (eds.), *Clinical Applications of Spectral Photon Counting Computed Tomography Technology*,
https://doi.org/10.1007/978-3-031-85078-3_9

9.1 Introduction

Spectral photon-counting CT (SPCCT) is widely known to provide high-resolution cross-sectional imaging for a wide range of clinical applications. Research continues to show clinically relevant advantages compared to conventional CT and other imaging modalities.

This chapter focuses on the MARS Extremity 5x120, which is a self-shielded system, designed specifically to bring advanced SPCCT to the point of care. Results from the recent clinical trial and pre-clinical research provide evidence of the clinical applications of this novel SPCCT.

9.2 MARS Spectral Photon-Counting Technology

The MARS Extremity 5x120 is a point-of-care upper extremity imaging system developed by MARS Bioimaging Limited (MBI) in 2020. The system is specifically designed for use in a community clinic or clinical office environment. It is fully shielded, with the exception of the entry port, and powered by standard mains power. Imaging is achieved with the patient in a seated position with their limb resting inside the gantry, as demonstrated in Fig. 9.1.

The system has a bore size of 123 mm with an available scan length of 35 cm. A low-power X-ray tube with a tungsten target is used to produce a polychromatic X-ray beam. The X-ray beam is attenuated by the limb during helical scanning and the transmission data is captured by the detector in the camera assembly. Typical exposure factors for a wrist scan are a tube voltage of 120 kVp and a tube current of 61 μA. The small focal spot of the X-ray tube and the 0.11 mm voxels in the camera provide images with high spatial resolution.

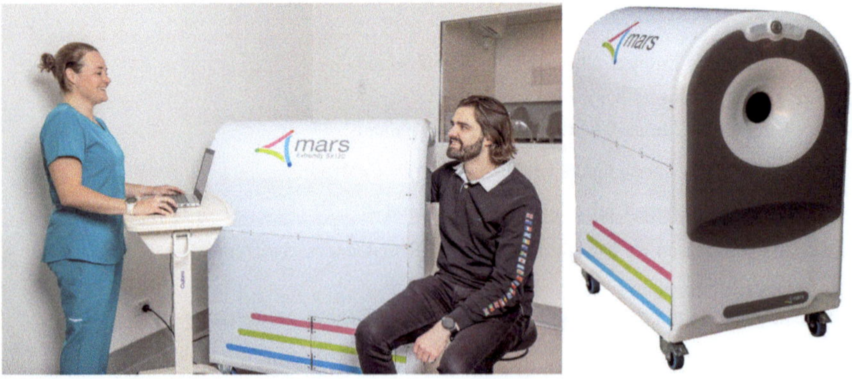

Fig. 9.1 MARS Extremity 5x120 scanner

MARS SPCCT uses the Medipix3RX photon counting detector [1]. The camera registers the energies of individual X-ray photons in the semiconductor sensor chips and allocates each photon to a protocol-defined energy bin. By registering the number of photons in each energy bin, the camera records different linear attenuation profiles for each projection. This approach allows attenuation images to be generated for up to 5 energy bins, using only one X-ray source and one scan.

The detector has two layers: a sensor layer and an application-specific integrated circuit (ASIC). The sensor layer converts the X-ray photon to an electron–hole charge cloud in the semiconductor. An electric field drives the charge cloud to electrodes on the top and bottom of the sensor chip. The electrodes convey the charge from each single X-ray photon as a voltage pulse to the second layer, the ASIC. The ASIC has multiple electrical pulse-shaping circuits to measure and record the size of the pulses. The Medipix3RX ASIC patented circuitry handles most of the inter-pixel charge sharing, measures each pulse height proportional to the X-ray photon energy, and allocates a count into the appropriate energy bins [2]. The MARS reconstruction engine transforms projection data into volumetric data of both the energy-dependent attenuation images and the material contents of each voxel.

Material decomposition algorithms are applied to convert energy images into material maps with associated material density information. This enables identification and quantification of materials such as bone, fat and water in clinically important anatomical and pathological structures [3]. Bone material maps, for example, are calibrated using known concentrations of hydroxyapatite (HA).

Proprietary software is used to perform multiplanar reformats to present images in axial, coronal, sagittal and oblique planes, as well as 3D volume renders. Specific regions of interest can be analysed and measured in detail using spectral or multi-energy scans [4, 5].

MARS systems provide high spatial resolution which enables the measurement of high-contrast objects such as bone trabeculae. Recent measurements have shown an average *x*/*y*-plane spatial frequency of 41.5 lp/cm for the 10% MTF. The energy resolution is less than 5 keV in 5 energy channels with charge summing mode. In addition, there are 3 further energy channels available without charge summing. This excellent spectral fidelity enables highly accurate measurements for calcium, fat and water, which are well suited to orthopaedic applications. It provides the potential to differentiate materials with similar attenuation profiles such as crystal deposits which are notoriously challenging to diagnose [6]. In addition, using MARS SPCCT, multiple contrast agents can be measured simultaneously, for example, hafnium, gadolinium and iodine, providing the possibility of molecular imaging in the future. These applications are currently being explored by researchers internationally.

Whilst the MARS imaging system focuses on high spatial resolution and excellent spectral fidelity, it compromises on high flux imaging such that acquisition time is similar to MRI. Other SPCCT implementations achieve short acquisition times to enable, for example, cardiac imaging. However, a focus on speed of acquisition could compromise both spatial resolution and spectral fidelity such that use in other clinical applications may be compromised.

While photon counting technology is still maturing, the clinical utility of scanners will vary depending on the specific implementation. In essence, scanner designers currently must make compromises between spatial resolution, energy resolution, and the number of energy thresholds versus acquisition time. In many clinical situations, high-quality diagnostic information is more important than fast acquisition.

9.3 Orthopaedic Applications of MARS SPCCT

9.3.1 Fracture Diagnosis

The first clinical trial of MARS SPCCT commenced in 2020 in New Zealand and is ongoing [7]. The aim is to determine whether, for orthopaedic applications, image quality is at least equivalent to conventional imaging modalities. Researchers are also exploring whether the technological advances of SPCCT result in clinically significant differences to support more accurate and efficient decision-making. Patients with known or suspected wrist fractures have been imaged with the MARS imaging system at various stages in their acute, pre and post-operative care.

For suspected wrist fractures, plain radiographs are typically the first imaging modality in the diagnostic pathway. They have high availability, low cost, low radiation dose and imaging is relatively fast. However, fractures can remain radiographically occult, due to the complex three-dimensional relationship between the carpal bones and the superimposition of bony structures [8, 9]. Up to 25% of suspected scaphoid fractures are not confirmed on initial and sometimes follow-up plain X-rays [10, 11]. Diagnosis often requires cross-sectional imaging within 1–2 weeks, including computed tomography (CT), cone-beam CT or magnetic resonance imaging (MRI) [12]. Whilst these modalities are able to detect and characterise fractures, they are expensive, and they are not always available in the acute phase. This has both financial and long-term health implications for the patient. A delayed fracture diagnosis is associated with an increase in the incidence of avascular necrosis, non-union and osteoarthritis, increased operative risk and a loss of productivity due to cast immobilisation [13–16].

SPCCT technology within compact, small-footprint scanners shows the potential to add value for acute and early follow-up imaging. If scanners are deployed in community clinics, high-quality, cross-sectional imaging could be achieved when the patient presents at the point of care. The goal is to provide early diagnosis, minimise unnecessary immobilisation and reduce the need for repeated clinical imaging.

The clinical trial of MARS SPCCT has focused primarily on scaphoid fractures. Figure 9.2 presents an example imaging. Figure 9.2a, b were acquired whilst the patient was in a fibre-glass cast. An oblique scaphoid waist fracture is demonstrated in the coronal and sagittal planes at a slice thickness of 90 μm. Figure 9.2c, d are examples of 3D volume renders created using material maps.

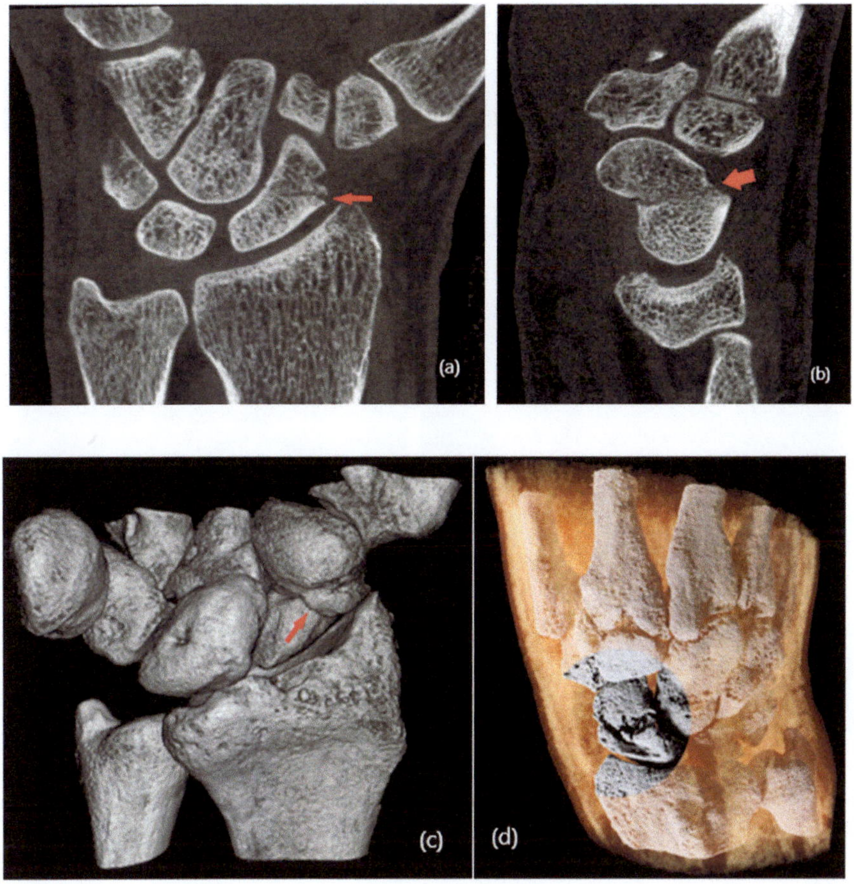

Fig. 9.2 MARS images of scaphoid fractures. (**a** + **b**) Coronal and sagittal oblique reformats to demonstrate a scaphoid waist fracture at a slice thickness of 0.09 mm. (**c**) 3D HA channel (**d**) 3D advanced volume render

A range of other carpal bone and wrist fractures have been detected using MARS SPCCT in the acute care setting. Examples are shown in Fig. 9.3. Figure 9.3a, b demonstrate a subtle distal radius fracture. This patient was otherwise booked for an MRI scan for diagnosis. Figure 9.3c, d show radiographically occult fractures of the triquetrum, which were not detected through the standard of care pathway, using plain radiographs.

The clinical trial has demonstrated that fracture diagnosis is easily achievable using MARS SPCCT. However, the scan duration is longer than that of conventional CT systems, similar to a limited-series MRI. Therefore, specific attention must be paid to patient comfort during positioning and effective use of immobilisation techniques. Furthermore, reconstruction times are longer than that of conventional CT to produce such high-resolution spectral data. Currently, the system is considered to

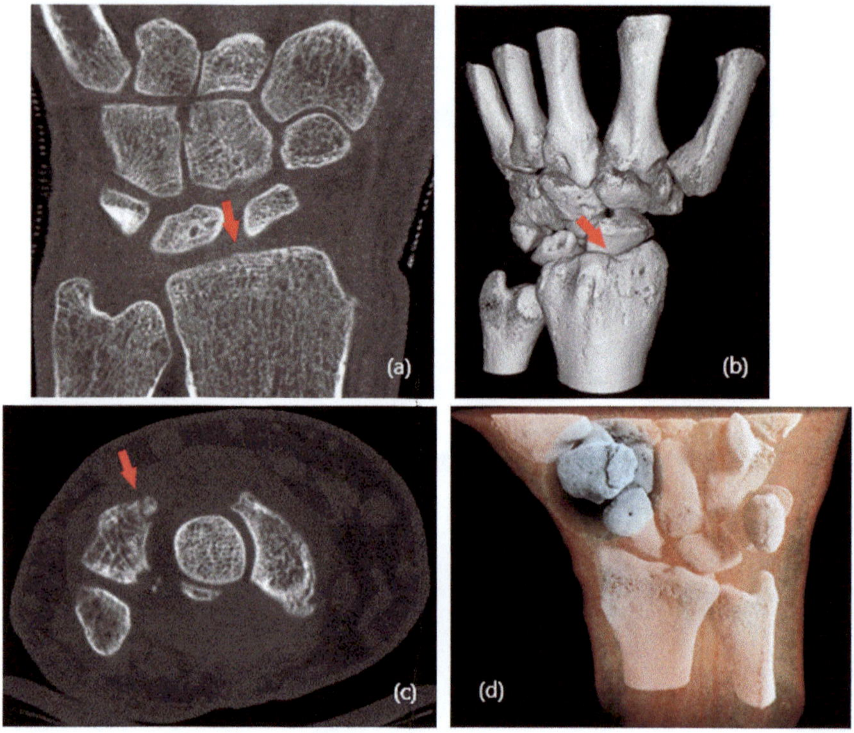

Fig. 9.3 MARS images of acute fractures (**a** + **b**) 2D and 3D imaging of a patient with a distal radius fracture. (**c**) Axial reformat demonstrating a triquetrum fracture. (**d**) 3D HA, Lipid and water channel, using advanced volume rendering techniques to demonstrate a triquetrum fracture

be better suited to pre- and post-operative imaging pathways rather than a diagnosis in a time-critical acute setting.

9.3.2 *Pre-operative Assessment of Fracture Healing*

Plain radiographs are typically used to evaluate fracture healing for minimally displaced fractures. They provide 2D qualitative information about the presence of callus or osseous bridging, using density changes to determine union [17]. Limitations include the inability to detect early calcification, which may take weeks to become visible [18]. Open, displaced or otherwise complex fractures typically warrant cross-sectional imaging such as CT to assess rotational alignment, estimate bone bridging and detect subtle changes in structure and density. Imaging must provide sufficient detail to differentiate fracture union from non-union, which is critical for deciding between surgical and conservative management. For scaphoid fractures, this is particularly important. Vascularity is often compromised when fractures are located at the proximal pole, resulting in bone loss or necrosis, which further complicates recovery and long-term health outcomes [19, 20].

For pre-operative planning, MARS SPCCT provides cross-sectional imaging of a fracture that is comparable to conventional CT. However, the smaller voxel size, charge summing and photon-counting capability should, in theory, provide higher-resolution imaging. In addition, the spectral data provides bone material maps which offer a quantitative component to the evaluation of fracture healing. These could be used to measure callus formation by density as well as volume and accurately monitor any changes in the proximal and distal poles. Researchers are investigating whether this level of information will lead to earlier detection of non-union and associated complications such as avascular necrosis.

Figure 9.4 is a case of scaphoid fracture non-union imaged using MARS SPCCT. The patient has sclerotic margins at the fracture site, resorption of bone at the distal pole, increased bone density at the proximal pole and cystic changes. The 35–45 keV energy channel was selected to create a series of multiplanar reformats. Figure 9.4a, b demonstrate high-resolution imaging of bone as well as good soft tissue contrast resolution. From the same scan, the bone (HA) material map is presented with a rainbow look-up table (Fig. 9.4c). Higher-density bone, such as cortical bone, appears red, whilst trabecular bone with a lower density is blue. Using this map, sclerosis at the fracture site and relative differences in bone density become obvious.

In Fig. 9.4c, the HA material map has been used to measure bone density in clinically useful regions of interest. In this slice, the distal pole (region 1) measures 0.20 g/mL, the bone bridging at the fracture site measures 0.60 g/mL (region 2), and the proximal pole measures 0.40 g/mL (region 3). These regions of interest can be extended across multiple slices to produce more accurate volumetric data. They are likely to be tracked over time by clinicians during follow-up scans.

9.3.3 Post-operative Imaging of Metal Implants

Metallic hardware, or metal implants, are commonly used in orthopaedics for internal fixation of fractures to restore stability and alignment and promote bone healing [21, 22]. Steel and titanium screws, plates and wires are common implant materials due to their mechanical strength, corrosion resistance and biocompatibility [23]. Post-operative imaging is important to assess implant positioning and osseous integration which is determined by the tightness of the implant and integrity of the surrounding tissue [24, 25]. The aim is early detection of implant loosening or infection to avoid lengthy delays, complex revision surgeries and significant costs for the patient and health service [26].

The imaging of metal implants is often affected by artefacts which make it difficult to evaluate the bone–metal interface. In CT systems, physical effects such as beam hardening, photon starvation, and partial volume effects can severely limit the assessment of any tissue in the vicinity of an implant [27, 28]. Beam hardening

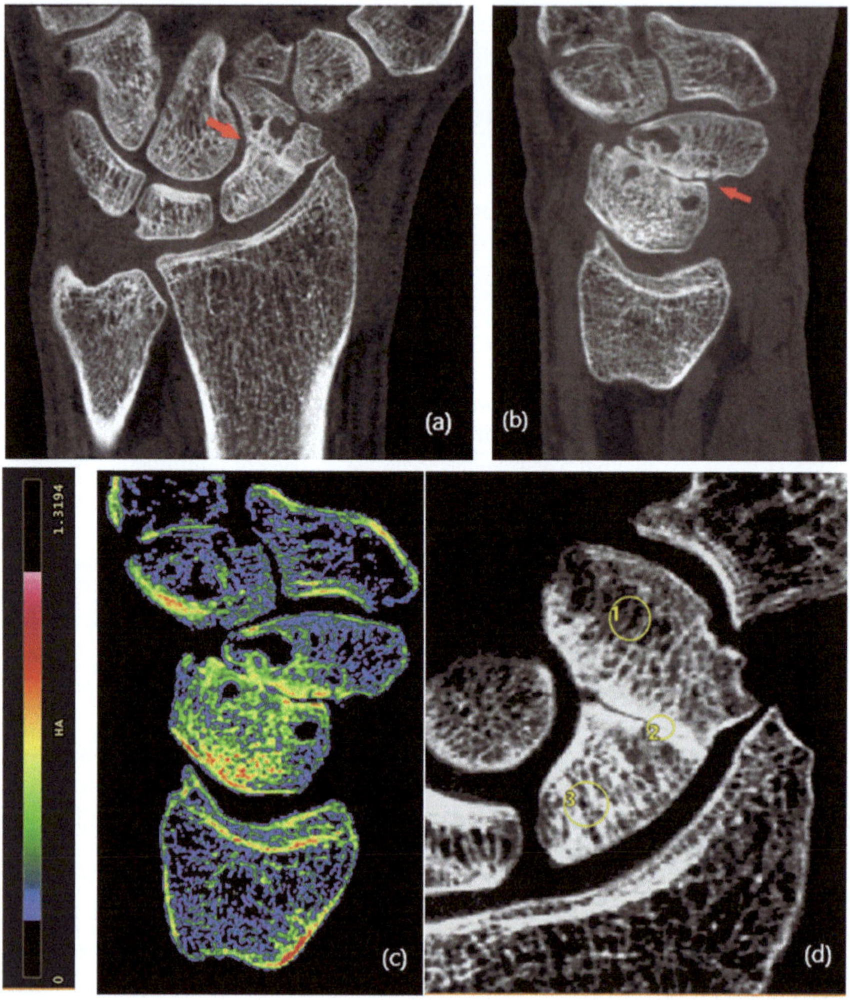

Fig. 9.4 MARS imaging of a scaphoid fracture non-union. (**a** + **b**) Coronal and sagittal oblique reformats at slice thickness of 0.15 mm. (**c**) MARS HA bone channel with a rainbow look-up table. (**d**) MARS HA bone channel with a greyscale look-up table. Material map analysis shows the differences in bone mineralisation for each region: 0.20 g/mL in ROI 1, 0.60 g/mL in ROI 2 and 0.40 g/mL in ROI3

occurs when a polychromatic X-ray beam passes through a dense material and increases the overall mean energy of the beam. Photon starvation produces bright and dark streaks around the metal implant in the direction of greatest attenuation. Partial volume effects occur when high-contrast material (metal) lies in the same voxel as low-contrast material (soft tissue), thereby producing attenuation values proportional to the average of these materials, resulting in blur. Correction of these artefacts using hardware-based approaches and algorithmic methods has been an active area of research since the invention of CT [29–31]. Dual-energy CT addresses this by scanning at two different energies and using image processing algorithms to correct [32–34]. MRI has limited diagnostic utility for this application, as the difference in magnetic susceptibility between the implant and soft tissue leads to low signal to noise ratio, geometric distortion, signal void, signal pile up and failed fat suppression [35, 36].

Spectral CT systems are able to use energy information to reduce artefacts associated with beam hardening [5, 37, 38]. The highest of 5 energy bins is used for optimal demonstration of the bone–metal interface (e.g. 65–120 kV). Simultaneously obtaining images in the lower energy bins means there is no compromise in terms of soft tissue differentiation (e.g. 35–45 keV) [37]. Further, the smaller voxel size and charge summing mode in MARS SPCCT should minimise partial volume effects.

MARS SPCCT has been used to scan patients with a range of metal implants including titanium screws, steel wires, distal radius plates and radio-ulnar joint replacement (Fig. 9.5).

MARS scans have been compared to the standard of care imaging, typically multislice or cone-beam CT. Figure 9.6 presents a MARS scan (left) and a comparison cone-beam CT (right) for a patient with a titanium screw in the scaphoid. In the MARS image the screw and surrounding bone are seen without blur or beam hardening artefact at the bone–metal interface.

For post-operative imaging, high spatial resolution and reduced metal artefact can be clinically significant. An example is presented in Fig. 9.7, where MARS SPCCT was able to demonstrate the end of the screw was positioned within and not outside of the cortical bone. For surgeons, this is critical information. Protrusion of the screw into the radio-carpal joint space can do irreversible harm and leads to costly revision surgery [21].

9.3.4 Assessment of Bone Quality

Bone mineral density (BMD) is the mineral content in bone tissue and is a crucial factor in determining bone strength and integrity. BMD testing is commonly performed to assess fracture risk, aid in the diagnosis of osteoporosis, and monitor the effectiveness of treatment [39, 40]. The clinical standard for assessing BMD is dual-energy X-ray absorptiometry (DXA) which measures areal BMD in g/cm2. However, DXA has limitations. It cannot simultaneously measure cortical and trabecular BMD, nor can it detect subtle variations in trabecular bone density. Additionally, DXA is population-based and may not be ideal for monitoring treatment response in individual patients [41, 42].

Standard single-energy or dual-energy CTs are advantageous as they provide volumetric bone density information (g/cm^3), which provides a more comprehensive assessment compared to DXA. Standard CT can be used to quantify bone micro-architectural changes, but with a low spatial resolution (350–1250 μm), it

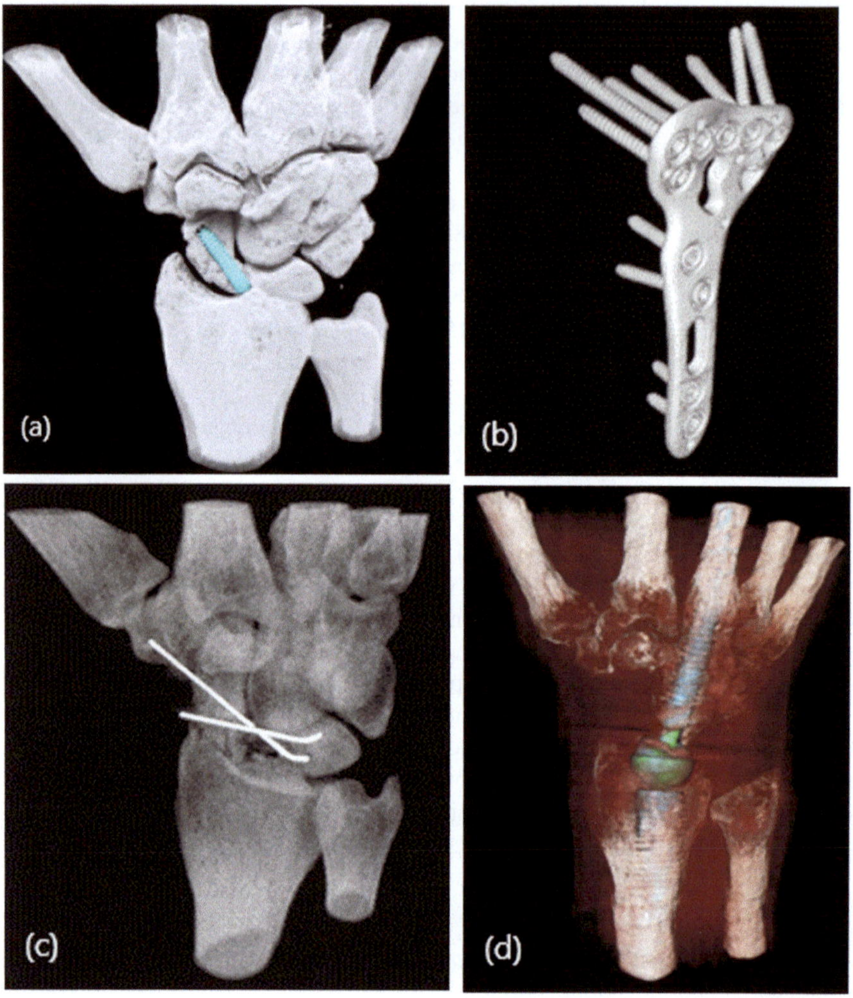

Fig. 9.5 MARS 3D imaging of a range of metal implants. (**a**) A scaphoid screw shown using 3D advanced volume rendering techniques. (**b**) A distal radius plate separated from the surrounding tissue using MARS proprietary software. (**c**) 3D MIP imaging of steel wires following a partial excision of the scaphoid. (**d**) A total wrist joint replacement demonstrated using the bone channel (white) and water channel (red)

presents challenges in terms of the accuracy of structural information. High-resolution peripheral quantitative CT (HR-pQCT) has the potential to overcome these limitations; however, at present it is used primarily in the pre-clinical space [43, 44].

MARS SPCCT provides simultaneous high-resolution images for structural assessment and material maps for evaluating the composition of bone. With a voxel size of 110 μm, MARS achieves a resolution closer to the required level for detailed

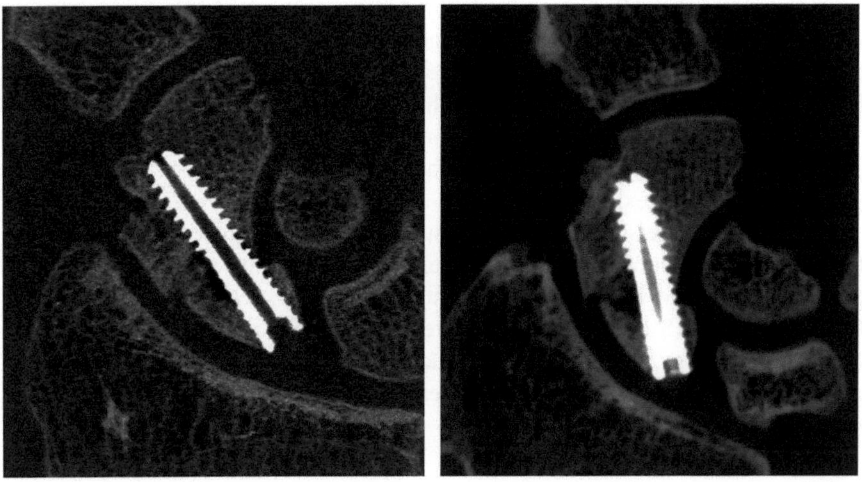

Fig. 9.6 (Left) A scaphoid screw demonstrated in the coronal plane using the MARS 65–120 kV energy channel. (Right) A comparison cone-beam CT scan of the same patient

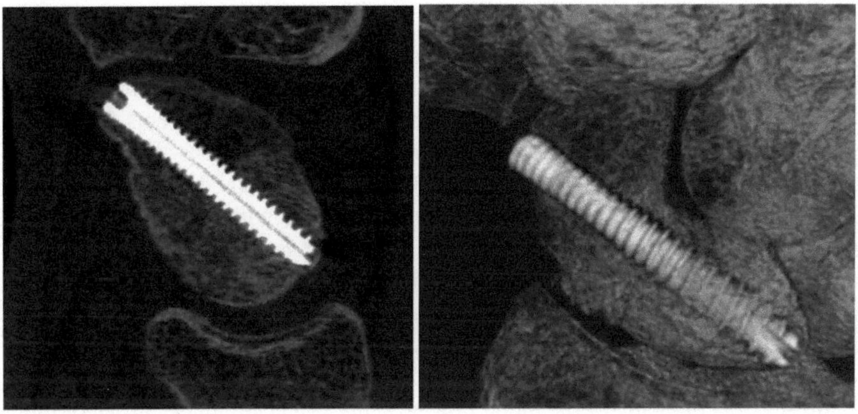

Fig. 9.7 (Left) A scaphoid screw demonstrated in the sagittal oblique plane using the MARS 65–120 kV energy channel. (Right) 3D MARS imaging of the scaphoid screw

visualization of trabecular bone. This higher resolution reduces partial volume averaging, leading to improved material identification and quantification.

The MARS material identification and quantification algorithm generates lipid, water, and bone maps, allowing for detailed assessment of bone composition [45, 46]. These maps are based on mass attenuation values obtained from the linear attenuation and density of materials. Various bone mineral parameters, including total bone mineral density (TBMD), cortical bone mineral density (CBMD), and trabecular bone mineral density (TBMD), can be derived from the hydroxyapatite material maps.

In pathological conditions, such as osteoporosis, changes in trabecular bone volume and shape occur, leading to inhomogeneity in the trabecular region. MARS is capable of capturing these pathological changes, providing valuable insights into bone health. Additionally, by assessing the proportional variation of bone marrow adipose tissue and bone marrow water content, MARS could facilitate quantitative evaluation of bone strength in various pathophysiological conditions, such as diabetes, obesity, and anorexia nervosa [7].

While MARS offers promise for assessing bone health, further research is needed to establish population norms and correlate measurements obtained from other extremity scans and with other prominent fracture sites such as the hip and spine. Overall, the MARS imaging system represents a significant advancement in bone imaging technology, offering detailed structural and compositional information that can enhance our understanding of bone health and aid in the diagnosis and management of bone-related conditions.

9.4 Imaging of Atherosclerotic Plaques and Diagnosis of CVD

The diagnosis and treatment of cardiovascular disease (CVD) is a significant challenge to clinicians, especially in the absence of significant clinical indication, with stroke, heart attacks and kidney failure being the most extreme events for the disease. Cardiovascular disease is the pathological growth of pus-like deposits of cholesterol-filled macrophages and smooth muscles within the wall of arteries called atherosclerotic plaques. These artery plaques can, with time, become significantly large enough to disrupt the flow of blood. The narrowing of the artery lumen by the plaque contributes to raising high blood pressure and reduced blood supply to various tissues such as the heart or limbs. The rupture or erosion of an atherosclerotic plaque exposes the blood to the plaque interior which triggers clot formation which may completely occlude the artery, thus starving the tissue downstream [47]. Clinically, this blocking of the blood flow presents as a heart attack, stroke or limping, depending on the organ affected.

Though a significant amount is known about the histology of such unstable plaques, the key structure features of the plaque structure are extremely challenging to image in situ [48]. Clinically significant carotid artery plaques generally only increase the overall thickness of the artery wall by only 2–4 mm, with the potentially unstable cap area overlying the top of the plaques being only 0.1–0.5 mm in thickness. The X-ray attenuation of the plaque tissue is low and similar to the blood and surrounding tissue. The low X-ray attenuation combined with the relatively small spatial size of the plaques means that they are generally not visible with the energy-integrating detectors used in standard clinical energy-integrating CT systems.

The narrowing caused by advanced plaques is observable with contrast agents by energy integrating CT and Ultrasound, but this imaging gives no histology-level information showing the structure or clinical state of the plaque. Intravascular ultrasound (IVUS) has been used to generate significantly improved plaque imaging, but this technique is invasive and is considered to be a relatively high-risk procedure.

Spectral photon counting-CT (SPCCT) using the MARS V5 and V6 small bore imaging systems have been successfully used to image key features of unstable plaque in excised human plaques taken from the carotid artery of stroke patients [49, 50]. Imaging has been optimised for soft tissues with energy bins between 20 and 120 KeV to identify lipid-, water- and calcium-rich regions. This has allowed the identification of the key features of unstable plaque in excised carotid plaque specimens where thinned plaque cap, extensive lipid-rich necrotic cores, possible plaque haemorrhage and extensive calcification have been imaged. The resulting imaging has been shown to agree with photographic histology of the plaques.

Lipid-rich necrotic cores with an overlying cap, usually made of collagen-rich tissue, are a key feature of unstable plaques (Fig. 9.8). The necrotic core regions of the plaques lack significant structure integrity, as they are made of dying cholesterol-rich cells and extracellular debris released by the death of these cells. Erosion of the overlying cap, seen as a thinning of this structure, is thought to be a precursor to plaque rupture. Both these features of unstable plaque are clearly seen in MARS SPCCT-generated images of excised plaque. Large sheets of calcified tissue can be

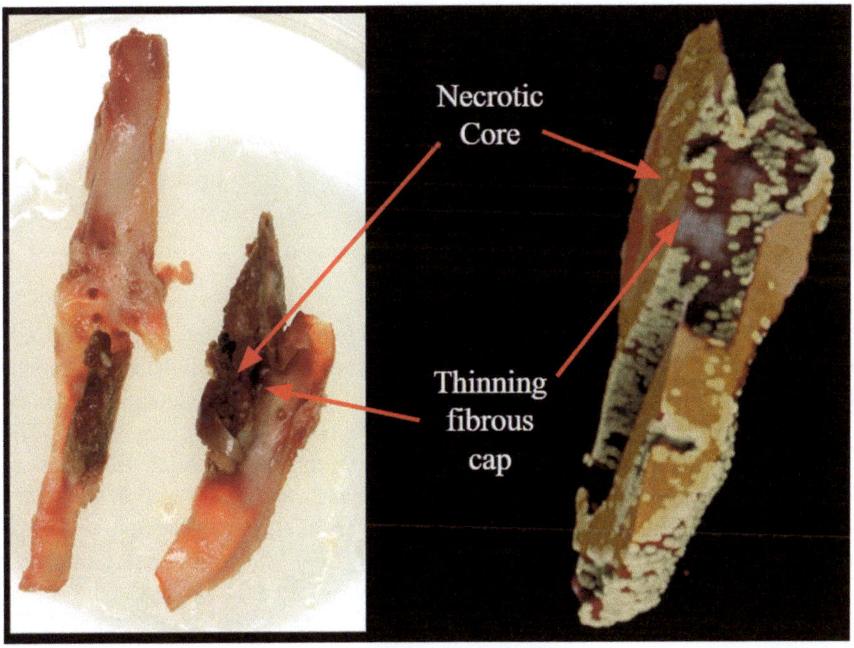

Fig. 9.8 Plaque structure as visualised by MARS SPCCT

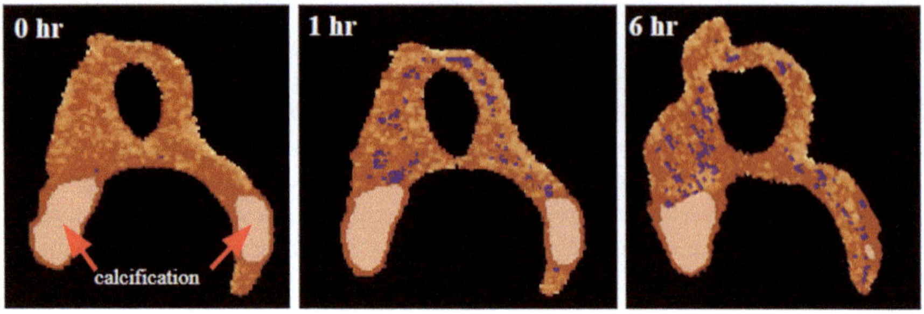

Fig. 9.9 LDL diffusion through live excised atherosclerotic plaque in tissue culture

visualised, as well as sub-voxel-sized spotty calcifications where the calcium X-ray absorbance dominates. The importance of calcification is not fully understood in relation to plaque rupture but does represent an advanced disease stage, which is unlikely to be reversible by drug therapy and requires surgical intervention [51].

The maturing of this imaging technology meant that we routinely use MARS SPCCT systems in our laboratory to non-destructively generate histology-level images of the plaque's internal structure for research. A key question for studies on live cultured plaque slices is how readily various chemicals and agents can diffuse into the tissue such as low-density lipoprotein (LDL). This circulating cholesterol-rich particle is the primary source of cholesterol, driving the formation of the cholesterol-filled cells which make up the necrotic core of the plaques. Its movement through the plaque can be imaged by labelling the LDL particle with iodine. Iodine has a distinct X-ray absorption which is easily visualised by the MARS imaging system (Fig. 9.9). As the imaging has no effect on the plaque tissue, the iodine-labelled LDL could be seen diffusing through the plaque showing that at least 20 nm particles can enter the plaque tissue.

9.5 In Situ Imaging of Artery Plaque

Though informative, imaging of excised plaque is but a stepping stone to SPCCT imaging of atherosclerotic plaque in situ for a patient's clinical diagnosis and management. Using a prototype large bore scanner [4], we have imaged the popliteal artery behind the knee of a patient with extensive peripheral vascular disease. The scan was carried out at a relatively low resolution to reduce the length of patient scanning time (Fig. 9.10). The resulting image clearly shows the wall of the artery and the extensive lipid and calcium deposition. This image was generated without a contrast agent and is possible due to the scanner and image processing systems' ability to distinguish small differences in the tissues' attenuation around the artery. Though the image is not of the highest quality, given the prototype status of the system, it does demonstrate SPCCT systems are capable of generating clinically informative images which could be superior to current energy-integrating CT. The

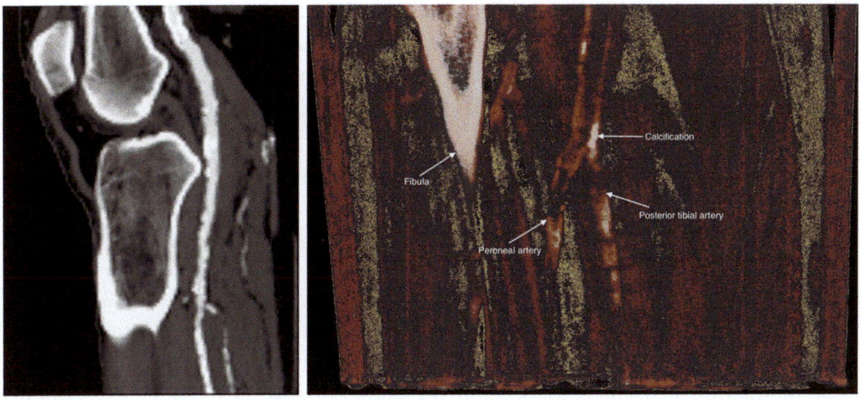

Fig. 9.10 Imaging of patient peroneal artery bifurcation

key feature of this capacity is the combination of submillimetre spatial resolution and good energy resolution into five energy bins. Application of improved camera architecture and processing as used in the latest MARS V6 limb scanners are expected to deliver much higher resolution artery images without contrast agents. This will allow clinicians to measure the progress or regression of plaques under treatment and make better clinical decisions for surgical intervention.

Atherosclerotic plaque obtained from carotid endarterectomy surgery on a stroke patient was sectioned longitudinally and photographed (left image). The right plaque section was imaged using a V5 small bore MARS SPCCT scanner with setting as described in Fig. 9.9. The yellow areas are lipid-rich, red is water-rich tissues and white is calcification.

Plaque tissue taken from a patient during a carotid endarterectomy was taken on ice to the laboratory and sliced into 3 mm thick sections and incubated in tissue culture media. After 24 h the tissue continues to generate lactic acid, indicating the tissue is still alive. LDL purified from pool serum from health donors by ultracentrifugation was labelled with iodine using the iodine monochloride method. The iodine labelled LDL was added to the culture media and incubated for the indicated times before washing in phosphate saline and imaging on a small-bore MARS V5 scanner with 7x CdZnTe-Medipix3RX SPCD, before being returned to the culture media. The blue regions are areas of iodine absorbance.

The X-ray tube current was 30 μA and voltage 118 kVp. Seven hundred twenty projections were taken per rotation with a 100 ms exposure time per projection. The energy bins were 18–26, 26–36, 36–46, 56–118 KeV. Material decomposition was performed using a phantom of water, calcium hydroxyapatite, triacylglycerol (olive oil) and sodium iodide.

A 70-year-old with heavily calcified multifocal stenosing disease of the right femoro-popliteal and crural vessel systems. The left image is a standard CT angiography in the sagittal plane of the patient's knee region. The right image shows the MARS material decomposition image in the coronal plane. The MARS image was obtained using a prototype large bore scanner containing a 14-chip linear array of Medipix3RX ASICs with 2 mm thick CdZnTe (CZT) sensors. The tube current was 345 μA with 50 ms exposure time and 360 projections per rotation. Energy bins were set at 7–25, 25–36, 36–49, 49–65, 65–120 keV [52].

9.6 Summary

MARS Bioimaging Ltd and its research partners have shown the clinical utility of the MARS SPCCT imaging systems. High-resolution imaging of cortical and trabecular bone is key for orthopaedic applications where an accurate visual assessment of fracture healing is significant for clinical decision-making. For patients with metal implants, MARS SPCCT provides excellent resolution at the bone–metal interface. This is critical for early identification of complications in an area, which is traditionally affected by beam hardening artefact.

Bone quality evaluation extends beyond visual assessment. Researchers and ultimately clinicians can use material maps for quantitative measures of bone health. The metrics have many applications including, for example, the evaluation of avascular necrosis, osteoporosis and monitoring response to treatment.

Pre-clinical research is underway to validate the use of MARS SPCCT for rheumatology, particularly gout and pseudogout, for vascular applications such as the identification and characterisation of carotid plaques and for metabolic imaging using nanoparticles.

In the future, the design of point-of-care MARS scanners for lower extremity and head and neck imaging will further extend these clinical applications.

Acknowledgments This project received funding from the Ministry of Business, Innovation and Employment (MBIE), New Zealand under contract number UOCX1404, by MARS Bioimaging Ltd and the Ministry of Education through the MedTech CoRE. The authors would like to acknowledge the Medipix 2, Medipix 3 and Medipix 4 collaborations. Atherosclerotic imaging was partially funded by the New Zealand Neurological Foundation under a project grant. Devyani Dixit was supported through a Callaghan Innovation R&D fellowship. Also we would like to acknowledge the work of the MARS collaboration.

MARS Collaboration: Maya R Amma[a], Marzieh Anjomrouz[a], Ali Atharifard[a], James Atlas[b], Anthony P H Butler[a, b, c, d, e], Philip H Butler[a, b, c, d, e], Pierre Carbonez[c, d], Krishna M Chapagain[a,c], Alexander I Chernoglazov[a, e], Jennifer A Clark[a,c,] Jonathan S Crighton[c, k], Shishir Dahal[c, h, i], Jérôme Damet[c, d, j], Lucia Gallego-Manzano[j], Yann Sayous[b], Theodorus Dapamede[c], Niels J A de Ruiter[a, b, c, e], Devyani Dixit[a,b], Karen Dombroski[a], Neryda Duncan[a], Steven P Gieseg[b, c, d], Joseph L Healy[a], Kevin Jonker[a,b], Tracy Kirkbride[f], V B H Mandalika[a,e], Mahdieh Moghiseh[a, c], Jereena S Sheeja[a], Aaron Smith[b], L, Vivek V S[a], E Peter Walker[c], Fatemah Agariomabad[b,] Emily Searle[b]

[a] MARS Bioimaging Limited, Christchurch, New Zealand

[b] University of Canterbury, Christchurch, New Zealand

[c] University of Otago Christchurch, Christchurch, New Zealand

[d] European Organisation for Nuclear Research (CERN), Geneva, Switzerland

[e] Human Interface Technology Laboratory New Zealand, University of Canterbury, Christchurch, New Zealand

[f] Ara Institute of Canterbury, Christchurch, New Zealand

[g] Lincoln University, Lincoln, New Zealand
[h] Ministry of Health, Nepal
[i] National Academy of Medical Sciences, Kathmandu, Nepal
[j] Institute of Radiation Physics, Lausanne University Hospital, Lausanne, Switzerland
[k] Te Whatu Ora—Health New Zealand Waitaha Canterbury.

References

1. Ballabriga R, Alozy J, Blaj G, Campbell M, Fiederle M, Frojdh E, et al. The medipix3RX: a high resolution, zero dead-time pixel detector readout chip allowing spectroscopic imaging. J Instrum. 2013;8(2) https://doi.org/10.1088/1748-0221/8/02/C02016.
2. Aamir R, Chernoglazov A, Bateman CJ, Butler APH, Butler PH, Anderson NG, et al. MARS spectral molecular imaging of lamb tissue: data collection and image analysis. J Instrum. 2014;9(2) https://doi.org/10.1088/1748-0221/9/02/P02005.
3. Moghiseh M, Clark JA, Amma MR, Chapagain KM, Dixit D, Lowe C, et al. MARS for molecular imaging and preclinical studies. In: Photon counting computed tomography: clinical applications, image reconstruction and material discrimination; 2023. https://doi.org/10.1007/978-3-031-26062-9_4.
4. Panta RK, Butler APH, Butler PH, De Ruiter NJA, Bell ST, Walsh MF, et al. First human imaging with MARS photon-counting CT. In: 2018 IEEE nuclear science symposium and medical imaging conference, NSS/MIC 2018—proceedings; 2018. https://doi.org/10.1109/NSSMIC.2018.8824513.
5. Willemink MJ, Persson M, Pourmorteza A, Pelc NJ, Fleischmann D. Photon-counting CT: technical principles and clinical prospects. Radiology. 2018;289(2) https://doi.org/10.1148/radiol.2018172656.
6. Becce F, Viry A, Stamp LK, Pascart T, Budzik JF, MARS Collaboration, et al. Winds of change in imaging of calcium crystal deposition diseases. Joint Bone Spine. 2019;86(6):665–8.
7. Clark JA, Chapagain KM, Amma MR, Moghiseh M, Lowe C, Butler APH. MARS for orthopaedic pathology. In: photon counting computed tomography: clinical applications, image reconstruction and material discrimination; 2023. https://doi.org/10.1007/978-3-031-26062-9_3.
8. Geijer M. Diagnosis of scaphoid fracture: optimal imaging techniques. Rep Med. Imaging. 2013;6(1) https://doi.org/10.2147/RMI.S48951.
9. Kaewlai R, Avery LL, Asrani AV, Abujudeh HH, Sacknoff R, Novelline RA. Multidetector CT of carpal injuries: anatomy, fractures, and fracture-dislocations. Radiographics. 2008;28(6):1771–84.
10. Dorsay TA, Major NM, Helms CA. Cost-effectiveness of immediate MR imaging versus traditional follow-up for revealing radiographically occult scaphoid fractures. Am J Roentgenol. 2001;177(6):1257–63.
11. Jenkins PJ, Slade K, Huntley JS, Robinson CM. A comparative analysis of the accuracy, diagnostic uncertainty and cost of imaging modalities in suspected scaphoid fractures. Injury. 2008;39(7):768–74.
12. Bhat M, McCarthy M, Davis TRC, Oni JA, Dawson S. MRI and plain radiography in the assessment of displaced fractures of the waist of the carpal scaphoid. J Bone Joint Surg Br. 2004;86(5):705–13.
13. Ahn JM, El-Khoury GY. Occult fractures of extremities. Radiol Clin North Am. 2007;45(3):561–79. ix
14. Gelberman RH, Wolock B, Siegel DB. Current concepts review: fractures and non-unions of the carpal scaphoid. J Bone Joint Surg Am. 1989;71(10):1560–5.
15. Gupta V, Rijal L, Jawed A. Managing scaphoid fractures. How we do it? J Clin Orthop Trauma. 2013;4(1):3–10.

16. Mallee WH, Wang J, Poolman RW, Kloen P, Maas M, de Vet HCW, et al. Computed tomography versus magnetic resonance imaging versus bone scintigraphy for clinically suspected scaphoid fractures in patients with negative plain radiographs. Cochrane Database Syst Rev. 2015;2015(6):CD010023.
17. Corrales LA, Morshed S, Bhandari M, Miclau T III. Variability in the assessment of fracture-healing in orthopaedic trauma studies. J Bone Joint Surg Am. 2008;90(9):1862.
18. Davis BJ, Roberts PJ, Moorcroft CI, Brown MF, Thomas PBM, Wade RH. Reliability of radiographs in defining union of internally fixed fractures. Injury. 2004;35(6):557–61.
19. Amrami KK, Frick MA, Matsumoto JM. Imaging for acute and chronic scaphoid fractures. Hand Clin. 2019;35(3):241–57.
20. Soufi M, See A, Hassan S. Scaphoid fractures and non-union: a review of current evidence. Orthop Trauma. 2021;35(4):198–207.
21. Dias JJ, Ring D, Grewal R, Clementson M, Buijze GA, Ho P-C. Acute scaphoid fractures: making decisions for treating a troublesome bone. J Hand Surg (European Volume). 2022;47(1):73–9.
22. Pawelec K, Planell JA. Bone repair biomaterials: regeneration and clinical applications; 2018.
23. Prasad K, Bazaka O, Chua M, Rochford M, Fedrick L, Spoor J, et al. Metallic biomaterials: current challenges and opportunities. Materials. 2017;10(8):884.
24. Shah FA, Omar O, Suska F, Snis A, Matic A, Emanuelsson L, et al. Long-term osseointegration of 3D printed CoCr constructs with an interconnected open-pore architecture prepared by electron beam melting. Acta Biomaterialia. 2016;36:296–309.
25. Wirth AJ, Goldhahn J, Flaig C, Arbenz P, Müller R, van Lenthe GH. Implant stability is affected by local bone microstructural quality. Bone. 2011;49(3):473–8.
26. Cyteval C, Bourdon A. Imaging orthopedic implant infections. Diagn Interv Imaging. 2012;93(6):547–57.
27. Boas FE, Fleischmann D. CT artifacts: causes and reduction techniques. Imaging Med. 2012;4(2) https://doi.org/10.2217/iim.12.13.
28. Rinkel J, Dillon WP, Funk T, Gould R, Prevrhal S. Computed tomographic metal artifact reduction for the detection and quantitation of small features near large metallic implants: a comparison of published methods. J Comput Assist Tomogr. 2008;32(4) https://doi.org/10.1097/RCT.0b013e318149e215.
29. Meyer E, Raupach R, Lell M, Schmidt B, Kachelrieß M. Normalized metal artifact reduction (NMAR) in computed tomography. Med Phys. 2010;37(10) https://doi.org/10.1118/1.3484090.
30. Meyer E, Raupach R, Lell M, Schmidt B, Kachelrieß M. Frequency split metal artifact reduction (FSMAR) in computed tomography. Med Phys. 2012;39(4) https://doi.org/10.1118/1.3691902.
31. Sun H, Qiu S, Lou S, Liu J, Li C, Jiang G. A correction method for nonlinear artifacts in CT imaging. In: Annual international conference of the IEEE engineering in medicine and biology—proceedings, 26 II; 2004. https://doi.org/10.1109/iembs.2004.1403407.
32. Bamberg F, Dierks A, Nikolaou K, Reiser MF, Becker CR, Johnson TRC. Metal artifact reduction by dual energy computed tomography using monoenergetic extrapolation. Eur Radiol. 2011;21(7) https://doi.org/10.1007/s00330-011-2062-1.
33. Wellenberg RHH, Hakvoort ET, Slump CH, Boomsma MF, Maas M, Streekstra GJ. Metal artifact reduction techniques in musculoskeletal CT-imaging. Eur J Radiol. 2018;107 https://doi.org/10.1016/j.ejrad.2018.08.010.
34. Yazdi M, Mansourian Z. Metal artifact reduction in spiral fan-beam CT using a new sinogram segmentation scheme. J X-Ray Sci Technol. 2017;25(5) https://doi.org/10.3233/XST-16224.
35. Aboelmagd SM, Malcolm PN, Toms AP. Magnetic resonance imaging of metal artifact reduction sequences in the assessment of metal-on-metal hip prostheses. Rep Med Imaging. 2014;7(1) https://doi.org/10.2147/RMI.S40052.
36. Hargreaves BA, Worters PW, Pauly KB, Pauly JM, Koch KM, Gold GE. Metal-induced artifacts in MRI. Am J Roentgenol. 2011;197(3) https://doi.org/10.2214/AJR.11.7364.

37. Rajendran K, Walsh MF, De Ruiter NJA, Chernoglazov AI, Panta RK, Butler APH, et al. Reducing beam hardening effects and metal artefacts in spectral CT using Medipix3RX. J Instrum. 2014;9(3):P03015. https://doi.org/10.1088/1748-0221/9/03/P03015.
38. Tortora M, Gemini L, D'Iglio I, Ugga L, Spadarella G, Cuocolo R. Spectral photon-counting computed tomography: a review on technical principles and clinical applications. J Imaging. 2022;8(4):112. https://doi.org/10.3390/jimaging8040112.
39. Felsenberg D, Boonen S. The bone quality framework: determinants of bone strength and their interrelationships, and implications for osteoporosis management. Clin Ther. 2005;27(1) https://doi.org/10.1016/j.clinthera.2004.12.020.
40. Srinivasan B, Kopperdahl DL, Amin S, Atkinson EJ, Camp J, Robb RA, et al. Relationship of femoral neck areal bone mineral density to volumetric bone mineral density, bone size, and femoral strength in men and women. Osteoporos Int. 2012;23(1) https://doi.org/10.1007/s00198-011-1822-8.
41. Briggs AM, Perilli E, Parkinson IH, Wrigley TV, Fazzalari NL, Kantor S, et al. Novel assessment of subregional bone mineral density using DXA and pQCT and subregional microarchitecture using micro-CT in whole human vertebrae: applications, methods, and correspondence between technologies. J Clin Densitom. 2010;13(2) https://doi.org/10.1016/j.jocd.2010.01.120.
42. Perilli E, Briggs AM, Kantor S, Codrington J, Wark JD, Parkinson IH, et al. Failure strength of human vertebrae: prediction using bone mineral density measured by DXA and bone volume by micro-CT. Bone. 2012;50(6) https://doi.org/10.1016/j.bone.2012.03.002.
43. Cheung AM, Adachi JD, Hanley DA, Kendler DL, Davison KS, Josse R, et al. High-resolution peripheral quantitative computed tomography for the assessment of bone strength and structure: a review by the Canadian bone strength working group. Curr Osteoporos Rep. 2013;11(2) https://doi.org/10.1007/s11914-013-0140-9.
44. Zhou B, Wang J, Yu YE, Zhang Z, Nawathe S, Nishiyama KK, et al. High-resolution peripheral quantitative computed tomography (HR-pQCT) can assess microstructural and biomechanical properties of both human distal radius and tibia: ex vivo computational and experimental validations. Bone. 2016;86 https://doi.org/10.1016/j.bone.2016.02.016.
45. Bateman CJ, Knight D, Brandwacht B, Mahon JM, Healy J, Panta R, et al. MARS-MD: rejection based image domain material decomposition. J Instrum. 2018;13(5) https://doi.org/10.1088/1748-0221/13/05/P05020.
46. de Ruiter NJA, Butler PH, Butler APH, Bell ST, Chernoglazov AI, Walsh MF. MARS imaging and reconstruction challenges. In: The 14th International meeting on fully three-dimensional image reconstruction in radiology and nuclear medicine, 14 June 2017.
47. Geovanini GR, Libby P. Atherosclerosis and inflammation: overview and updates. Clin Sci (Lond). 2018;132(12):1243–52.
48. Saba L, Saam T, Jäger HR, Yuan C, Hatsukami TS, Saloner D, et al. Imaging biomarkers of vulnerable carotid plaques for stroke risk prediction and their potential clinical implications. Lancet Neurol. 2019;18(6):559–72.
49. Dahal S, Raja AY, Searle E, Colgan FE, Crighton JS, Roake J, et al. Components of carotid atherosclerotic plaque in spectral photon-counting CT with histopathologic comparison. Eur Radiol. 2023;33(3):1612–9.
50. Healy J, Searle E, Panta RK, Chernoglazov A, Roake J, Butler P, et al. Ex-vivo atherosclerotic plaque characterization using spectral photon-counting CT: comparing material quantification to histology. Atherosclerosis. 2023;378:117160.
51. Saba L, Nardi V, Cau R, Gupta A, Kamel H, Suri JS, et al. Carotid artery plaque calcifications: lessons from histopathology to diagnostic imaging. Stroke. 2022;53(1):290–7.
52. Dahal S. Characterization of atherosclerotic plaque with MARS imaging. Dunedin: University of Otago; 2021.

Chapter 10
Photon-Counting Computed Tomography: Clinical Applications, Detector Technology, and Artificial Intelligence

Devon Richtsmeier, Xinchen Deng, Pierre-Antoine Rodesch, Joanna Nguyen, James Day, Krzysztof Kris Iniewski, William Siu, and Magdalena Bazalova-Carter

Contents

D. Richtsmeier · X. Deng · J. Day · M. Bazalova-Carter
University of Victoria, Victoria, BC, Canada

P.-A. Rodesch
University of Victoria, Victoria, BC, Canada

Lausanne University Hospital and Lausanne University, Lausanne, Switzerland

J. Nguyen
University of Victoria, Victoria, BC, Canada

GE HealthCare, Waukesha, WI, USA

K. K. Iniewski (✉)
Redlen Technologies, Saanichton, BC, Canada

W. Siu
Royal Columbian Hospital, New Westminster, BC, Canada

S. A. Si-Mohamed, K. K. Iniewski (eds.), *Clinical Applications of Spectral Photon Counting Computed Tomography Technology*,
https://doi.org/10.1007/978-3-031-85078-3_10

10.1 Introduction

Photon-counting computed tomography (PCCT) marks a significant leap forward in medical imaging, presenting considerable improvements over traditional CT technologies. PCCT utilizes photon-counting detectors (PCDs) to count individual photons, significantly enhancing the image resolution and reducing the patient's radiation exposure [1]. These advancements allow for more precise diagnostics and the ability to capture high-quality images at lower doses than energy-integrating detectors (EIDs) used in conventional CT scanners [2–8].

PCCT is a diagnostic imaging system with a photon-counting detector that can identify multiple material compositions. It is expected to improve diagnostic accuracy by providing superior images quantitatively. In addition, with the higher resolution of PCCT, the detectability of lesions should be improved at even lower exposure doses than with conventional systems. Based on these advanced system capabilities, PCCT is expected to become a next-generation CT imaging technique capable of providing outstanding clinical value.

The underlying technology of PCCT leverages PCDs to distinguish each photon's energy level. This specificity can be crucial for differentiating between tissues and materials within the body. Furthermore, reducing radiation exposure is a critical advancement, considering the potential risks associated with higher doses of radiation to patients over time. By minimizing the dose required for high-quality diagnostic imaging, PCCT addresses one of the primary concerns in medical imaging practices, offering a safe diagnostic tool for patients. Moreover, the enhanced image resolution and quality afforded by PCCT can be particularly beneficial in detecting and diagnosing complex medical conditions, including cancers, cardiovascular diseases, and skeletal disorders [3]. PCCT can help clinicians identify diseases at earlier stages and with greater accuracy, improving treatment outcomes.

In summary, PCCT, through its innovative use of photon-counting detectors, significantly improves medical imaging technology. It provides high-resolution images and reduces patients' radiation exposure, offering a more precise and safer diagnostic tool. This advancement holds the promise of revolutionizing medical imaging, with wide-ranging implications for patient care and diagnostics.

10.2 Photon-Counting Technology

X-ray detectors are the major and arguably most important component of a clinical CT system and have a substantial impact on image quality and patient radiation dose, consequently affecting a wide range of clinical applications. Compared to the conventional scintillator-based CT detectors using an indirect conversion technology that integrates photon energy deposited by all received x-ray photons, photon-counting detectors possess many inherent advantages, largely attributed to their direct conversion technology and energy discrimination capability. Although PCD technology has been widely utilized in nuclear medicine in the last several years, its adoption in X-ray CT systems has long been hampered by the high photon flux effects encountered in clinical CT imaging. The recent dramatic advances in CZT (Cadmium Zinc Tellurium) technology have overcome these problems, and CZT-based Photon Counting Computed Tomography (PCCT) is now a clinical reality!

In this chapter, we will discuss the potential clinical benefits of PCD-CT that have been demonstrated in a series of recent studies. These include:

- High spatial resolution and ability to distinguish plaque in cardiovascular applications in particular.
- Increased iodine contrast-to-noise ratio (CNR), increased dose efficiency and reduced electronic noise.
- Reduced beam-hardening and metal artifacts, enabling more certain clinical diagnoses.
- K-edge contrast imaging simultaneous multi-contrast agents for future emerging clinical applications.

In the following sections, we will then discuss how these benefits can impact CT applications in each clinical area, such as cardiovascular, neuro and musculoskeletal imaging.

10.2.1 High Spatial Resolution

Detectors with high spatial resolution enable clinicians to see smaller details like plaque in coronary arteries much better, and this is one of the most important selling features of PCCT scanners. Spatial resolution is much better in photon counting detectors than in energy-integrated detectors due to much smaller pixels, typically by a factor of 10 in the pixel area, used in PCCT. Conventional EID detectors incorporate a reflective layer (septa) of a finite width thickness between each scintillating detector pixel along the detector row and channel directions. Since scintillating detectors create visible light in the process of X-ray detection, the septa are designed to prevent leakage of visible light between adjacent detector pixels. Figures 10.1 and 10.2 illustrate the contrast between EID and PCCT detectors.

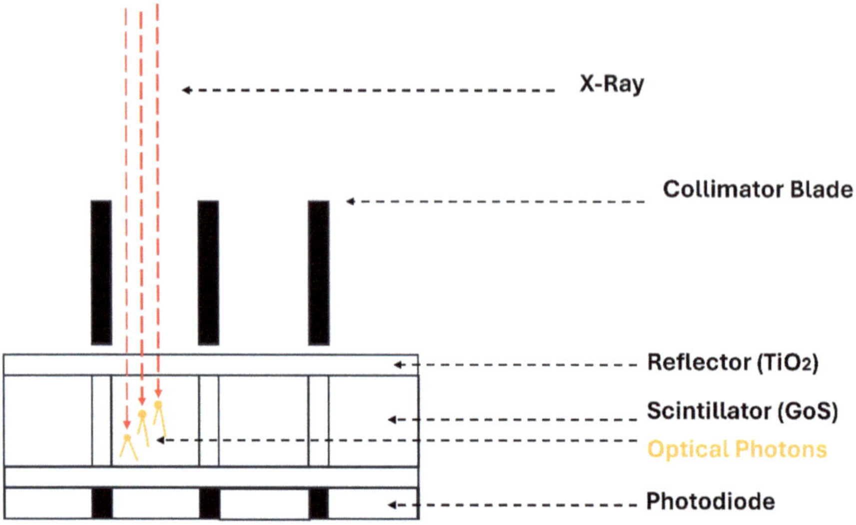

Fig. 10.1 Schematic cross-section though EID scintillation indirect detector

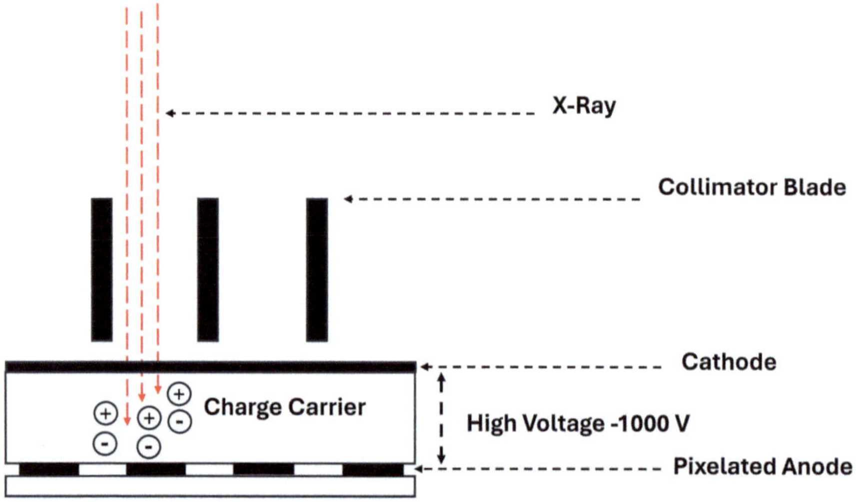

Fig. 10.2 Schematic cross-section through PCCT direct conversion detector

While it is possible to create smaller EID pixels, there is a physical limit to the width of the septa. Smaller detector pixels with finite-width septa lead to reduced geometric dose efficiency due to the reduced detector fill factor. Today's EID scanners typically utilize a 1 mm pixel pitch, translating into roughly 580 μm spatial resolution at the ISO center. PCCT scanners typically utilize 330 μm pixels, which translates into roughly 190 μm spatial resolution at the ISO center.

Further improvement in EID scanner spatial resolution can be obtained using comb filters that are placed near the detector to reduce the pixel aperture, resulting in smaller pixel sizes for ultra-high-resolution imaging. Though this approach is routinely used for specific imaging tasks such as the CT scan of the inner ear, it is a highly dose-inefficient approach as the comb filter is placed in front of the detector array, X-rays blocked by the filter have already passed through the patient, contributing to patient dose.

This spatial resolution limitation is overcome in direct conversion PCCT technology, where no wall between detector pixels is necessary. The schematic shown in Fig. 10.2 shows a cross-section of a PCCT direct-conversion detector. This design allows PCCT detectors to contain smaller pixel pitches, ranging between 300 and 500 μm, without losing fill factor or dose efficiency.

10.2.2 Reduced Electronic Noise and Better Signal-to-Noise Ratio (SNR)

High electronic noise is a permanent feature of the conventional EID technology as the energy-integrated detectors integrate all noise into the signal by their principle of operation. The first obvious advantage of PCCT detectors is lower electronic noise. Lower noise leads to higher image quality and lower radiation dose.

Image noise in CT mainly originates from two sources: quantum noise and electronic noise. The quantum noise, determined by the number of detected photons, can be traced back to the random nature of X-ray photon interactions. The electronic noise comes from the detector's electronic circuits within the X-ray detection system and is, therefore, to a first degree, independent of the number of photons reaching the detector. The relative significance of quantum and electronic noise is determined by the incident photon flux and clinical parameters. In practice, this is affected by the patient's size; larger patients result in a very small number of photons reaching the detector.

Photon counting CT has a distinct advantage over EID CT when handling electronic noise. As illustrated in Fig. 10.3, in the PCD case, directly counting from the electronic noise can be avoided by setting an electronic counting threshold (typically at 20 keV, but it can be adjusted by a clinician) just above the level of electronic noise, making PCCT almost "noise-free." In contrast, this thresholding is not possible in the EID case, where the low-energy electronic noise is integrated together with the useful signal, thereby reducing the signal-to-noise ratio (SNR).

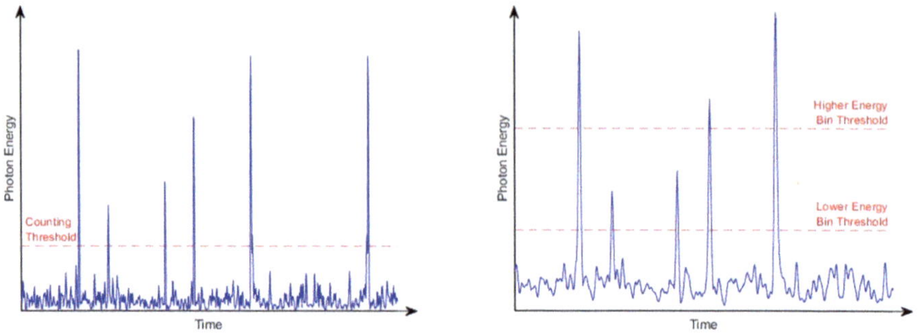

Fig. 10.3 Schematic illustration of photon electron pulses processed by PCCT detector. With one threshold to eliminate electronic noise (left) or two (or more) thresholds to create energy bins for additional material discrimination

10.2.3 Lower Radiation Dose and Iodine Contrast

Computed tomography tests frequently utilize contrast agents to increase contrast to noise ratio (CNR). Contrast agents increase CNR through the increased attenuation at the K-edge of the material located within the applied X-ray spectrum. The most common contrast agent is Iodine, which has a K-edge position at 33.2 keV. Iodinated contrast, such as iohexol, is a water-soluble, intravenous radiocontrast agent containing iodine, which enhances the visibility of vascular structures and organs during radiographic procedures. Many pathologies, such as cancer, have remarkably improved visibility with iodinated contrast due to blood pooling from poorly formed vasculature near the tumor volume. The radiodensity of iodinated contrast is 25–30 Hounsfield units (HU) per milligram of iodine per milliliter at a tube voltage of 100–120 kVp.

PCCT detectors show great advantages over EIDs in iodine detection for the following reasons. In the diagnostic X-ray energy range, the X-ray photons primarily interact with scanned objects through two physical mechanisms: the photoelectric effect and Compton scattering. Theoretically, the attenuation due to the photoelectric effect is proportional to the effective atomic number of a material and is inversely proportional to the energy of the incident X-ray photon. On the other hand, the Compton scattering has little energy dependence. As a result, its attenuation coefficient remains constant throughout the diagnostic energy range. Consequently, the photoelectric effect contributes more to the attenuation of low-energy photons, while Compton scattering is the dominant effect accounting for high-energy photons. The photoelectric effect is not solely decreasing with energy for all materials; this mechanism is also affected by the number of participating electron shells in the atom; for high Z materials, one or more energy thresholds may exist, resulting in another shell participating in attenuation. For iodine, the k-shell begins participating at 33.2 keV, resulting in a sudden spike in the attenuation coefficient. This is known as the K-edge. As a result, high-Z materials, such as iodine, can have a higher photon attenuation in the low-energy range due to the photoelectric effect at the k-edge and, therefore, demonstrate higher signal or contrast in CT images.

In conventional EID detectors, the detector signal is proportional to the total energy of all detected X-ray photons. Due to the energy-integrating nature of EID, lower-energy photons contribute less to the detector signal than higher-energy photons. This creates what is known as an energy-fluence-weighted image. However, higher-energy photons have less contrast information content from high-Z materials such as iodine, which has the greatest contrast information in the lower energies for the 100–120 kVp spectra used in CT. Compared to a fluence-weighted image, the energy-fluence-weighted image from an EID has lower dose efficiency when generating iodine contrast. PCCT detectors operate on fluence-weighted imaging; they count each individual photon equally regardless of the photon energy, without the energy weighting as in the case of EID. Hence, the low-energy photons have a greater contribution to the image contrast for PCD-CT than EID-CT, improving the iodine contrast and CNR. For similar reasons radiation dose in PCCT can be reduced because of the improved dose efficiency. This advantage of PCCT is sometimes referred to as the better Swank factor and has been shown to produce 30–40% gains in CNR values for the same dose and can be translated to a similar reduction in radiation dose. The following section discusses test cases of detailed studies of PCCT vs. EID dose comparison.

10.2.4 Managing Beam Hardening and Metal Artifacts

Beam hardening is a well-known effect in X-ray imaging. As photons pass through the scanned object, low-energy photons are preferentially attenuated compared to high-energy photons. Since poly-energetic beams are used in CT, the effective photon energy is shifted toward the higher end of the spectrum. This introduces artifacts, which are typically dark areas adjacent to highly attenuating objects such as cortical bone, affecting the image appearance and CT number accuracy for nearby soft tissues. In the presence of foreign objects such as metal implants, severe image streaking artifacts together with dark/bright regions can be observed across the image.

These metal artifacts have a characteristic appearance and are caused by multiple physical mechanisms, including X-ray scattering, photon starvation, and beam hardening effect. The severity of metal artifacts can vary according to the atomic number, density, and shape of the metal, as well as the surrounding anatomy and scan acquisition/reconstruction parameters. In many cases, the resulting metal artifacts can dramatically obscure critical structures, significantly reducing diagnostic confidence in various clinical tasks, such as distinguishing pathologic findings from normal structures.

Since each individual photon is sorted according to its energy in a PCCT acquisition, an energy-bin image can be reconstructed using only higher-energy photons. Compared to the conventional EID-CT image or the low-energy threshold image in a PCCT acquisition, the high-energy bin image is more immune to the beam hardening effect in areas around dense bones. In the follow-up test case section, we are reporting on a research study of metal artifact reduction.

10.3 Clinical Applications

Due to extensive research over the last few years, the clinical benefits of photon-counting CT (PCCT) are becoming more evident. As photon-counting detectors (PCDs) directly record individual photons and their associated energy, the technology offers multiple benefits over CT imaging using conventional energy-integrating detectors. Clinically, PCCT affords clinicians higher contrast (clearer images) and improved spatial resolution (finer details) at a reduced dose (less radiation). In this section, we will glimpse the clinical benefits of PCCT technology powered by CdTe/CZT photon counting detectors [9–23].

10.3.1 Cardiovascular Imaging

Conventional CT systems can rule out coronary artery disease in relatively large normal vessels. Still, their specificity, the ability to determine the artery is not diseased when no disease is present, remains low, as precisely measuring the degree of stenosis requires very high spatial resolution. In addition, calcified plaques are often affected by blooming artifacts. The increased spatial resolution of PCCT allows for increased detectability, more accurate coronary artery calcium scoring (CAC), and the minimization of blooming artifacts.

There are specific technical issues when imaging coronary stents with current CT scanners in clinical practice. These include metallic, blooming, and beam-hardening artifacts, in addition to partial volume effects due to relatively low spatial resolution. With conventional CT scanners, the lumen of the coronary stent can be visualized accurately in a still unsatisfactory percentage of cases. This problem is resolved when using PCCT technology.

PCCT's superior image quality may produce new diagnostic capabilities and enable novel clinical applications. Examples include a clear assessment of stenosis, determination of aortic stent endo leaks, and surveillance after aortic endovascular aneurysm repair. PCCT imaging can also improve patient management by lowering Coronary Artery Disease Reporting and Data System (CAD-RADS) scores for patients with dense calcifications and more efficiently identifying cerebrospinal fluid leaks. As cardiovascular health is arguably the number one issue facing modern humanity, many improvements in cardiovascular imaging using PCCT can be expected in years to come.

10.3.2 Pulmonary Imaging

PCCT has the potential to improve the detection of pulmonary nodules and the assessment of other lung structures. In particular, the high spatial resolution of PCCT, compared to conventional EID scanners in current clinical use, leads to a more precise assessment of nodules and the smallest pulmonary structures, such as the terminal divisions of the bronchial tree. PCCT guarantees a strong improvement in the detectability of both lesions with low contrast compared to surrounding parenchyma and lesions with high contrast.

PCCT can revolutionize the role of imaging in detecting interstitial lung disease and pathological key signs, including intra-lobular reticulations, bronchiectasis, and honeycombing. Small pulmonary features could be visualized more distinctly with PCCT.

10.3.3 Musculoskeletal Imaging

PCCT technology has many potential applications in the domain of musculoskeletal imaging. For example, its material decomposition capabilities can be used to reconstruct calcium-free images to evaluate bone edema. PCCT can detect and characterize the deposition of calcium crystals within the articular cartilage with high sensitivity and good specificity. PCCT can also distinguish between calcium pyrophosphate and hydroxyapatite deposits, providing further insight into rheumatic diseases and aiding in identifying new potential therapeutic targets.

PCCT is also very useful for cartilage imaging. Due to its avascular nature and limited capability for regeneration, an acute joint injury may induce post-traumatic osteoarthritis with a loss of proteoglycans, breakdown of fibers, and increased water content. By estimating the concentration of two contrast agents, PCCT could provide information directly indicative of the water and proteoglycan content in the articular cartilage. Therefore, contrast-enhanced PCCT imaging has the potential use for the assessment of articular cartilage health and monitoring.

PCCT is also helpful in Osteoarthritis treatment. Osteoarthritis represents the most common cause of disability, and it can be managed by arthroplasty, especially for the knee. Establishing the cause of symptoms and pain is fundamental to proper management, as revision surgery with an uncertain diagnosis has a poor outcome. The currently available imaging techniques are important, but each modality presents its own limitations. PCCT could clearly identify polyethylene inserts and metallic tray wear as the causes of prosthesis failure, and that would offer significant improvement over current imaging practices.

10.3.4 Abdominal Imaging

Diagnostic imaging of liver disease is very common in clinical practice, often requiring a dedicated multiphase CT acquisition protocol for both benign and malignant lesions. PCCT offers the ability to depict characteristic arterial and portal venous enhancement with a single scan, allowing for streamlined lesion detection and characterization with additional benefits regarding reduced radiation exposure dose.

Although CT imaging offers a high spatial and temporal resolution, its main drawback is the lack of contrast resolution. Therefore, PCCT, with its super contrast detection, offers important advantages for peritoneal imaging. If pre-clinical results translate to clinical practice, the higher sensitivity provided by PCCT could avoid unnecessary "open-close" surgical procedures.

10.3.5 Neck Imaging

PCCT can potentially improve the quality of carotid and intracranial angiography imaging compared to conventional EID CT. By minimizing electronic noise and beam-hardening artifacts, spectral material decomposition is feasible for vascular imaging, also offering a reduction in overall radiation dose.

PCCT also offers important improvements in the head and neck district, such as the staging of laryngeal and hypo-pharyngeal cancer, for which it is essential to identify the degree of invasion of the laryngeal cartilage. Since the density of non-ossified laryngeal cartilage is similar to that of tumors in conventional CT imaging, greater spatial resolution, material decomposition, and differentiation techniques allowed by PCCT may lead to improved lesion local staging.

10.4 PCCT Imaging Test Cases

Test cases illustrate PCCT technology's advantages in laboratory and real-life examples. In this section, we will illustrate some of the reported pre-clinical and clinical test cases obtained using Redlen's PCCT detectors (www.redlen.com).

10.4.1 Spatial Resolution and Dose Reduction Study

The following section provides examples of a comparative study between Canon's CZT-based PCCT system and scintillator-based EID-CT systems [9, 10]. Phantom scans using the setup shown in Fig. 10.4 were acquired using 120 kVp from 50 to

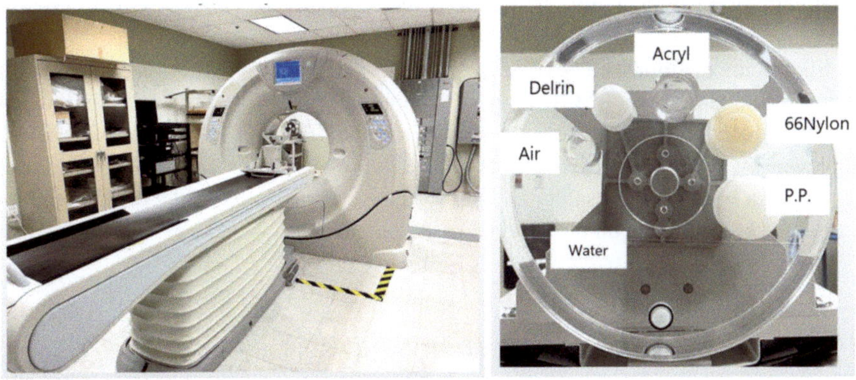

Fig. 10.4 Canon's first engineering prototype CZT-based PCCT system (left) and image of one of the phantoms used in this study (right) [10]

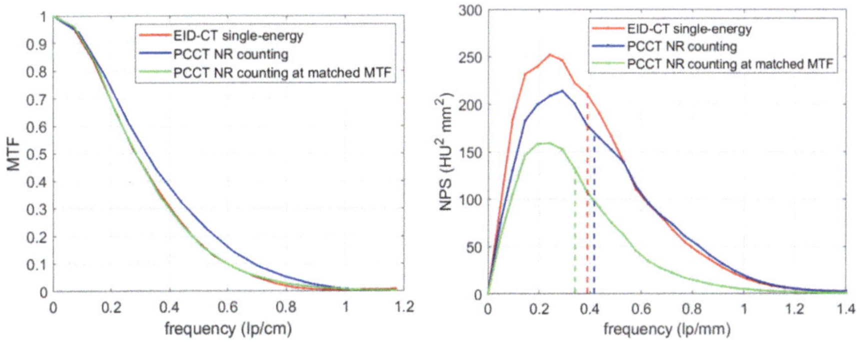

Fig. 10.5 MTF (right) and NPS (left) measurements of EID-CT and PCCT systems [10]

400 mAs to assess the imaging performance on CT number accuracy, uniformity, noise, spatial resolution, and material differentiation and quantification. This prototype imaging system supports a 500 mm field-of-view (FOV) and a slice thickness of 10 mm at the isocenter.

The quantitative evaluations demonstrate that PCCT has superior imaging performance with lower noise and improved spatial resolution compared to conventional energy integrating detector (EID)-CT under the tested conditions. For example, Fig. 10.5 shows the results of the Modular Transfer Function (MTF) and Noise Power Spectrum (NPS). Using a projection domain material decomposition approach with multiple energy bin measurements, PCCT virtual monoenergetic images (VMIs) have lower noise and good accuracy in quantifying iodine and calcium concentrations. These results lead to an increased contrast-to-noise ratio (CNR) for both high and low-contrast phantom objects compared to EID-CT at matched dose and spatial resolution. PCCT can also generate super-high resolution

(SHR) images using a much smaller detector pixel size than EID-CT, greatly improving image spatial resolution.

Significance Improved spatial resolution and quantification accuracy with reduced image noise in PCCT images can potentially lead to more precise diagnoses at reduced radiation doses than conventional EID-CT. The increased CNR achieved by PCCT suggests a potential reduction in iodine contrast media load, resulting in better patient safety and reduced cost. Please refer to the original publication for more details on the original research papers.

10.4.2 Metal Artifacts Reduction Study

Metal artifacts have been an outstanding issue in computed tomography (CT) since their first use in clinics. Metal artifact reduction (MAR) methods continue to be developed, and photon-counting detectors have recently been the subject of research for this purpose. PCCT detectors can distinguish the energy of incident X-rays and assign them to a set number of energy bins. High-energy data captured using PCDs have been shown to reduce metal artifacts in reconstructions due to reduced beam hardening.

However, high-energy reconstructions using PCCT have drawbacks, such as reduced image contrast and increased noise. Here, we demonstrate the recently developed MAR algorithm, called "trace replacement MAR" (TRMAR), in which the data corrupted by metal artifacts in full-energy spectrum projections are corrected using the high-energy data captured during the same scan. The resulting reconstructions offer similar MAR to high-energy reconstructions but with improved image quality.

To illustrate the power of PCCT technology and the TMAR algorithm for metal artifact reduction, we summarize the recent research performed using the PCCT330 detectors described earlier. Experimental data were collected using a benchtop PCCT readout system supplied by Redlen, as shown in Fig. 10.6. Simulations were performed to determine the optimal high-energy threshold and to test TRMAR in simulations using the XCAT phantom source (DOI: 10.1118/1.3480985) and a biological sample.

The first step in TRMAR, outlined in Fig. 10.7, is identifying the corrupted data in the original sinogram. First, an uncorrected CT image is reconstructed, and the metal is segmented in a high-energy bin using thresholding. The algorithm allows for the selection of the appropriate threshold value based on the uncorrected images. Here, the center of the images was excluded from metal thresholding due to the presence of ring artifacts. Then, a number of iterations of binary dilation are applied to account for any errors in thresholding. From there, the metal-only images were forward projected to obtain the location of the metal traces in the original sinogram. Once the metal traces have been identified in the sinogram, a mask can be created of the metal traces from which the original sinograms can be manipulated. Using the mask, the metal trace data in the original sinogram are replaced with the equivalent data from a high-energy bin from the PCD. The corrected sinogram is then

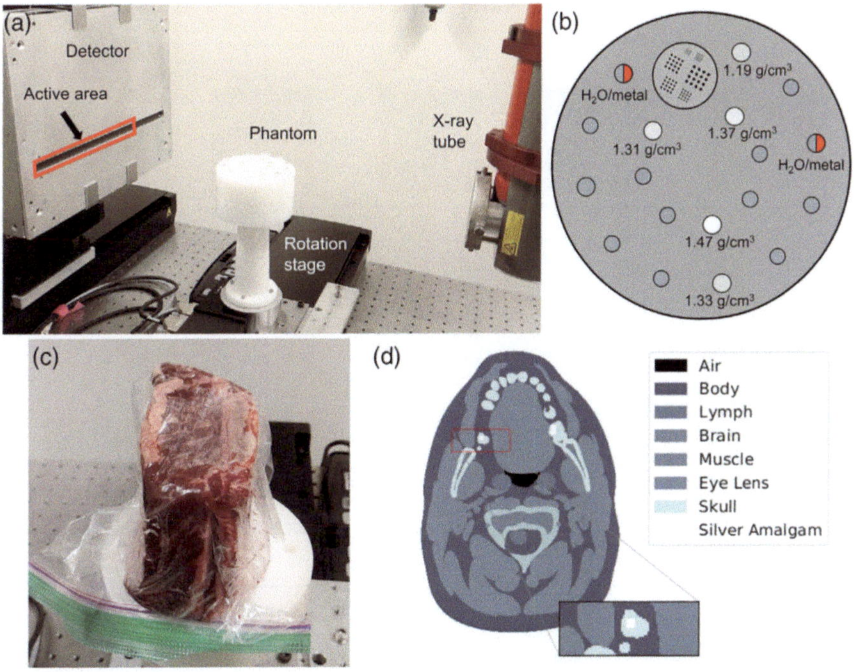

Fig. 10.6 (**a**) Benchtop photon-counting detector PCCT test setup. (**b**) The phantom layout. Dark grey circles represent vials that contained water only, while the lighter gray vials contained various concentrations of $Ca(ClO_4)_2$ at the densities shown. The vials that were half red were either water vials or metal screws, depending on whether metal artifacts were to be present or not. (**c**) An image of the ex-vivo bovine tissue sample on the rotation stage. (**d**) A schematic of the XCAT head phantom used in simulations shows the various tissue types and materials [11]

reconstructed with reduced metal artifacts. Also shown in Fig. 10.7 is the modified scheme for normalized metal artifact reduction (NMAR), a common MAR method, which was used for comparative purposes.

For experiments, a 100-mm diameter cylindrical phantom containing vials of water, two screws, vials with various densities of $Ca(ClO_4)_2$, and a spatial resolution phantom was imaged with and without the screws. The screws were segmented in the initial reconstruction and forward projected to identify them in the sinogram space to perform TRMAR. The resulting reconstructions were compared to the control and to reconstructions corrected using NMAR. Additionally, a beef short rib was imaged with and without metal to provide a more realistic phantom. Some of the results are shown in Fig. 10.8 and indicate NMAR offers better performance compared to TRMAR in streak reduction and image quality metrics.

However, TRMAR is less susceptible to metal segmentation errors and can closely approximate the reduced streak metal artifacts seen in NMAR at 1/3 the computation time. With the recent introduction of PCCT technology into clinics worldwide, TRMAR offers notable potential for fast, effective MAR. For more details of the original research papers, please refer to the following publication [11].

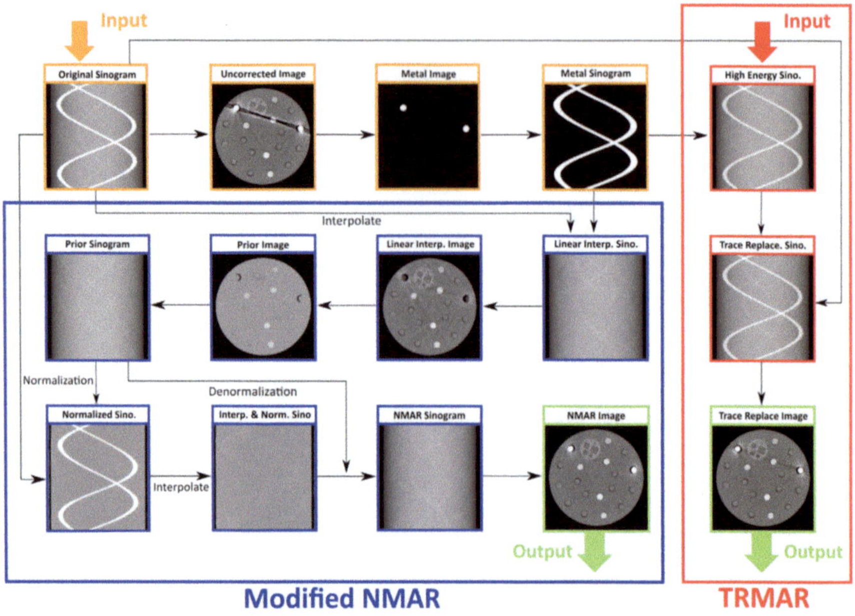

Fig. 10.7 Scheme of trace replacement metal artifact reduction (TRMAR) and modified normalized metal artifact reduction (NMAR) [11]

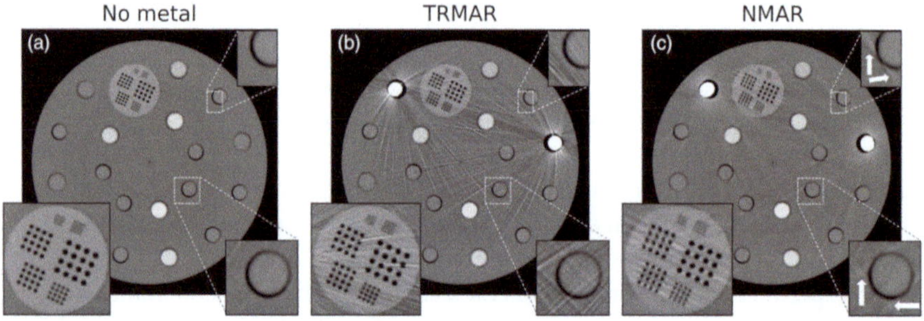

Fig. 10.8 Computed tomography (CT) reconstructions (**a**) without metal (**b**) with the 100–110 keV trace replacement metal artifact reduction (TRMAR) in the 30–120 keV energy bin, and (**c**) normalized metal artifact reduction (NMAR) in the 30–120 keV energy bin [11]

10.4.3 Stent Imaging Study

CT angiography (CTA) is a common imaging assessment used to diagnose vascular diseases and monitor potential complications before and after treatment. Vascular issues such as stenoses can often be treated with the implantation of a stent, which can prevent the blood vessels from narrowing. However, monitoring of the area in which the stent is implanted can be hindered due to the stent itself. As discussed in the section above, stents are usually manufactured from various metals, which cause metal artifacts to form blooming artifacts due to their high atomic number and density. These artifacts can reduce the diagnostic value of CTA scans and need to be mitigated as much as possible. The artifacts obscure the visualization of the stent lumen, which must be examined to assess whether any blockages have developed within the stent. Stent imaging with CT is especially problematic with stents with lumen diameters under 3 mm, as the lumen covers a smaller area, and visualization is more readily impaired by artifacts. Several CT advances are currently approaching the point where they could be used in the clinic to address some of the issues of stent imaging, including ultra-high-resolution (UHR) CT and photon-counting detector (PCD) CT (Fig. 10.9).

In summary, in this study, the performance of the laboratory benchtop PCCT scanner was evaluated and compared with respect to two clinical EID-CT scanners for stent imaging, as shown in Fig. 10.10. Three stents were imaged: a Medtronic Protégé stent, a Cordis Precise stent, and a Cordis S.M.A.R.T. Control stent. The researchers have found that the apparent strut, or individual wire thickness, was closest to the physical measurement using the CZT-based PCCT system for all stents. The significant improvement in the measurement of the strut thickness with PCCT can be largely attributed to the higher inherent spatial resolution of CZT

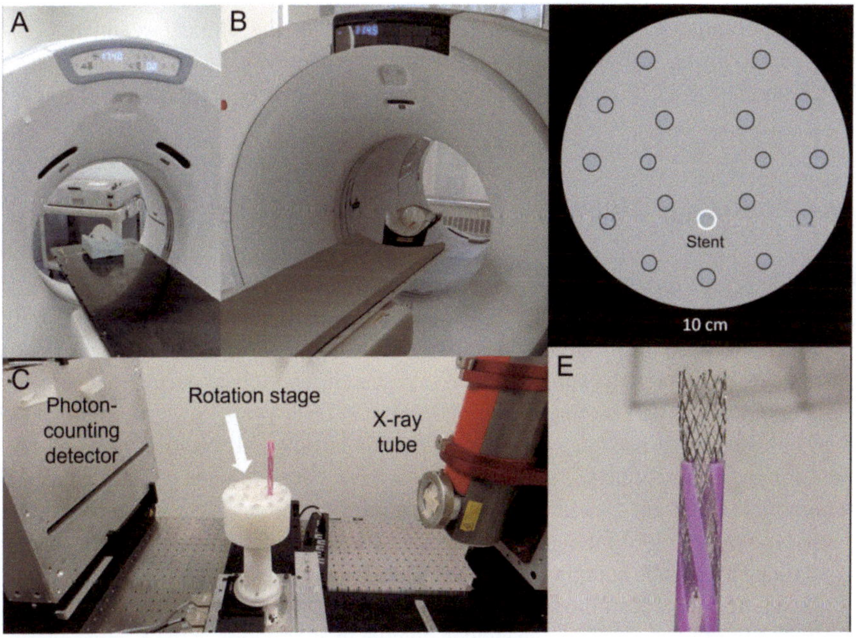

Fig. 10.9 Setup images of the (**a**) GE Optima 580, (**b**) GE Discovery IQ, and (**c**) the benchtop PCD-CT system. (**d**) Phantom layout with water vials placed in each of the circular locations. The stents were inserted into a straw, which was then filled with water and inserted in the indicated location in the phantom. (**e**) Image the Cordis Precise stent inside the straw [12]

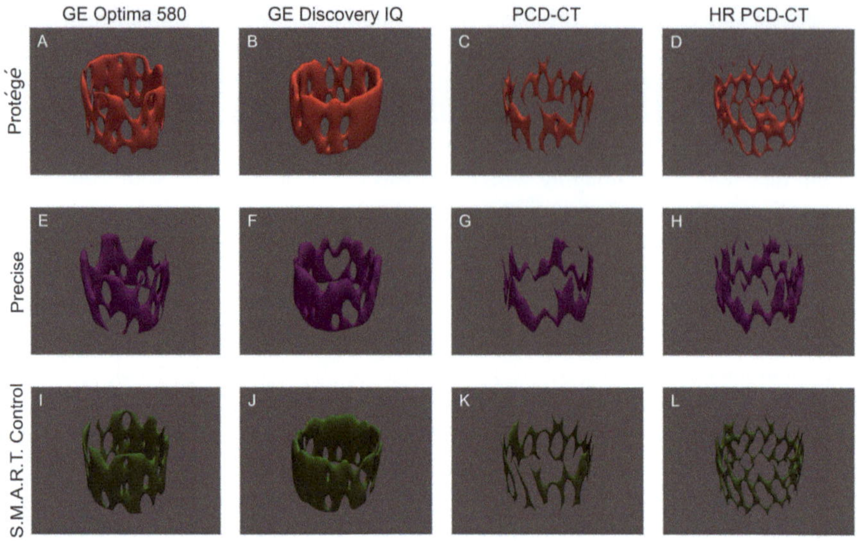

Fig. 10.10 3D volume renderings of all three stents. Volume renderings created from scans of (**a–d**) the Medtronic Protégé stent, (**e–h**) the Cordis Precise stent, and (**i–l**) the Cordis S.M.A.R.T. Control stent with each of the four imaging methods [12]

detectors compared with the EIDs detectors used in the clinical scanners. This reduced blooming artifacts in image reconstructions, even though the laboratory PCCT and clinical CT scans were reconstructed with nearly identical voxel sizes.

10.4.4 K-Edge Molecular Imaging Study

PCCT detectors provide an opportunity to discriminate multiple tissue types and contrast materials simultaneously. Their spectral signatures provide an opportunity to develop new contrast agents and imaging techniques. Nanoparticle-based contrast agents have been the subject of active research, and it is expected that some of these applications will hit the clinical market in several years. Several reports have been published that use heavy metal-based nanoparticles like gold, ytterbium, and gadolinium as potential CT contrast agents have been published. Individualized nanoparticles that are functionalized and antibody-conjugated to target a specific tissue type or region of interest have been used in combination with PCCT systems to facilitate molecular imaging.

The ability to detect multiple contrast agents relies on K-edge imaging, a concept illustrated in Fig. 10.11 where the sudden change in attenuation at the K-edge energy for the various contrast agents used in the study is shown. PCCT detectors, due to their ability to discriminate photon energy, can measure the position of K-edges of those high-Z materials that fall within the diagnostic CT energy range. The

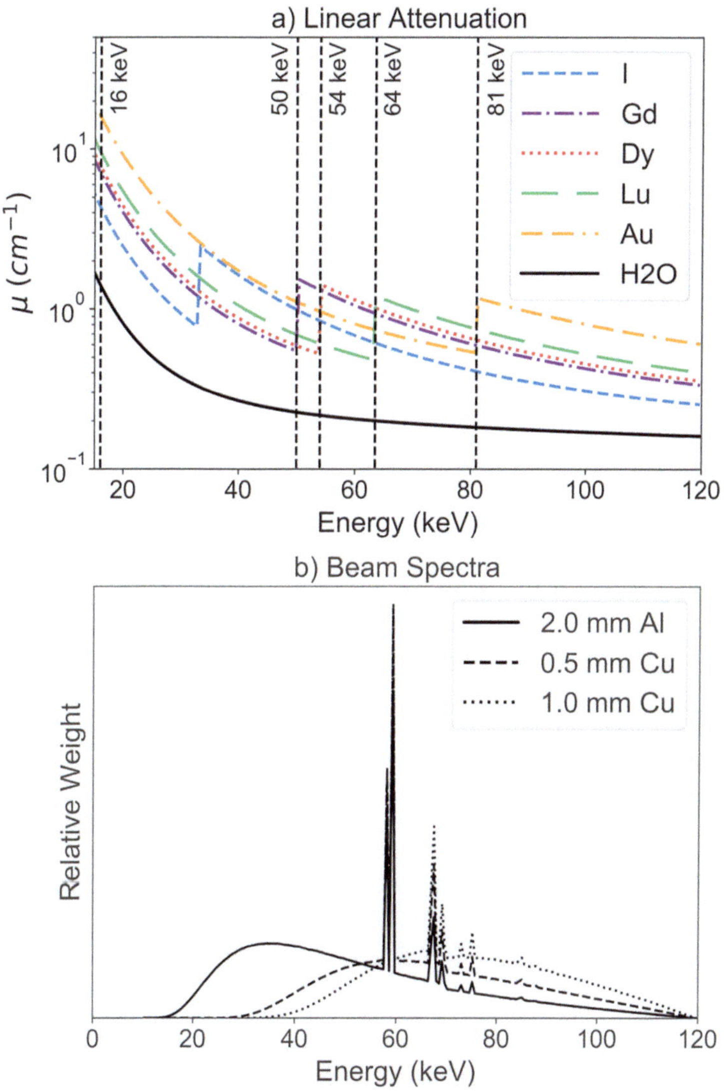

Fig. 10.11 Linear attenuation coefficients of the contrast agents and beam spectra that were used in the study. (**a**) Semi-log plot of the linear attenuation coefficients of water and all contrast agents included in the study, over the relevant energy range. (**b**) Relative number of photons with respect to energy of the three filtered beam spectra. Each spectrum was scaled by the tube current used for that filtration to give an equal total flux

user-defined energy thresholds in PCDs can be placed close to the K-edge energy of high-Z contrast materials in order to capture the discontinuity in the attenuation profile. This helps distinguish the K-edge material from other materials (e.g., bone, soft tissue, second contrast agent) and, more importantly, quantify the concentration of the contrast materials in a given target site using material decomposition techniques. This approach has been mainly tested on rodents imaged using small-animal PCCT scanners and is expected to be tested on humans in the future.

In the study reported here [13], a tabletop PCCT system with a Redlen PCCT 330 detector capable of separating six energy bins was used to image multiple contrast agents in a small phantom. The contrast agents were separated, and the concentration was quantified using K-edge subtraction. The researchers investigated three parameters to increase system performance: beam filter type and thickness, projection acquisition time, and energy bin width. The results from the parameters were compared based on the PCCT signal and contrast-to-noise ratio (CNR) or noise in K-edge images. The concentrations of the contrast agents were quantified in K-edge images and compared to known concentrations.

Three different filter setups were investigated for this study: 2 mm Al, 0.5 mm Cu, and 1 mm Cu. To obtain similar noise levels in the final images using each filter, the tube current was varied to acquire a similar total number of photons in the air scan compared to the total number using a 1.0 mA tube current with 2 mm Al filtration. This yielded tube currents of 2.25 mA and 4.75 mA for 0.5 mm Cu and 1.0 mm Cu, respectively. The spectra for each beam filtration can be seen in Fig. 10.11. The unfiltered spectrum was calculated using a validated Monte Carlo model and filtered according to Lambert-Beer's Law. The three filters were chosen due to the use of aluminum and copper in many CT and PCCT studies and to cover a lower, middle, and higher weighted energy spectrum.

Three phantom layouts (the Lanthanide phantoms), shown in Fig. 10.12, were scanned using the tube currents listed above. Bin thresholds were set to 16, 50, 54, 64, 81, and 120 keV to accommodate the K-edges of Gadolinium (50.2 keV), Dysprosium (53.8 keV), Lutetium (63.3 keV), and Gold (80.7 keV). This translated to bins with energy ranges of 16–50 keV, 50–54 keV, 54–64 keV, 64–81 keV, and 81–120 keV.

To facilitate the modification of bin width, only two contrast agents were investigated and placed in the second (AuGd) phantom layout, as shown in Fig. 10.12. 0.5 mm Cu filtration and a tube current of 2.25 mA were used for image acquisitions of the second layout as it offered a compromise between the other two filters detailed above. The energy thresholds were shifted to create different bin widths on either side of the K-edge of the two contrast materials. The two energy thresholds that remained the same were those set at 50 and 81 keV, corresponding to the K-edges of gadolinium and gold. Four scans of the AuGD phantom were taken corresponding to bin widths of 5, 8, 10, and 14 keV for gadolinium and bin widths of 5, 10, 14, and 20 keV for gold.

K-edge images were created by subtracting the CT image of the energy bin below each respective K-edge from the CT image above the K-edge. All K-edge images can be found in the original publication [13]. Figure 10.13 demonstrates how bin width affects the CNR of gadolinium and gold K-edge images, respectively. The data was fit with a quadratic curve to show how the K-edge CNR would likely behave over the range of bin width data that was collected. The ideal bin width must be found for every contrast material separately, as the bin width that results in the highest K-edge CNR can vary, as demonstrated here. For Gadolinium,

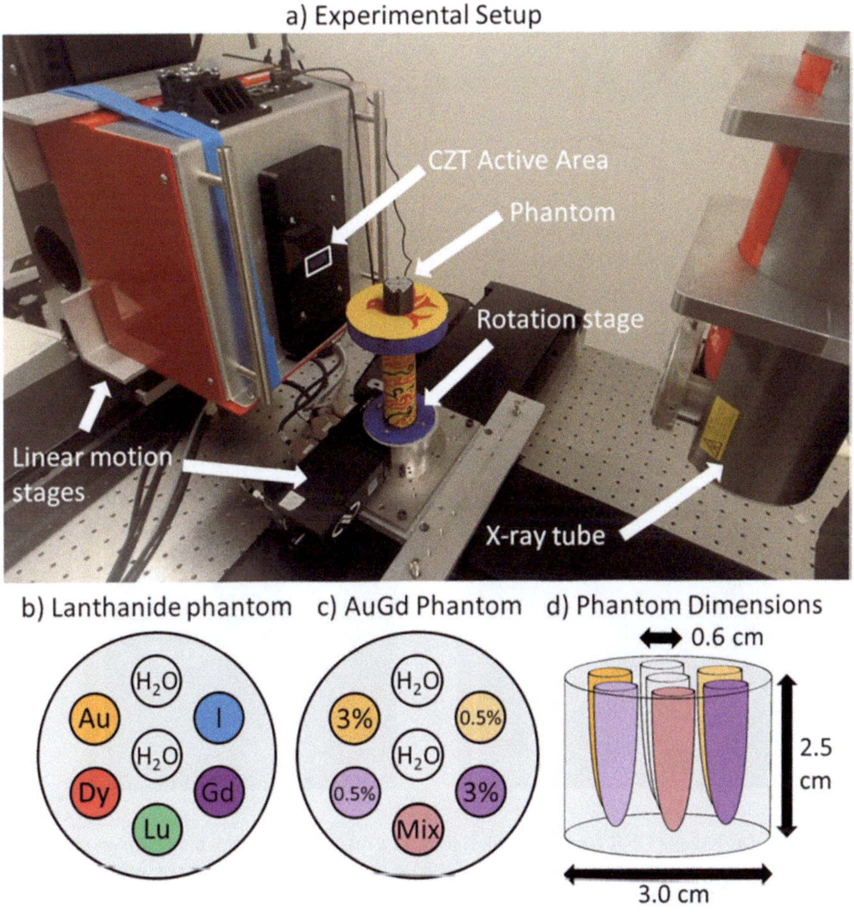

Fig. 10.12 (**a**) Experimental setup showing the x-ray tube, detector, phantom, and movement stages. (**b**) Contrast material layout for the three Lanthanide phantoms, one for each concentration of 5%, 3%, and 1% of all contrast materials. (**c**) The layout for the AuGd phantom shows the percent concentration in each vial with Au (orange) and Gd (violet). The mix contained 0.5% concentration of both materials. (**d**) The phantom dimensions. From [13]

peak K-edge CNR was achieved with a bin width of 10.7 keV, while for gold, the optimal bin width was 15.8 keV.

In conclusion, it has been demonstrated how parameters in an experimental PCCT benchtop system can be varied to determine the configuration that offers the best imaging performance for a certain contrast agent. For research details, the reader is referred to the original publication [10]. Redlen's CZT-based PCCT system could separate four different contrast agents in a range of Z-values (64–79) at varying concentrations using K-edge subtraction, even without ideal parameter selection. This demonstrates the power of CZT-based PCCT technology for K-edge clinical imaging applications in the future.

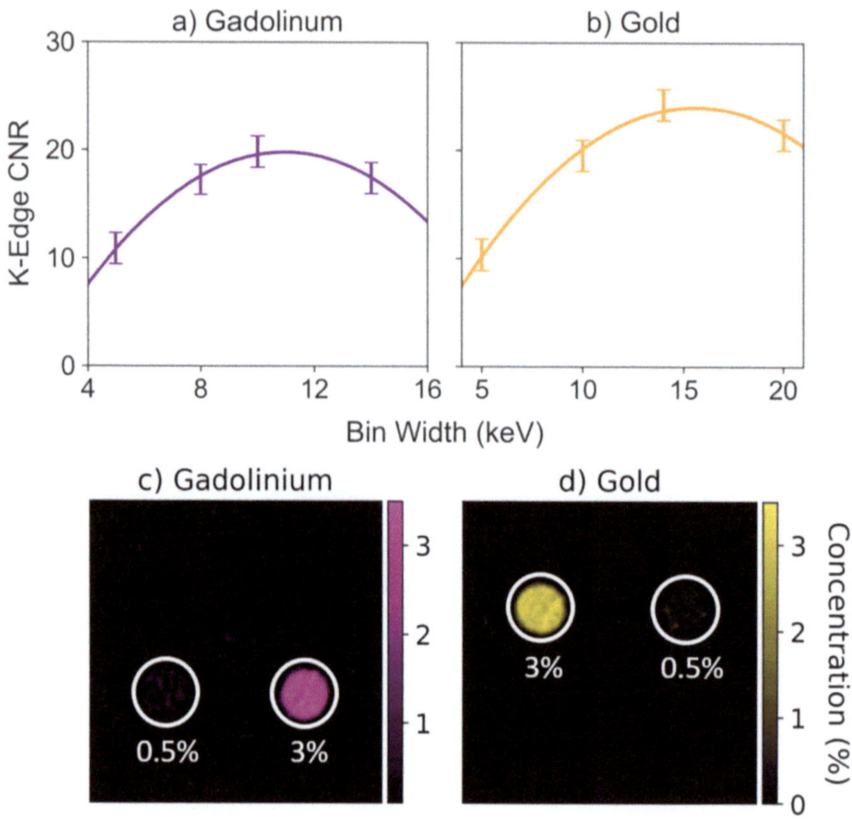

Fig. 10.13 (**a**) K-edge CNR curve for gadolinium as a function of energy bin width on either side of the K-edge energy. (**b**) K-edge CNR curve for gold as a function of energy bin width. (**c**) K-edge image of gadolinium reconstructed with 10 keV bin widths. (**d**) K-edge image of gold reconstructed with 14 keV bin widths

10.4.5 Bone Density Study

Osteoporosis is a huge inviable issue as one out of two Caucasian women over 50 years old will develop osteoporosis. Most of the cases are undiagnosed and untreated, that is, about 20 million women in the USA only. Breaking hips in old age can be deadly, with 25% mortality after hip fracture. As a result, the total cost to US healthcare for osteoporosis is larger than for breast cancer.

Osteoporosis is low Bone Mass and micro-architectural deterioration of bone tissue, leading to enhanced fragility and consequent increase in fracture risk. Bone Mass Density (BMD) is determined by Bone Mass Weight (in kg) divided by the area (in cm^2). An example of normal and osteoporotic bone is shown in Fig. 10.14.

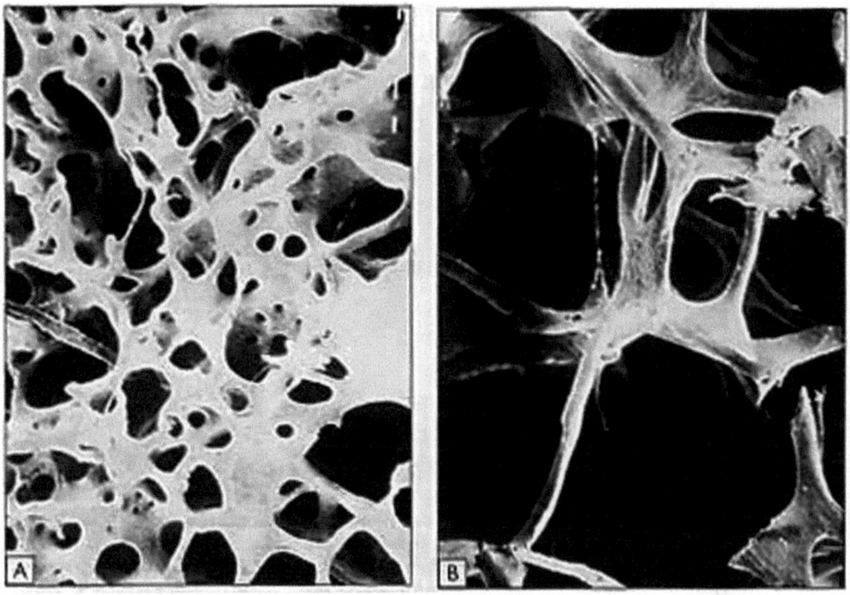

Fig. 10.14 Illustration of the normal (left) and osteoporotic bone (right) (www.bonehealthandosteoporosis.org/patients/what-is-osteoporosis/)

Standard assessment of bone mass density is currently done using a Dual Energy X-ray Absorption (DEXA) scan. As explained below, DEXA suffers from numerous problems and can be dramatically improved using multi-energy photon counting.

In today's clinical practice, Bone Mass Density (BMD) is measured using DEXA, Dual-Energy X-ray Absorptiometry, sometimes called DXA. DEXA is an X-ray imaging technique primarily used to derive the mass of one material in the presence of another through knowledge of their unique X-ray attenuation at different energies. Two images are made from the attenuation of low and high average X-ray energy. DEXA is a special X-ray imaging modality not typically available with general-use X-ray systems because of the need for special beam filtering and near-perfect spatial registration of the two attenuations.

DEXA's primary commercial application has been to measure BMD to assess fracture risk and to diagnose osteoporosis; the X-ray energies used are optimized for bone density assessment. The lumbar spine, proximal hip, and, sometimes, the distal forearm are scanned for osteoporosis diagnosis. The region of interest (ROI) corresponding to these parts is used, and the diagnostic criteria are well-defined. The whole body can also be scanned to measure bone mass and soft tissue body composition. In image areas that contain only soft tissue, lipid and lean tissue can be assessed, from which percent lipid mass can be calculated. In contrast, areas that contain bone use an estimated percent lipid from the surrounding tissue. Reference populations have been scanned and defined by sex, ethnicity, and age. Diagnosis of the disease is typically undertaken by comparing individuals to their peer group or a young, healthy population. Currently, there are estimated to be over 60,000 whole-body DEXA systems in use worldwide.

In all DEXA systems on the market, the X-ray tubes are standard tungsten anode tubes with focal spot sizes of 0.5–1 mm^2. However, there are differences in how the dual-energy images are created. The two methods in use are:

- kVpp voltage switching systems
- K-edge filtering systems

In the kVpp voltage switching system, two X-ray tube voltage settings create low and high-energy images. The X-ray tube power supply switches between a low (70 kVpp) and high (140 kVpp) voltage setting during alternate half cycles of the power supply. The resulting pulses are short (8.33 ms for 60 Hz.) Copper or brass is used to pre-harden the high-energy beam, removing the low-energy part of the spectrum and minimizing the overlap between low and high X-ray spectra. The filter, kVpp voltage switching, and X-ray detectors are all electronically and mechanically synchronized to collect information for each position of the X-ray gantry sequentially.

In the K-edge filter system, the X-ray tube is operated in a steady direct current mode. A K-edge absorption filter splits a single X-ray spectrum into low and high-energy components. The X-ray tube kVpp is set so the K-edge places a notch in the X-ray spectrum; for the given X-ray voltage, one of several rare earth filters can be used between the patient and the X-ray tube for those purposes. Examples include Samarium filtering (K-edge = 46.8 keV) for kVpp of 100 keV or Cerium (K-edge = 40 keV). Examples of the X-ray spectra for both systems are shown in Fig. 10.15.

Despite its widespread use, DEXA suffers from several fundamental problems, making its application in clinical care problematic. First, DEXA is a relative measurement! World Health Organization (WHO) defines severe osteoporosis as "A T-score more than −2.5 standard deviations below the young adult female reference mean in the presence of one or more fragility fractures." In other words, the DEXA assessment relies on a comparison against an imaginary average person representing your weight, height, and ethnicity. You can have low BMD from birth and be assessed as having osteoporosis, or already have osteoporosis and be assessed normal—many false positives and negatives. Absolute measurement is needed!

Second, the reference group is typically Caucasian females. Not enough data has been collected for children, men, and some ethnicities. Special considerations are involved in using DEXA to assess bone mass in children. Specifically, comparing the bone mineral density of children to the reference data of adults will underestimate the BMD of children because children have less bone mass than fully developed adults. This would lead to an overdiagnosis of osteopenia in children.

The third problem is that DEXA makes body composition assumptions and can only solve for two materials simultaneously. Specifically soft tissue composition can only be solved in areas exclusive of bone. In some cases, accurate estimates cannot be made (head, hands, feet and upper torso) as there is inadequate soft tissue. Bone mass can only be determined by assuming the soft tissue composition

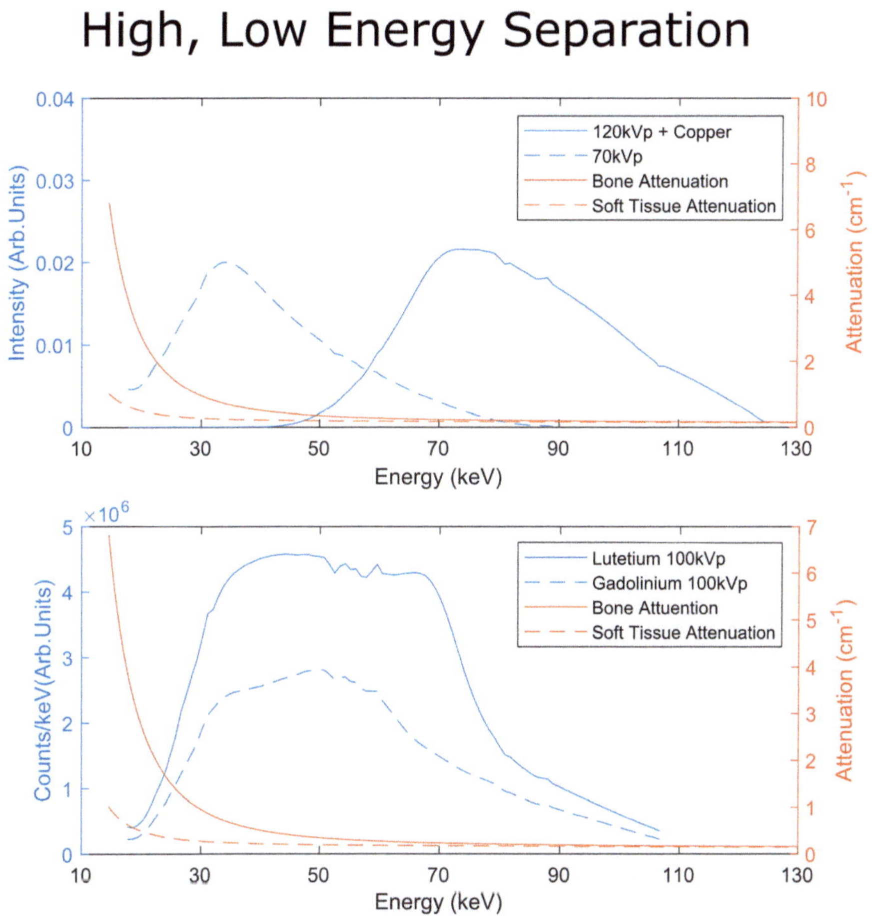

Fig. 10.15 Typical DEXA X-ray spectra obtained using kVpp switching (above) and K-edge filtering (below). Bone and soft tissue attenuation curves are superimposed

overlaying the bone. As bone is typically 40% or more of the body image pixels, the soft tissue composition has to be estimated from surrounding tissue.

These simplifying assumptions had to be made as DEXA, by definition, uses two effective X-ray energies and tries to solve for at least 3-unknowns as the human body has to be represented by at least three unknowns (more components than three is better for accurate human body representation as discussed below). This leads to unreliable measurements that must be repeated on the same machine each time, or at least a machine from the same manufacturer. Errors between DEXA scanners or trying to convert measurements from one manufacturer's standard to another can introduce errors large enough to wipe out the sensitivity of the measurements, making the BMD test completely useless.

To resolve DEXA problems and provide an absolute measure of bone mass density, we are proposing to use a photon counting detector that uses multiple energy bins. To understand why many energy bins are required, Appendix A provides models of human body decomposition first. DEXA effectively relies on the Two-Component Model as it only has two measured data points: attenuation at low and high energies, meaning the human body is considered bone and no-bone mass. By making certain approximations, DEXA attempts to consider fat by analyzing areas where bone is absent. As the human body comprises at least four components: bone, fat, protein, and water, the ideal way to decompose the contributing elements is to use four energy bins in X-ray imaging. The most effective way of accomplishing that is through the use of multi-energy bin photon counting.

We have proposed [14] using a multi-energy photon counting detector in the DEXA configuration, where the X-ray source is collimated to be a pencil beam or a fan beam. Fan beam leads to faster X-ray scans and is typically preferred to the pencil beam configuration. In this work, the PCD used five energy bins ranging from 38 to 109 keV with energy thresholds of 38, 42, 68, 72, 98, and 102 keV, Given that a DEXA scan typically utilizes the ratio of two X-ray energies at 40 keV and 70 keV, the energy bins were chosen such that the average X-ray energy of the first and third energy bins in the PCD imaging system was equivalent to 40 keV and 70 keV, respectively. Using a series of five bone phantoms of known densities, a linear calibration curve was established by plotting the ratio of the sum of counts at 40 keV and 70 keV as a function of bone density. Five additional bone phantoms of unknown densities were estimated using this calibration curve, resulting in a mean absolute error of 3.8%. This suggests that a CZT PCD can mimic a DEXA scan without relying on dual-exposure imaging or dual-layer detector.

10.5 Artificial Intelligence (AI) and PCCT

AI encompasses a broad range of technologies that enable computers to perform tasks that typically require human intelligence [24–35]. Machine learning, a subset of AI, focuses on developing algorithms that enable computers to learn from and make predictions based on data [24]. In the broader field of computed tomography (CT), machine learning can optimize scanning parameters, improve image quality, and reduce noise. By learning from a dataset of annotated images, machine learning models can enhance the efficiency of image processing and analysis, leading to faster and more accurate diagnostics. Deep learning, a more advanced subset of machine learning, utilizes neural networks with multiple layers to analyze and generate complex data [25]. Deep learning algorithms excel at image classification, segmentation, and enhancement tasks in medical imaging [24, 25].

This review section will survey four areas of applying AI techniques to PCCT. Noise reduction is a critical area where AI can substantially improve PCCT. In PCCT, noise can obscure anatomical features despite the high resolution of images, making it challenging to achieve accurate diagnoses. AI algorithms are adept at identifying and filtering out noise, thus preserving the integrity of the image details [26, 27]. Material decomposition in PCCT is an advanced application that benefits from AI's capabilities. By differentiating between various materials based on the energy levels of counted photons, AI can provide unprecedented insights into the composition of tissues and other bodily substances [28, 29]. Metal artifact reduction is one area where AI-driven methodologies can help to mitigate the impact of metal artifacts on CT imaging, thus substantially elevating both the diagnostic precision and the visual clarity of images [30]. Image reconstruction, another domain where AI significantly contributes, holds immense potential. Traditional image reconstruction methods may only partially exploit the potential of the data collected by PCCT. However, AI-driven techniques can reconstruct images to maximize resolution and clarity, opening exciting possibilities for more precise interpretations [31]. These areas of research are beneficial in detecting and characterizing diseases. Enhancing overall image quality is one of the most straightforward yet profound benefits of applying AI in PCCT. Through deep learning, AI can improve images' clarity, contrast, and detail, making it easier for clinicians to diagnose, monitor disease progression, and plan treatments.

In essence, the combination of AI and PCCT has the potential to revolutionize medical imaging. By addressing key challenges and significantly enhancing image quality and interpretation, AI empowers PCCT to unlock new potentials in clinical research, diagnostics, and patient care, setting a new standard in the precision and efficacy of medical imaging practices.

10.5.1 Noise Reduction

The intersection of deep learning and PCCT has led to significant advancements in noise reduction techniques, enhancing image clarity and diagnostic value. Studies have explored various network architectures and their effectiveness across different domains, indicating a potential for broader application and improved image quality in medical diagnostics.

One study published by Baffour et al. investigates the application of deep learning for denoising photon-counting CT images, focusing on whether denoising is more effective in the projection or image domain. Three network architectures (U-Net, Dilated U-Net, and ResNet) were examined across both domains. The performance was assessed based on image quality metrics such as contrast-to-noise ratio and noise power spectrum. Results indicated that U-Net and Dilated U-Net architectures performed best in the image domain for the specific phantoms trained on. However, denoising in the projection domain generally yielded better results for different phantoms, suggesting a potential for broader generalization. The study suggests that the choice of domain for implementing deep learning denoising can significantly impact the quality of photon-counting CT images, with implications for optimizing CT imaging protocols [26].

Further expanding on the utility of PCCT when combined with deep learning, another study focused on noise reduction for detecting multiple myeloma lesions, benchmarking it against the conventional energy-integrating detector (EID) CT. Conducted at the Mayo Clinic, it involved 27 adult participants with multiple myeloma, using both PCD CT in ultra-high-resolution mode and EID CT for whole-body scans. The findings indicated that PCD CT, especially with deep learning denoising, provided superior visualization of multiple myeloma lesions, including lytic lesions, intramedullary lesions, and pathological fractures, without increasing radiation dosage. This suggests that PCD CT could significantly enhance the detection and diagnosis of multiple myeloma, potentially influencing treatment and management strategies [27].

10.5.2 Material Decomposition

Deep learning also plays a crucial role in advancing material decomposition capabilities in PCCT, enabling more precise differentiation of materials based on photon energy levels. This precision is critical for diagnosing and characterizing diseases, further underscoring AI's transformative impact on medical imaging.

The study conducted by Eguizabal et al. introduces a deep learning post-processing approach to enhance the maximum likelihood estimate of material decomposition in photon-counting spectral CT. It leverages convolutional residual blocks, blending model-based and data-driven strategies for noise reduction and accuracy improvement. This method shows promising results in simulations using Shepp-Logan phantoms and human numerical phantoms, demonstrating its potential to improve CT image quality efficiently. The approach suggests a balance between computational efficiency and the quality of material decomposition, indicating its promise for clinical application [28].

Further advancing the field, the research detailed in "Multi-material spectral photon-counting micro-CT with minimum residual decomposition and self-supervised deep denoising," authored by Trapani et al., unveils an innovative cone-beam setup optimized for spectral micro-CT imaging. This setup utilizes a CdTe photon-counting detector to address charge-sharing effects and incorporates advanced image processing techniques for material decomposition and noise reduction. The study demonstrates improved differentiation of materials with close K-Edge separations and quantitative analysis of spectral imaging systems, suggesting significant implications for functional and multi-organ imaging applications [29].

10.5.3 Metal Artifact Reduction

AI-driven methods have shown promising results in reducing artifacts caused by metal objects in CT images, improving diagnostic accuracy and image clarity. These advancements highlight the potential of integrating spectral information and sophisticated algorithms to enhance CT imaging quality.

A study by Schmidt et al. presents a photon-counting normalized metal artifact reduction (PCNMAR) method to enhance diagnostic CT imaging by minimizing distortion caused by metal objects. It introduces a technique that leverages spectral information intrinsic to photon-counting CT alongside a sophisticated prior-based metal artifact reduction algorithm. By integrating these approaches, PCNMAR effectively reduces metal artifacts across various bin images and combinations, showcasing a notable improvement over the conventional method. This advancement could significantly enhance the clarity of CT images, offering promising implications for clinical diagnostics [20].

10.5.4 Image Reconstruction Methods

Deep learning-based image reconstruction techniques offer advantages over traditional methods, including improved image quality, resolution, and reduced computation time. These innovations represent a significant leap forward in utilizing PCCT data more effectively, enhancing the diagnostic process.

A study titled "Photon-Counting CT Reconstruction With a Learned Forward Operator" addresses the challenge of efficiently modeling cross-talk between detectors using a deep convolutional neural network trained to model this cross-talk. The study shows that incorporating a learned cross-talk model into the reconstruction process improves image quality (with a notable increase in peak signal-to-noise ratio) and reduces computation time by half compared to traditional complete cross-talk modeling methods. Furthermore, this approach generalizes well across different data and detector settings, suggesting that partially learned forward operators could be a computationally advantageous method for modeling data generation in Photon-Counting CT [21].

Another pioneering work introduces a Joint reconstruction and Spectrum Refinement (JoSR) algorithm tailored for photon-counting detector CT. JoSR addresses the challenge of accurate spectrum estimation by combining spectral reconstruction with spectrum refinement, critical for reducing noise and enhancing image quality. The algorithm uses non-negative matrix factorization to reduce the dimensionality of unknowns, making the problem more manageable. Through simulation studies and real experiments, JoSR demonstrates its effectiveness in enhancing image quality and quantitative accuracy, suggesting its potential for clinical applications in CT angiography and quantitative examinations [22].

Furthermore, an investigation into using machine learning (ML) models to augment the quality of photon-counting CT images reveals promising outcomes. Three ML techniques—dictionary learning, U-Net, and Noise2Noise—were utilized on simulated and experimental PCCT images. These models were trained using low- and high-dose image pairs to improve the peak signal-to-noise ratio (PSNR) and other image quality metrics. The application of these ML models to PCCT images showed promising results, suggesting their potential for improving diagnostic imaging quality, especially in situations with limited photon statistics [23].

10.5.5 AI Challenges

Data Requirements: The prerequisite of large datasets for training deep learning models poses a significant challenge, amplified by concerns over privacy and restrictions on data sharing. Unfortunately, this issue has been prevalent in medical imaging in general, and PCCT cannot be excused from these concerns. These hurdles complicate the data acquisition process and raise ethical issues that must be navigated carefully. Ensuring data diversity to train robust and generalizable models across various demographics and conditions adds another layer of complexity. Innovations in synthetic data generation offer potential pathways to mitigate these challenges, enabling the use of AI in settings where data are scarce or sensitive [23–25].

Interpretability: The opaque nature of deep learning models, often called "black boxes," poses significant challenges to clinician trust and technology adoption for PCCT. The inability to understand the decision-making process of AI models for medical imaging has constantly been discussed in clinical research and practice, where explainability and accountability are paramount. Developing explainable AI methods that provide insights into the models' reasoning processes is critical to overcoming this barrier.

10.5.6 AI Future Directions

Algorithm and model development is crucial in overcoming the barriers of model complexity and data scarcity that currently hinder broader AI integration in medical imaging, including PCCT. Current research endeavors can be directed toward creating powerful and understandable deep-learning frameworks that are both data and resource-efficient. Innovations in algorithm design focus on models that can learn from smaller datasets without compromising their predictive accuracy, thus democratizing AI technology for a broader range of healthcare institutions, regardless of size or resource availability. Moreover, advancements in making these models more interpretable aim to bridge the gap between AI capabilities and clinical applicability, ensuring that healthcare professionals can understand and trust AI-generated insights.

Clinical validation is another critical area of research in applying AI techniques to PCCT. Integrating deep learning-enhanced PCCT into routine clinical practice requires rigorous validation through extensive clinical trials. One pioneering example is using a PCCT simulator to generate projection images of computational phantoms, given the geometry and physics of the scanner and imaging parameters for specific clinical tasks [24]. Clinical validation studies are essential to ascertain the diagnostic precision and effectiveness of AI-augmented PCCT across a spectrum of medical fields, including but not limited to oncology and cardiology. By systematically evaluating the performance of these AI-enhanced imaging modalities in diverse clinical scenarios, researchers can identify areas of improvement, adapt AI algorithms to meet clinical needs better, and ultimately demonstrate the practical value of this technology in enhancing patient outcomes. The successful clinical validation of PCCT diagnostic accuracy will pave the way for its adoption in standard medical protocols, revolutionizing patient care with cutting-edge precision diagnostics.

10.6 Conclusions

CdTe and CZT-based PCCT detectors present a variety of advantages over conventional EID detectors. In this white paper, we have highlighted the primary benefits of this technology for Computed Tomography applications: higher spatial resolution, absence of electronic noise, lower radiation, and contrast dose, and the possibility of using advanced contrast agents. The great value of photon counting technology in CT is that it incorporates all of these elements into a single acquisition platform. For many routine applications, PCCT scanners using CZT detectors will greatly improve clinical care. Scans with metal hardware will become incrementally easier to read because of better artifact suppression. Scans with iodine contrast will show better detectability, leading to lower imaging and contrast doses, especially at high kVp.

The new capabilities of PCCT scanners include simplifying and consolidating a standardized universal imaging protocol and the ability to classify small, high-contrast lesions such as lung tumors and kidney stones. The radiation dose will be reduced for examinations of high-resolution structures such as the temporal bone or sinus. Truly transformative capabilities may occur when targeted molecular contrast agents become clinically available. PCCT technology is only in its infancy, with the first human scanners being announced and deployed. The next several years will reveal a wealth of information on the potential applications of PCCT.

Photon-counting CT technology continues to improve. New detectors from various detector manufacturers have already been developed and are being evaluated. An example of recent detector technology is showing high-quality images in Fig. 10.16 (cardiac stent and deer teeth). PCCT has the potential to dramatically

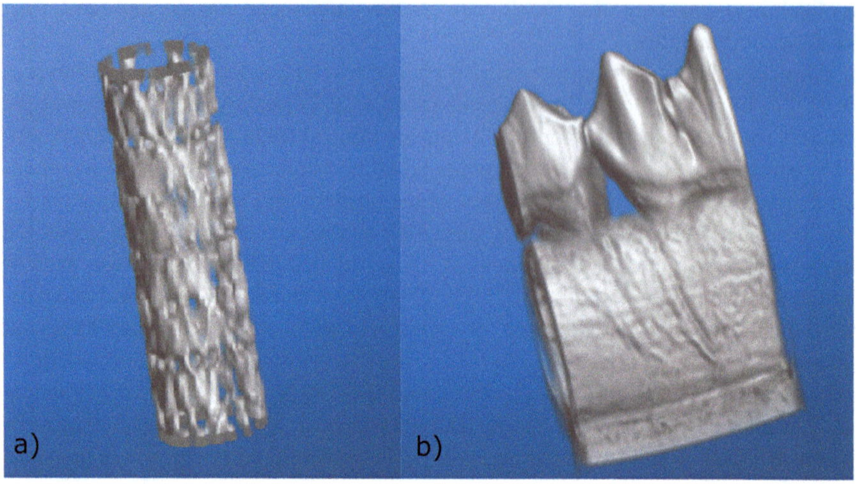

Fig. 10.16 Sample images from the latest CZT technology from Redlen. (**a**) A medical stent at 120 kVp and (**b**) a set of deer teeth at 60 kVp

alter the clinical use of CT in the coming years. Using energy-resolving detectors instead of EID, these new CT systems can substantially reduce image noise, increase the spatial resolution, and use K-edge imaging to measure the concentration of specific elements with reduced radiation doses compared to current standards. These properties can be useful in evaluating multiple anatomical districts, where it is necessary to identify small physiological and pathological structures, as well as for using different contrast agents, even concurrently, as shown in our review.

In addition, the synergy between AI and PCCT holds immense promise for revolutionizing medical imaging. The fusion of these cutting-edge technologies enables overcoming longstanding limitations in traditional imaging methods, offering clearer, more detailed images and facilitating the early detection and accurate characterization of a wide range of medical conditions. Although powerful, applying AI techniques to PCCT still needs to be navigated carefully. Again, this overview of the current research trends and challenges between AI and PCCT does not intend to cover all the ongoing research in the field. As the field continues to evolve, the continued exploration and integration of AI within PCCT promises to redefine the boundaries of medical imaging and enhance patient care and treatment outcomes significantly.

References

1. Hsieh J, Flohr T. Computed tomography recent history and future perspectives. J Med Imaging. 2021;8(5):052109. https://doi.org/10.1117/1.JMI.8.5.052109.
2. Flohr T, Petersilka M, Henning A, Ulzheimer S, Ferda J, Schmidt B. Photon-counting CT review. Phys Med. 2020;79:126–36. https://doi.org/10.1016/j.ejmp.2020.10.030.
3. McCollough CH, Leng S, Yu L, Fletcher JG. Dual- and multi-energy CT: principles, technical approaches, and clinical applications. Radiology. 2015;276(3):637–53. https://doi.org/10.1148/radiol.2015142631.
4. Tanguay J, Kim HK, Cunningham I. The role of X-ray swank factor in energy-resolving photon-counting imaging. Med Phys. 2010;37:6205. https://doi.org/10.1118/1.3512794. https://pubmed.ncbi.nlm.nih.gov/21302777/
5. Taguchi K, Iwanczyk J. Vision 20/20: single photon counting X-ray detectors in medical imaging. Med Phys. 2013;40(10):100901. https://www.ncbi.nlm.nih.gov/pmc/articles/PMC3786515/
6. Willemink M, Persson M, Pourmorteza A, Pelc N. Photon-counting CT: technical principles and clinical prospects. Radiology. 2018;00:1–20. https://pubmed.ncbi.nlm.nih.gov/30179101/
7. Taguchi K, Blevis I, Iniewski K, editors. Spectral, photon counting computed tomography technology and applications. London: CRC Press/Routledge; 2020. ISBN 9780367490119. https://www.routledge.com/Spectral-Photon-Counting-Computed-Tomography-Technology-and-Applications/Taguchi-Blevis-Iniewski/p/book/9780367490119.
8. Hsieh S, Iniewski K, editors. Photon counting computed tomography: clinical applications, image reconstruction and material discrimination. Cham: Springer; 2023. https://link.springer.com/book/10.1007/978-3-031-26062-9.
9. Start of clinical research of the first japanese produced next-generation photon-counting computed tomography. Canon Medical. www.global.medical.canon/News/PressRelease/Detail/131535-834.
10. Zhan X, Zhang R, Niu X, Hein I, Budden B, Shuoxing W, et al. Comprehensive evaluations of a prototype full field-of-view photon counting CT system through phantom studies. Phys. Med. Biol. 2023;68:175007. https://doi.org/10.1088/1361-6560/acebb. https://iopscience.iop.org/article/10.1088/1361-6560/acebb3
11. Richtsmeier D, O'Connell J, Rodesch P-A, Iniewski K, Bazalova-Carter M. Metal artifact correction in photon-counting detector computed tomography: metal trace replacement using high-energy data. Med Phys. 2023;50(1):380–96. https://doi.org/10.1002/mp.16049. https://pubmed.ncbi.nlm.nih.gov/36227611/
12. Richtsmeier D, Rodesch P-A, Iniewski K, Siu W, Bazalova-Carter M. The feasibility of accurate stent visualization with photon-counting detector CT and K-edge imaging. IEEE Trans Radiat Plasma Med Sci. 2023;7(7):712–8. https://doi.org/10.1109/TRPMS.2023.3287566. https://pubmed.ncbi.nlm.nih.gov/31882698/
13. Richtsmeier D, Dunning C, Iniewski K, Bazalova-Carter M. Multi-contrast K-edge imaging on a bench-top photon-counting CT system: acquisition parameter study. J Instrum. 2020;15 https://doi.org/10.1088/1748-0221/15/10/P10029. https://iopscience.iop.org/article/10.1088/1748-0221/15/10/P10029
14. Nguyen J, Richtsmeier D, Iniewski K, Bazalova-Carte M. Estimating bone densities using a CZT PCD, AAPM Conference; 2021.
15. Meloni A, Cademartiri F, Positano V, Celi S, et al. Cardiovascular applications of photon-counting CT technology: a revolutionary new diagnostic step. J Cardiovasc Dev Dis. 2023;10(9):363. https://doi.org/10.3390/jcdd10090363Xxx.
16. Zanon C, Cademartiri F, Toniolo A, Bini C, Clemente A, Colacchio EC, et al. Advantages of photon-counting detector CT in aortic imaging. Tomography. 2024;10(1):1–13. https://doi.org/10.3390/tomography10010001Xxx.

17. Remy-Jardin M, Hutt A, Flohr T, Faivre J-B, Felloni P, Khung S, et al. Ultra-high-resolution photon-counting CT imaging of the chest: a new era for morphology and function. Invest Radiol. 2023;58(7):482–7. https://doi.org/10.1097/RLI.0000000000000968Xxx.
18. Gaillandre Y, Duhamel A, Flohr T, Faivre JB, Khung S, Hutt A, et al. Ultra-high resolution CT imaging of interstitial lung disease: impact of photon-counting CT in 112 patients. Eur Radiol. 2023;33(8):5528–39. https://doi.org/10.1007/s00330-023-09616-x.
19. Van Ballaer V, Dubbeldam A, Muscogiuri E, Cockmartin L, Bosmans H, Coudyzer W, et al. Impact of ultra-high-resolution imaging of the lungs on perceived diagnostic image quality using photon-counting CT. Eur Radiol. 2024;34(3):1895–904. https://doi.org/10.1007/s00330-023-10174-5.
20. Huflage H, Hendel R, Kunz AS, Ergün S, Afat S, Petri N, et al. Investigating the small pixel effect in ultra-high-resolution photon-counting CT of the lung. Invest Radiol. 2024;59(4):293–7. https://doi.org/10.1097/RLI.0000000000001013.
21. Esquivel A, Ferrero A, Mileto A, Baffour F, Horst K, Rajiah PS, et al. Photon-counting detector CT: key points radiologists should know. Korean J Radiol. 2022;23(9):854–65. https://doi.org/10.3348/kjr.2022.0377.
22. Graafen D, Halfmann MC, Emrich T, Yang Y, Kreuter M, Düber C, et al. Optimization of the reconstruction settings for low-dose ultra-high-resolution photon-counting detector CT of the lungs. Diagnostics (Basel). 2023;13(23):3522. https://doi.org/10.3390/diagnostics13233522.
23. Booz C. Innovations in chest imaging: how can patients benefit? Diagnostics. 2024;14(2):141. https://doi.org/10.3390/diagnostics14020141.
24. Ongsulee P. Artificial intelligence, machine learning and deep learning. In: 2017 15th international conference on ICT and knowledge engineering (ICT&KE); 2017. p. 1–6. https://doi.org/10.1109/ICTKE.2017.8259629.
25. Castiglioni I, Rundo L, Codari M, Di Leo G, Salvatore C, Interlenghi M, et al. AI applications to medical images: from machine learning to deep learning. Phys Med. 2021;83:9–24. https://doi.org/10.1016/j.ejmp.2021.02.006.
26. Karlsson H, Moro V, Eguizabal A, Eriksson M, Agren A, Akerström D, et al. Deep-learning-based denoising for photon-counting CT: image domain or projection domain? In: Medical imaging 2022: physics of medical imaging, vol. 12031. SPIE; 2022. p. 698–715. https://doi.org/10.1117/12.2612480.
27. Baffour FI, Huber NR, Ferrero A, Rajendran K, Glazebrook KN, Larson NB, et al. Photon-counting detector CT with deep learning noise reduction to detect multiple myeloma. Radiology. 2022;306(1):229–36. https://doi.org/10.1148/radiol.220311.
28. Eguizabal A, Persson MU, Grönberg F. A deep learning post-processing to enhance the maximum likelihood estimate of three material decomposition in photon counting spectral CT. In: Medical imaging 2021: physics of medical imaging, vol. 11595. SPIE; 2021. p. 1080–9. https://doi.org/10.1117/12.2581044.
29. Trapani VD, Brombal L, Brun F. Multi-material spectral photon-counting micro-CT with minimum residual decomposition and self-supervised deep denoising. Opt Express. 2022;30(24):42995–3011. https://doi.org/10.1364/OE.471439.
30. Schmidt TG, Sammut BA, Barber RF, Pan X, Sidky EY. Addressing CT metal artifacts using photon-counting detectors and one-step spectral CT image reconstruction. Med Phys. 2022;49(5):3021–40. https://doi.org/10.1002/mp.15621.
31. Ström E, Persson M, Eguizabal A, Oktem O. Photon-counting ct reconstruction with a learned forward operator. IEEE Trans Comput Imaging. 2022;8:536–50. https://doi.org/10.1109/TCI.2022.3183405.
32. Shen L, Xing Y, Zhang L. Joint reconstruction and spectrum refinement for photon-counting-detector spectral CT. IEEE Trans Med Imaging. 2023;42(9):2653–65. https://doi.org/10.1109/TMI.2023.3261999.
33. Toyoda T, Sato S, Kiji H, Kataoka J, Kotoku J, Taki M. Application of machine-learning models to improve the image quality of photon-counting CT images. J Instrum. 2021;16(05):05021. https://doi.org/10.1088/1748-0221/16/05/P05021.

34. Abadi E, McCabe C, Harrawood B, Sotoudeh-Paima S, Segars WP, Samei E. Development and clinical applications of a virtual imaging framework for optimizing photon-counting CT. Proc SPIE Int Soc Opt Eng. 2022;12031:120311. https://doi.org/10.1117/12.2612079.
35. Wang S, Yang Y, Yin Z, Wang AS. Noise2Noise for denoising photon counting CT images: generating training data from existing scans. In: Medical imaging 2023: physics of medical imaging, vol. 12463. SPIE; 2023. p. 15–9. https://doi.org/10.1117/12.2653975.

Index

S. A. Si-Mohamed, K. K. Iniewski (eds.), *Clinical Applications of Spectral Photon Counting Computed Tomography Technology*,
https://doi.org/10.1007/978-3-031-85078-3

MIX
Papier aus verantwortungsvollen Quellen
Paper from responsible sources
FSC® C105338

If you have any concerns about our products,
you can contact us on
ProductSafety@springernature.com

In case Publisher is established outside the EU,
the EU authorized representative is:
Springer Nature Customer Service Center GmbH
Europaplatz 3, 69115 Heidelberg, Germany

Printed by Libri Plureos GmbH
in Hamburg, Germany